Boston Studies in the Philosophy and History of Science

Volume 270

More information about this series at http://www.springer.com/series/5710

Rivka Feldhay • Jürgen Renn
Matthias Schemmel • Matteo Valleriani
Editors

Emergence and Expansion of Preclassical Mechanics

Springer

Editors
Rivka Feldhay
Cohn Institute for the History and Philosophy of Science and Ideas & Minerva Humanities Center
University of Tel Aviv
Tel Aviv-Yafo, Israel

Matthias Schemmel
Max Planck Institute for the History of Science
Berlin, Germany

Jürgen Renn
Max Planck Institute for the History of Science
Berlin, Germany

Matteo Valleriani
Max Planck Institute for the History of Science
Berlin, Germany

Technische Universität Berlin
Berlin, Germany

University of Tel Aviv
Tel Aviv, Israel

ISSN 0068-0346 ISSN 2214-7942 (electronic)
Boston Studies in the Philosophy and History of Science
ISBN 978-3-319-90343-9 ISBN 978-3-319-90345-3 (eBook)
https://doi.org/10.1007/978-3-319-90345-3

Library of Congress Control Number: 2018945536

This Springer imprint is published by the registered company Springer Nature Switzerland AG
The registered company address is: Gewerbestrasse 11, 6330 Cham, Switzerland

Foreword: The Historical Epistemology of Mechanics

Jürgen Renn

The historical epistemology of mechanics studies the long-term development of mechanical knowledge. Mechanical knowledge concerns material bodies in time and space, their motions, and the forces that cause or resist such motions. Mechanical knowledge enables us to predict how bodies change their position with time as long as we know their current state and the forces acting upon them. Mechanical knowledge of this kind played a special role in the process of transformation from natural philosophy to modern science. Natural philosophy from its very inception in the works of Aristotle constructed conceptual systems to represent pictures of the world as a whole. But, in contrast to such global intentions, the origins of mechanical knowledge have to be sought in the much more down-to-earth practical activities of achieving the specific tasks of everyday life.

Over a long historical period, the development of mechanical knowledge and its transmission from one generation to the next remained an inherent dimension of such activities, unrelated to any cognitive endeavors aimed at constructing a mechanical worldview. It was only after the first attempts in classical antiquity to include mechanical knowledge in the conceptual systems of natural philosophy that its assimilation to them and the corresponding accommodation of such systems to mechanical concepts led to conflicts between mechanical knowledge and knowledge about nature as a whole. It was only after the growing body of mechanical knowledge became a vital resource of early modern societies that mechanical knowledge within its own conceptual systematization started to compete with natural philosophy by constructing its own worldviews. This finally resulted in early modern times in what has been called the "mechanization of the world picture."

The main goal of the series *The Historical Epistemology of Mechanics*, conceived in analogy to the four-volume set on *The Genesis of General Relativity*, is to explain the development and diffusion of mechanical knowledge in terms of historical-epistemological concepts. The studies presented within the series are based on a research project centered at the Max Planck Institute for the History of Science in Berlin. While the emphasis of the research has been on the period of the Scientific Revolution, the analysis also takes into account the long-term development of mechanical knowledge without which neither its emergence nor the consequences of this period can be adequately understood. Just as the reconstruction of the relativity revolution in *The Genesis of General Relativity* takes Einstein's work as the point of reference for a thorough contextualization of his achievements, the reconstruction of the transformation of mechanical knowledge during the Scientific Revolution similarly refers to Galileo's work as a point of departure for outlining a historical epistemology of mechanics.

The development of an adequate theoretical framework provides a common basis for the investigations constituting *The Historical Epistemology of Mechanics*. The longevity of mechanics makes it particularly clear that large domains of human knowledge accumulated by experience are not simply lost when theories are revised, even if this knowledge does not explicitly appear in such theories. Thus formal logic is of little use for a description of the multi-layered architecture of the scientific knowledge that allows both the continuous and the discontinuous aspects of the transmission of mechanical knowledge to be accounted for. In order to explain structural transformations of systems of knowledge, it is furthermore necessary to take into account the collective character and the historical specificity of the knowledge being transmitted and transformed, as well as to employ sophisticated models for reconstructing processes of knowledge development. Concepts such as 'mental model', 'shared knowledge', 'challenging object', and 'knowledge reorganization' have turned out in our work to be pivotal for such explanations.

We conceive of mental models as knowledge representation structures based on default logic, which allow inferences to be drawn from prior experiences about complex objects and processes, even when only incomplete information on them is available. Mental models relevant to the history of mechanics either belong to generally shared knowledge or to the shared knowledge of specific groups. Accordingly, they can be related either to intuitive, to practical, or to theoretical knowledge. They are, in any case, characterized by a remarkable longevity—even across historical breaks—as becomes clear when considering examples such as the mental models of an atom, of a balance, of the center of gravity, or of positional weight. Their persistence in shaping the shared knowledge documented by the historical sources becomes particularly apparent in the consistency of the terminology used, a consistency that offers one important element for an empirical control in the reconstruction of mental models and their historical development. The concept of mental model is particularly suited to study the role of practical knowledge for the transformation of mechanics in the early modern period.

Conceiving a body in terms of the intuitive mental model of gravitation, for instance, implies that a heavy body falls down in "natural motion" if its motion is not inhibited or deflected by a force, that it makes an impact when it falls, that the force of this impact is greater the longer it falls, but also—at a later stage of development—that the body has a weight that can be measured with a balance. Whenever the question of a quantitative measure of impact arises—as it does in the early modern period—this model of intuitive and practical knowledge could be and was further extended in reaction to new experiences with the challenging objects of contemporary technology.

More generally, we conceive of challenging objects as historically specific material objects, processes or practices that enter the range of application of a system of knowledge without the system being capable of providing a canonical explanation for them. Examples run from mechanical devices challenging Aristotelian dynamics, via artillery challenging early modern theories of motion, to black body radiation challenging classical radiation theory. In reaction to such challenges, knowledge systems are typically further elaborated, occasionally to the extent that they give rise to internal tensions and even inconsistencies. Such explorations of their limits may then become starting points for their reorganization where often previously marginal insights take on a central role in an emerging new system of knowledge. Such processes of reorganization may be exemplified by the emergence of theoretical mechanics from Aristotelian natural philosophy in ancient Greece, the transformation of preclassical into classical mechanics in early modern times, or the emergence of quantum theory from classical physics at the turn of the last century.

The investigations constituting *The Historical Epistemology of Mechanics* build on this theoretical framework, centering on the role of shared knowledge, of challenging objects, and of knowledge reorganization. The first study, Matthias Schemmel's *The English Galileo: Thomas Harriot's Work on Motion as an Example of Preclassical Mechanics*, investigates the shared knowledge of preclassical mechanics by relating the work of Thomas Harriot on motion, documented by a wealth of manuscripts, to that of Galileo and other contemporaries. Harriot and Galileo indeed exploited the same shared knowledge resources in order to approach the same challenging objects. The study of Harriot's parallel work thus allows one to explore the structure of the shared knowledge of early modern mechanics, to perceive possible alternative histories, and to distinguish between individual peculiarities and shared structures of early modern mechanical reasoning.

The second study of the series, Matteo Valleriani's *Galileo Engineer*, looks more closely at the role of Galileo as a practical mathematician and engineer-scientist. It focuses on his intellectual development in the frame of the interaction between natural philosophy and the challenging objects provided by technological developments. It analyzes Galileo's contribution to the practical science of machines as well as his role as a teacher involved in the contemporary art of war. The results of this analysis highlight Galileo's profile as a military engineer. The book develops a model according to which new scientific knowledge was generated on the basis of the interaction between theoretical knowledge—basically Aristotelian—and the practical knowledge Galileo shared with his contemporaries. Galileo's work is reinterpreted in its entirety against the background of a historiographical investigation concerning the early modern figure of the engineer-scientist.

The third contribution to this series, Jochen Büttner's study *Galileo's Challenges: The Genesis of a New Theory of Motion,* looks more closely at the reorganization of mechanical knowledge that took place in the course of Galileo's research process, stimulated by contemporary challenging objects. It reconstructs in detail Galileo's research process extending over a period of more than thirty years, pursued within a network of exchanges with his contemporaries, and documented by a vast collection of research notes. Analyzing the challenging objects that motivated and shaped

Galileo's thinking, it shows, for example, how the problem of reducing the properties of pendulum motion to the laws governing naturally accelerated motion on inclined planes became a mainspring for the formation of Galileo's comprehensive theory of naturally accelerated motion.

This fourth and conclusive volume articulates more extensively the theoretical foundations of a historical epistemology of mechanics, still with a focus on preclassical mechanics, its prehistory, and its contexts. The framework presented in this volume makes it possible to reconstruct the long-term development of mechanical knowledge from its anthropological origins via the formation of a mechanical worldview to the transformations of classical mechanics in modern physics. It is based on the results of a long-term collaboration between two groups of scholars, one centered at the Max Planck Institute for the History of Science in Berlin, the other at the Cohn Institute of Tel-Aviv University. In the final stages of this long journey, we were assisted by a research grant endowed by the German Israeli Foundation (GIF) during the years 2006–2011, for which we would like to extend our sincere thanks. In the course of this collaboration, the framework of a historical epistemology of mechanics was not only enriched by further case studies, covering a wide range of early modern protagonists, but also further developed in order to capture the specific cultural and societal conditions under which mechanics flourished in the early modern period. The account presented here emphasizes and makes explicit the connections among diverse forms of knowledge that have hitherto been mostly treated in isolation from each other. It also investigates the societal conditions fostering the integration of this knowledge in the emergence and expansion of preclassical mechanics, and the far-reaching consequences it had on the development of physical knowledge.

Introduction

Rivka Feldhay, Jürgen Renn, Matthias Schemmel, and Matteo Valleriani

This book aims to introduce a systematic framework for the historical representation and analysis of preclassical mechanics from a historical-epistemological perspective. Preclassical mechanics is understood here as a heterogeneous knowledge system emerging in the period roughly between the fifteenth and the seventeenth centuries, before classical mechanics was formulated, in continuation of Newton's work, as a coherent and comprehensive mechanical theory. Preclassical mechanics is characterized by the investigation of problems, often resulting from disparate technological challenges, that were at first isolated from each other. In the course of their exploration by a growing community of engineer-scientists, the solutions to these problems increasingly cohered within a broad intellectual framework. The distinctive cognitive and social architecture of this intellectual framework of preclassical mechanics is analyzed in the contributions to this volume. Classical mechanics, as we shall argue, emerged from the tensions within this framework in a process of the reorganization of the knowledge accumulated in the period of preclassical mechanics, while the community of its practitioners became part of a new social world.

The first part of the volume "The Historiographical Framework" comprises three introductory essays, which reflect tendencies in contemporary historiography of science, but also open it up to new horizons.

The first chapter is concerned with the epistemic structures of mechanical knowledge. It focuses on mental models as cognitive instruments that function as mediators between the realm of experience and practice on the one hand, and conceptual systems on the other. With the help of the concept of mental model, the chapter discusses how mechanical knowledge has emerged from experience in practical contexts and how it was transformed into theoretical and mathematically formalized knowledge systems. Focusing on one particular mental model, which describes the cognitive structure conceptualizing motion as being caused by forces, the chapter then follows its transformations in the long-term history of mechanical thinking. This so-called *motion-implies-force* model is rooted in intuitive, non-written mechanical knowledge. Over the course of history, the model was recruited, complemented, and transformed in the context of the use of mechanical tools and articulated in the work of practitioners dealing with machines, arms, ships, buildings, fortifications, and the like. Eventually, under specific cultural circumstances, this and other mental models were elaborated and integrated into mathematically formalized systems that were used, for example, in the explanation of terrestrial and celestial motions in early modern natural philosophy and the mathematical disciplines of European universities.

While sharing interest with histories of mechanics such as Bertoloni Meli's *Thinking with Objects* (2006), or histories of practical mathematics such as Jim Bennet's "Practical Geometry and Operative Knowledge," (*Configurations* 6.2 (1998), 195–122), this first chapter presents a historical-epistemological perspective anchored in cognitive science that justifies breaking the dichotomy between thought processes and empirical experiences. It points at interactions and feedback patterns between the different levels that constitute mechanical bodies of knowledge, from challenging objects, such as inclined planes, projectiles, or pendulums, and the empirical knowledge shared by practitioners about their functioning, up to the theories formulated by the mathematicians who were in close contact with practitioners. Such theories were then disseminated to generations of students through college and university courses as well as through printed texts.

The second introductory chapter is concerned with the socio-political context of preclassical mechanical knowledge between the mid-sixteenth and mid-seventeenth centuries. The chapter identifies a political structure that emerged and became predominant in early modern Europe, namely the sovereign state, as the relevant context in which technological developments of the Renaissance were enhanced and acquired political relevance. The knowledge of preclassical mechanics was recruited to the service of the state. This took place either through the newly organized "state universities" of the monarchs, through the imperial and religious colleges that were strongly supported by sovereigns throughout Europe, through the courtly patronage of scientists, or through academies under the patronage of sovereign rulers. The centralization of political power, with its specific military and economic needs, thus conditioned the transformations and reorganization of mechanical knowledge. Mechanical knowledge was also advanced by the general approval of its utility and its diffusion through religious schools—especially through the broad network of Jesuit colleges in Europe and even in China, Japan, and South America. Finally, the environment of the sovereign state also reinforced the social position of the carriers of mechanical knowledge, their authority, and even their identity.

The chapter provides the context in which feedback between technology, empirical knowledge, and theoretical knowledge was enhanced in the political environment of the sovereign state. It shares common ground with historiographies that emphasize patterns of patronage in the early modern era, such as Schaeffer and Shapin's *Leviathan and the Airpump* (1985), Biagoli's *Galileo Courtier* (1993), and Galluzzi's *Scienziati a Corte* (2001). It focuses on the specific political structure—formulated in the contemporary flourishing literature on sovereignty—within which patronage functioned. The structure of a sovereign regime revolves around a monarch claiming absolute power, which is of course beyond the reach of any human being. To compensate, the sovereigns tended to rely on military and economic forces to sustain their claims, on scientists and artists to provide technological development and a narrative of grandeur, and on the religious educational system to provide collective identity and to discipline the population, namely, shaping each individual as a subject in order to satisfy the monarch's centralistic ambitions. The thesis relies not only on political contemporary texts such as Jean Bodin's *Six Books on the Commonwealth* (1576) and Botero's *Della Ragion di Stato* (1589), but also on modern historical studies such as Kalmo and Skinner's (eds.) *Sovereignty in Fragments* (2010), Paci's *La Guerra nell'Europa del Cinquecento* (2005), and Alfani's *Calamities and the Economy in Renaissance Italy* (2013).

The third introductory chapter sets our understanding of preclassical mechanics in the context of the historiography of science in the twentieth century by focusing on the crucial importance of feedback between empirical, practical, and theoretical knowledge for the development of a science of mechanics. This importance was recognized early on by authors such as John Bernal, Henryk Grossmann, Boris Hessen, Leonhard Olschki, and Edgar Zilsel. However, these authors tended to oversimplify the relation between the economic structure of early modern societies and the emerging new science. Later historiography in contrast focused on the role of cultural contexts, often at the expense of an adequate account of the larger societal dynamics, including the class struggles at the focus of the pioneers of materialist historiographies. Preclassical mechanics, however, can only be understood as being shaped by both the economic dynamics fostering new technologies and by a rapid expansion of knowledge and its role as an instrument in the ideological struggles of the time. The theories and aspirations of preclassical mechanics could be exploited as models of rational science, and even as models of a rational organization of society. Accordingly, its protagonists, the engineer-scientists, were not yet "organic intellectuals" in the sense of Gramsci, that is, they did not embody the political aspirations of large social groups who reinforced their interests by means of cultural and educational programs, nor were they simply servants of political and economic powers—the two alternatives canonically represented by the two versions of Brecht's drama on Galileo Galilei. Rather, the protagonists of preclassical mechanics were, for the most part, faute de mieux, social explorers in search of a position in society, both in the sense of striving to secure their ever precarious individual careers, and in the sense of pursuing an intellectual agenda that unavoidably interfered—and often clashed—with the dominant political and social constellations. In some significant cases, such as that of Giovanni Battista Benedetti, discussed in Chap. 7, these social explorers identified themselves with the ruling elite of the early modern court society, thus embodying the ideals and interests of new centralized territorial states. Another characteristic case of a preclassical engineer-scientist, related to a different institutional context, is that of the Jesuit professor of mathematics at the college of Parma, Paolo Casati, which is discussed in Chap. 10.

The second part of the volume "The Ancient Roots and Emergence of Preclassical Mechanics" deals with core themes of preclassical mechanics such as the functioning of the balance and the lever and their long-term history from ancient science to their early modern transformations.

Chapter 4 describes the early history of weighing from the perspective of historical epistemology. It extends the analysis of mental models as bridging practical and theoretical knowledge with a detailed investigation of the interaction between technical innovations of weighing instruments and the emergence of certain abstract concepts that are key to theoretical mechanics, such as 'center of gravity.' In particular, technological artifacts are considered as external representations of mental structures, and technological innovation is conceived as resulting from an interplay between representation and reflection. While the analysis focuses on the ancient Mediterranean, the case of mechanical technology in ancient China is used to discuss the possibility of alternative developments in the history of mechanical knowledge.

Chapter 5 discusses the origins of theoretical mechanics in ancient Greece on the basis of an analysis, also within the framework of historical epistemology, of the *Mechanical Questions*, a text traditionally ascribed to Aristotle. This text served as a key reference of preclassical mechanics, which is characterized by attempts to reconcile the heterogeneous legacy of ancient science, in particular the Aristotelian and the Archimedean traditions. The chapter details the emergence of a theoretical science from practical experience and analyzes the simple machines that constituted a basic model for the discussion of complex technology in the period of preclassical mechanics.

Chapter 6 focuses on the concept of 'center of gravity.' This concept was not only related to the mechanical works of Archimedes and Pappus, but also included a cosmological dimension, beginning with Aristotle's discussion of the shape and immobility of the Earth in *De caelo*, via the Medieval reformulation by Johannes de Sacrobosco, to the centrobaric theories of preclassical authors such as Francesco Maurolico, Christophorus Clavius, and Giovanni Biancani. The early modern discussions about the implications of a new science of mechanics for cosmology thus had a long pre-history which is discussed in this contribution.

The third part of this volume "The Expansion of Preclassical Mechanics" deals with the social, political, and cultural positions of the protagonists of preclassical mechanics and with its expansion through the work of the members of the order of the Jesuits.

Chapter 7 opens this discussion with a case study of Giovanni Battista Benedetti who is interpreted as a prototypical court scientist of the sixteenth century. It illustrates how deeply the writings of this still neglected protagonist of preclassical mechanics were shaped by the socio-political structures of the time.

Chapter 8 is the result of the analysis of a recently discovered manuscript that testifies to the integration of Galilean mechanics within the mathematical curriculum of the Jesuit college in Naples. This chapter investigates the important role of teaching activities in the expansion process of preclassical mechanics. In particular, it demonstrates how Galilean mechanical ideas were disseminated and popularized through the Jesuit's elaborated educational system in the seventeenth century, even immediately after the trial of Galileo. The chapter thus critically reexamines a traditional historiography that perceived a dichotomy between "Jesuit science" and "the new science." It does so by showing, first, that Jesuit teachers, although bound not to deviate from the Aristotelian textual corpus by the *Ratio studiorum* (their official program of studies) were nevertheless open to new ideas, including those of Galileo and his school, incorporating them into their teaching materials; second and more generally, that Jesuits cultivated a genuine interest in transforming the traditional mathematical disciplines into physico-mathematical sciences.

Chapters 9 and 10 further elaborate this line of argumentation. Paolo Casati, for instance, is shown to have continued the tendency to physicalize the mathematical science of machines, combining Archimedes' lever principle and theory of buoyancy with the concept of impetus originating in Aristotelian medieval commentaries and modernized by Honoré Fabri and Franciscus Eschinardus. Both chapters show the dissolution of major elements of the traditional Aristotelian framework (such as the concept of 'levity'), and the engagement of Jesuit scholars with the experimental-mathematical discourse of the time. They thus illustrate the ways in which the scientific traditions pursued by Jesuits contributed to the emergence and expansion of preclassical mechanics.

Contents

Foreword: The Historical Epistemology of Mechanics v

Introduction ix

Contributors xvii

Part I The Historiographical Framework

1 **Mental Models as Cognitive Instruments in the Transformation of Knowledge** 3
Jürgen Renn, Peter Damerow, Matthias Schemmel, Christoph Lehner, and Matteo Valleriani

2 **Preclassical Mechanics in Context: Practical and Theoretical Knowledge Between Sovereignty, Religion, and Science** 29
Rivka Feldhay

3 **Socio-Political Coordinates of Early-Modern Mechanics: A Preliminary Discussion** 55
Pietro D. Omodeo

Part II The Ancient Roots and Emergence of Preclassical Mechanics

4 **The Early History of Weighing Technology from the Perspective of a Theory of Innovation** 81
Jochen Büttner, Jürgen Renn, and Matthias Schemmel

5 **The Balance, the Lever and the Aristotelian Origins of Mechanics** 111
Jürgen Renn and Peter McLaughlin

6 **A Journey to the Center of the Earth: Cosmology and the Centrobaric Theory from Antiquity to the Renaissance** 139
Elio Nenci

Part III The Expansion of Preclassical Mechanics

7 The Social Position and Intellectual Identity of the Renaissance Mathematician-Physicist Giovanni Battista Benedetti: A Case Study in the Socio-Political History of Mechanics 181
Pietro D. Omodeo

8 A Treatise on Galileian Mechanics: *Exercitationes in Mechanicis Aristotelis* by the Jesuit Giovan Battista Zupi 215
Romano Gatto

9 Heaviness, Lightness and Impetus in the Seventeenth Century: A Jesuit Perspective 255
Rivka Feldhay and Ayelet Even-Ezra

10 Jesuit Conceptions of Impetus After Galileo: Honoré Fabri, Paolo Casati, and Francesco Eschinardi 285
Michael Elazar and Rivka Feldhay

Correction to: Emergence and Expansion of Preclassical Mechanics C1

Correction to: Emergence and Expansion of Preclassical Mechanics C3

Index 325

Contributors

Jochen Büttner Max Planck Institute for the History of Science, Berlin, Germany

Peter Damerow (1939–2011) Max Planck Institute for the History of Science, Berlin, Germany

Michael Elazar Minerva Humanities Center, Tel-Aviv University, Tel-Aviv, Israel

Ayelet Even-Ezra History Department, The Hebrew University, Jerusalem, Israel

Rivka Feldhay Cohn Institute for the History and Philosophy of Science and Ideas & Minerva Humanities Center, University of Tel Aviv, Tel Aviv-Yafo, Israel

Romano Gatto Università degli Studi della Basilicata, Potenza, Italy

Christoph Lehner Max Planck Institute for the History of Science, Berlin, Germany

Peter McLaughlin Philosophy Department, University of Heidelberg, Heidelberg, Germany

Elio Nenci University of Milan, Milan, Italy

Pietro D. Omodeo Ca' Foscari University of Venice, ERC EarlyModernCosmology, Venice, Italy

Jürgen Renn Max Planck Institute for the History of Science, Berlin, Germany

Matthias Schemmel Max Planck Institute for the History of Science, Berlin, Germany

Matteo Valleriani Max Planck Institute for the History of Science, Berlin, Germany

Technische Universität Berlin, Berlin, Germany

University of Tel Aviv, Tel Aviv, Israel

The original version of this book was revised. A correction to this book is available at https://doi.org/10.1007/978-3-319-90345-3_11 and https://doi.org/10.1007/978-3-319-90345-3_12

Part I
The Historiographical Framework

Chapter 1
Mental Models as Cognitive Instruments in the Transformation of Knowledge

Jürgen Renn, Peter Damerow, Matthias Schemmel, Christoph Lehner, and Matteo Valleriani

Abstract The chapter is concerned with the epistemic structures of mechanical knowledge in its historical transformations. It describes these structures using the concept of mental models as cognitive instruments, which function as mediators between the realm of practice and experience on the one hand, and conceptual systems on the other. With the help of the concept of mental model, the chapter discusses how mechanical knowledge has emerged from experience in practical contexts and how it was transformed into theoretical and mathematically formalized knowledge systems. Focusing on one particular mental model, which describes the cognitive structure conceptualizing motion as being caused by forces, the chapter then follows its transformations in the long-term history of mechanical thinking. This so-called "motion-implies-force" model is rooted in intuitive, non-written mechanical knowledge. Over the course of history, the model was recruited, complemented, and transformed in the context of the use of mechanical tools and articulated in the work of practitioners dealing with machines, arms, ships, buildings, fortifications, and the like. Eventually, under specific cultural circumstances, this and other mental models were elaborated and integrated into mathematically formalized systems that were used, for example, in the explanation of terrestrial and celestial motions in early modern natural philosophy and the mathematical disciplines of European universities.

Keywords Mental model · Practical knowledge · Reflection · Experience · Causation · Motion · Force · Aristotelian dynamics · Classical physics · Impetus · Projectile trajectory · Galileo Galilei · Thomas Harriot · Isaac Newton

---

J. Renn · M. Schemmel (✉) · C. Lehner · M. Valleriani
Max Planck Institute for the History of Science, Berlin, Germany

P. Damerow (1939–2011)
Max Planck Institute for the History of Science, Berlin, Germany

R. Feldhay et al. (eds.), *Emergence and Expansion of Preclassical Mechanics*, Boston Studies in the Philosophy and History of Science 270,
https://doi.org/10.1007/978-3-319-90345-3_1

1.1 Mental Models

The object of the following study is the question as to how knowledge, especially the technical knowledge of antiquity, is transferred and transformed, with an emphasis on the historical transformation of this knowledge into preclassical mechanics first and then into classical mechanics of the modern age. Technical knowledge is to be understood in the broadest sense, as knowledge about the way to manipulate material objects, regardless of whether or not it has been represented as scientific knowledge in written language.

This study proceeds from a certain form of the internal, that is, mental, representation of knowledge, which we designate *mental model*, a concept borrowed from the field of *cognitive science*. What we mean with this concept is to be explained first using a paradigm from the area of reading comprehension.[1] Let us assume that a story in a children's book begins with the following words:

> It was Paul's birthday. Jane and Alexander went to get presents. 'Oh, look!' Jane said, 'I'll get him a kite.' 'No, don't,' Alex responded, 'He already has one. He'll make you take it back.'

Let us presume further that questions like the following were posed to the reader:

Why are Jane and Alex buying presents?
Where did Jane and Alex go?
What did Alex refer to by the word 'it' in his last sentence?

Most people from cultures in which birthdays are celebrated will be able to answer these questions spontaneously: Jane and Alexander bought gifts because it was Paul's birthday; to do so they went to a shop, and the word "it" refers to the kite that Jane would have to return. Most people would also be convinced that these answers are contained in the paragraph they read.

Both of these facts are astonishing to information theorists. For at the very least it has proven difficult to program a computer such that it is able to derive these answers, supposedly so obvious, from the text. None of the answers is contained explicitly in the text; they have to be inferred from the information given. The form this reasoning takes, however, does not correspond to the traditional understanding of deduction, for in a formal logical sense, the answers cannot be concluded from the information contained in the text without making far-reaching additional suppositions. Rather, the answers are obtained from the information contained in the

[1] The paradigm goes back to Eugene Charniak's dissertation (1972) about the question as to how children understand stories. Marvin Minsky, his thesis advisor, used the paradigm in 1974 in a memo for the Department for Artificial Intelligence at MIT to explain the concept of a *frame* (published electronically: http://web.media.mit.edu/~minsky/papers/Frames/frames.html). The paradigm became known through the later publication of the memo, Minsky (1975, 241–247). See also Minsky (1985, 261–265), German: Minsky (1990, 261–265). Here we use the version and the interpretation of the paradigm by Robert Davis (1984, 45–48).

text by linking it with previous experiences. Using a theory of mental models,[2] in which knowledge obtained from experience is represented, the course of the conclusions can be described as follows:

1. A catchword prompts the reader to call up the mental model of a birthday.
2. The model contains free variables that must be *instantiated* using information from the given situation before conclusions can be drawn.
3. In this the model raises questions that have to be answered, and the paragraph is searched for indications of such answers: Whose birthday is it? Who are the guests? What are the gifts?
4. If a suitable answer is found, it is inserted in the corresponding variable of the model.
5. If no suitable answer is found, what is known as a *default assumption,*[3] a standard assumption based on previous experiences, replaces the missing information.
6. When values are allocated to the essential variables in the model, the model is instantiated and available for information processing, for instance, to answer the questions posed.
7. If the model cannot be instantiated, however, it is discarded. Then the paragraph cannot be interpreted as a text about a birthday party.

Based on this example, several main features of the theory of mental models can be explained.

1. Mental models are instruments for drawing conclusions in the context of given knowledge. The available information is transferred into a model before such conclusions can be drawn. In our current example, the thinking on the basis of which the posed questions can be answered does not proceed directly from the

[2] The concept of the mental model as it is used here harks back to two theoretical traditions that focus on the structure of the representation of knowledge; see the section on *frame-based systems* and *mental models* in Wilson and Keil (1999, 324–326, 525–527). In the mid-1970s the concept of *frame* became a central concept of a new theory of meanings in both cognitive science and linguistics; see Minsky (1975) and Fillmore (1976), on this see also Davis (1984), Fillmore (1985) and Lehrer and Kittay (1992). Somewhat less than a decade later, two books appeared with the title *Mental Models, completely independent of each other*: Gentner and Stevens (1983) and Johnson-Laird (1983), see also Johnson-Laird (1989) as well as the web site http://www.tcd.ie/Psychology/Ruth_Byrne/mental_models/. Both traditions address the structures of cognition and language, but not the way these have changed throughout history. Here we observe mental models as special frames and use this concept for the theoretical definition of the historical change to the structure of knowledge systems.

[3] The concept of the default assumption became the foundation of a non-monotonic logic of reasoning. Classical logic is *monotone*, meaning that a theorem derived from certain assumptions can still be derived even after additional assumptions are added. In this sense classical logic is deterministic. The introduction of default assumptions, by contrast, makes the validity of a conclusion dependent on the available information, and in this manner makes possible the theoretical definition of the change in deductive methods. On this, see Besnard (1989) and Brewka (1991) as well as the section on *Nonmonotonic Logics* in Wilson and Keil (1999, 608–611).

information in the text, but from the model of a birthday party instantiated with this information.
2. Mental models are context-specific and not universally valid. Using mental models it is also possible to reach conclusions that presume the knowledge of certain objects and contexts. In order to decipher the meaning of the text in the current example, the reader has to be familiar with a culture in which birthdays are celebrated and associated with certain actions like bringing presents.
3. Mental models can be adapted to new experiences. Through such new experiences it is thus possible to conceive changes in conclusions as a consequence of changes to the context of experience. If, for instance, the age-specific preferences of children shift toward electronic toys, the "birthday" model can remain unchanged while the variable "birthday present" undergoes such a change that the text appears obsolete and no longer appropriate to the period.
4. Mental models link present with past experiences by embedding new experiences in a cognitive network of previous experiences. So they allow conclusions to be drawn from incomplete information. In our current example the sparse information of the text is integrated into a complex image of the course of a birthday party representing past experiences, and only then does this information become meaningful.
5. Mental models bridge various levels of knowledge that represent the same object in various forms of knowledge, from the level of practical knowledge by practitioners up to the level of scientists' theories. Therefore they allow us to grasp even, and indeed especially, implicit conclusions that are embodied in the logic of practitioners' actions, which are not recorded explicitly in the form of language or writing. In our current example it is sufficient to have taken part in birthday parties—without having ever thought about the meaning of the rituals associated with the celebration—to decipher the meaning of the short text.

Because of these characteristics, the idea of mental models, which is here presented by the paradigm of reading comprehension, is better suited to describing and explaining theoretically the transformation of knowledge independent of its given, shifting form of representation, than are theories based on a strict differentiation between thought processes and empirical experiences. In fact, the theory of mental models was developed in opposition to the widely held assumption that human thought could ultimately be traced back to the laws of a universal, formal logic or linguistic theory.

At least with regard to the history of knowledge and thinking, the theory of mental models has two decisive advantages. First, it allows the theoretical description and explanation of object-specific deductive methods and their changes through history, which were the subject of traditional philology and *metaphysical* logic[4] but,

[4] A comprehensive overview of the historical development of logic before its formalization is presented by Friedrich Ueberweg (1882). The most important attempt to formulate a logic that incorporates the changes in logical structures by reflecting on their application is Hegel's *System of Logic*. However, because of the importance that formal logic took on for deductive theories in mathematics and the natural sciences, his system enjoyed only short success; on this see Damerow and Lefèvre (1980).

due to the successes of formal logic at the beginning of the twentieth century, increasingly came to be regarded as irrelevant for natural sciences or even fell into oblivion. Second, this theory also allows us to grasp deductions and their changes which are based on incomplete information, so that conventional, anachronistic evaluations of the correctness or faultiness of findings in the history of science, along with the image of a linear and cumulative cognitive development they construct, can be replaced by a more adequate model of the development and transformation of knowledge.[5]

1.2 Mechanical Knowledge

1.2.1 What Is Mechanical Knowledge?

What does the term mechanical knowledge designate? Mechanical knowledge concerns material bodies in time and space, their motions, and the forces that cause or resist such motions. Mechanical knowledge allows us to predict how bodies change their position with time if only we know their current state and the forces acting on them.

Mechanical knowledge of this kind played a special role in the process of development from natural philosophy to modern science. Natural philosophy from its very beginnings in the works of Aristotle constructed conceptual systems representing pictures of the world as a whole, referred to in the following as "world views." In contrast to such global intentions, the origins of mechanical knowledge have to be sought in the much more down-to-earth activities of realizing specific aims of practical life.

Over a long historical period, the development of mechanical knowledge and its transmission from one generation to the next remained an inherent dimension of such activities unrelated to any cognitive endeavors aimed at constructing a mechanical worldview. It was only after the first attempts in classical antiquity to include mechanical knowledge into the conceptual systems of natural philosophy that its assimilation to them and the corresponding accommodation of such systems to mechanical concepts could lead to conflicts between mechanical knowledge and knowledge about nature as a whole. It was only after the growing body of mechanical knowledge became a vital resource of early modern societies that mechanical knowledge within its own conceptual systematization started to compete with natural philosophy by constructing its own world views. This finally resulted in early modern times in what has been called the "mechanization of the world picture"(Dijksterhuis 1956, 1986).

[5]A turn against the anachronistic historiography of the history of science fed by the nineteenth century belief in progress was ushered in by Thomas Kuhn (1962) with his work *The Structure of Scientific Revolutions*. Kuhn successfully drew attention to the autonomy of knowledge systems, which makes it impossible to integrate them seamlessly into the model of a continuous and cumulative development of knowledge. According to his theory, fundamental structural changes in knowledge are caused by paradigm shifts. Just how these shifts come about, however, is not convincingly explained by this theory.

In order to make understandable the long-term transformation of mechanical knowledge from specific knowledge inherent in practical activities to a seemingly universal fundament of modern science, the architecture of this knowledge has to be briefly reviewed.

1.2.2 Intuitive Mechanical Knowledge

One essential element of all human activities is the goal-oriented application of human forces to spatially rearrange the objects of our environment in such a way that this environment better suits human needs. The orientation toward the realization of such specific goals requires the application of past experiences to the situation at hand. Mechanical knowledge in this general sense is based on universal biological preconditions and is developed by every human being in their own individual development.

Building on the underpinnings of such universal resources, the cultural environments of human societies enforce the development of specific forms and contents of mechanical knowledge that is represented enactively in the use of mechanical tools and techniques.

Such incipient types of mechanical knowledge share one particular characteristic: The knowledge is to a great extent intuitive knowledge, that is, context-dependent, implicit knowledge of practical activities. Such knowledge is usually not explicitly written down in any form, nor is it encoded in descriptive or theoretical language. It is rather implicitly represented by conceptual schemes serving to control and transmit mechanical activities, which we have described above as "mental models." When a mental model is successfully instantiated, it allows content-specific inferences to be drawn about possible transformations of this situation, even without any explicit conceptual representation of such inferences encoded in spoken or written language. This, in particular, is the way in which intuitive mechanical knowledge gained in prior experience is used to draw conclusions that implicitly apply this experience to new situations.

1.2.3 Theoretical Reflection on Mechanical Knowledge

In contrast to intuitive mechanical knowledge, theoretical knowledge representing mechanical experience is based on generalized concepts encoded in language. Such knowledge results from reflecting on the mental models of intuitive mechanical knowledge. This reflection represents the intuitive inferences in language-encoded mechanical rules and statements. It furthermore constructs from the interaction of the mental models an integration of such rules and statements into higher-order systems of generalized concepts, systems that finally represent in theoretical form the structured experiences embodied in intuitive mechanical knowledge.

Within such forms of cognitive representation, multivalent terms such as place, motion, or force develop into technical terms with restricted mechanical meanings and standardized relations to one another. These terms and relations reflect directly or indirectly the integration of various mental models that are structures of intuitive mechanical knowledge. The models now ensure the applicability of the generalized mechanical concepts to the specific contexts of mechanical experiences from which they were originally derived, which thus become examples of general theoretical statements on mechanical experiences.

Theoretical mechanical knowledge thus emerges with technical terms related to specific forms of mechanical tools and techniques which are then used to elaborate general relations between them by discursive descriptions and, finally, deductive systems based on implicit or explicit meta-cognitive rules of operating with formalized language representations.

1.2.4 Mechanical Knowledge Without Theory: Mechanical Tools

On the foundation of intuitive mechanical knowledge, which is built up in more or less the same way in all cultures and by all individuals growing up in these cultures, new mechanical techniques are developed, the knowledge of which is culture-specific and, within a culture, not usually shared by all members. The reason for this differentiation is that these techniques no longer solely depend on the biologically determined capabilities of human individuals to perform certain actions, but additionally on the use of mechanical tools. Such tools serve to increase mechanical forces beyond the biologically limited human capabilities. The use of such tools demands special experience and knowledge, which is usually not achieved to the same extent by all members of a community, thus leading to processes of specialization. Once the prerequisite tools for gaining such experience and for building up such knowledge have been invented, the knowledge of how to produce and use them adequately is communicated among individuals and subsequently transmitted from one generation to the next.

Human mechanical knowledge, in contrast to the mechanical knowledge inherent in the behavior of animals, has its roots not only in biologically determined activities but also in the use of historically transmitted mechanical tools. A necessary consequence of this is that the development of tool-based mechanical knowledge is determined by social contexts. Since the skillful use of mechanical tools requires accumulated experience, which can best be acquired by specialization on the handling of these tools, the development of tool-based mechanical knowledge is to a great extent the result of a social division of labor leading to the emergence of professions and the transmission of professional knowledge from one generation to the next. This transmission is predominantly realized by participation in the process of work during a phase of apprenticeship. The tool-based mechanical knowledge is represented in mechanical tools and their use so that its development, application, and historical transmission does not require any symbolic representation in written descriptions or graphical depictions. This development leaves traces in the semantic topologies of languages, in particular in the emergence of professional terminologies representing a semantic derived from the tools, skills, and practices to which they are related.

Thus, building on the underpinnings of such universal, biologically determined resources, the cultural environments of human societies enforce in human minds the emergence of specific systems of context-dependent mechanical knowledge by turning the accumulated experiences of individuals into bodies of knowledge related to the mechanical tools available in these environments. These bodies of knowledge are historically transmitted by material tools, by shared practices and by oral communication using terminologies related to these tools and practices.

One consequence of the social preconditions of the historical occurrence of mechanical tools is that they are developed predominantly in those areas of social activities that are important for the reproduction of the community. Thus, main areas for which civilizations on early stages of human development developed mechanical tools were hunting and gathering, food processing, and tribal warfare. In such areas mechanical tools for enhancing human forces turned out to be favorable for the community so that incentives emerged for their systematic production and their transmission to the following generations.

1.2.5 Architecture of Mechanical Knowledge

These brief remarks on the nature of mechanical knowledge may be sufficient to elucidate the basic architecture of mechanical knowledge, which we will use here as an analytical tool for understanding the processes of the historical development of mechanical techniques and representations of knowledge. Given the broad definition of mechanical knowledge presented above, at least three different layers of mechanical knowledge may be distinguished:

1. Intuitive mechanics built up in ontogeny;
2. Practical and professional mechanical knowledge depending on familiarity with the use of specific mechanical tools and techniques;
3. Explicit mechanical theories based on reasoning codified in natural language, including deductive theories based on the representation of reasoning in rule-based systems of formalized language.

These three layers of mechanical knowledge represent not only different kinds of knowledge but moreover a developmental sequence that is closely connected to the social organization and control of human activities.

1.3 The Mental Model of the Causation of Motion

1.3.1 The Intuitive Motion-implies-force *Model*

The possibility of understanding the historical development and transformation of technical knowledge in terms of the change in mental models will first be elucidated on the basis of a simple example. Studies in cognitive theory have shown that, even in early childhood, the experience that objects can be moved by exerting force upon them results in the inverse conception that every perceived movement of an object is linked with the idea that it must have been caused by a mover exerting a force upon it.[6] At the same time the conviction develops that, under otherwise equal conditions, greater force must also cause stronger motion.

This inversion of the experience of forces as a cause of motions, into an interpretation of perceived motions as caused by force exerted by a mover, can be viewed as a typical example of the formation of a mental model—which is to be designated here, in keeping with the literature, as the *motion-implies-force model* (Clement 1983). The experiences that are represented by this model are so general that the *motion-implies-force* model is probably acquired by people in each and every culture over the course of their ontogenetic development. Experiences with the causation of motions through the exertion of forces result in the mental construction of a model of motions, the essential variables of which are the *mover*, the *moved object* and the *motion performed*. Whenever this intuitively acquired and applied model is retrieved by the perception of changes, it requires identifying the *mover*, the *moved object* and the *motion* of the object in the given situation. If any of these three variables cannot be instantiated, the application of the model fails and the perceived changes are not recognized as the motions of an object.

In the course of its application the model is differentiated more finely. It receives a number of additional variables:

- The *causing force* exerted by the mover.
- The *potential force* the moved object is able to exert on others.
- The *velocity* of the moved object.
- The *path* traversed by the object.
- The *time* the object needs for the motion.
- The *resistance* that opposes the motion.
- The *medium* that causes the resistance to the motion.

Such variables can be instantiated by assigning elements of perception to the given situation or through default assumptions, that is, through plausible default assumptions that represent previous experiences with moved objects. Through the application of this motion-implies-force model of an intuitive comprehension of motions differentiated in this way, the perception of a motion is linked with the

[6] On the connection between cognitive theoretical and historical studies, see Piaget (1978), Piaget and Garcia (1989), McCloskey (1983).

structured knowledge stemming from previous experiences. This knowledge allows conclusions to be drawn about the given situation that extend far beyond what was directly perceived.

1.3.2 The Mental Model of Aristotelian Dynamics

The philosophers of Greek antiquity explicated elementary models of intuitive thinking like the motion-implies-force model conceptually and in written form. In so doing they availed themselves of a meta-language that no longer relates to the perceived reality itself, but rather to the thought directed toward this reality. Implicit mental models of intuitive reasoning thus became generalized assumptions of an explicit natural philosophy with a claim to universal validity.

The determination of the relationship between force and motion addressed by Aristotle in his *Physics* offers a telling example for this consequence of a change in media. Aristotle made the motion-implies-force model of intuitive thinking the core of his theory of local motion, which defined philosophers' theories of motion well into the Late Middle Ages and the early modern period.[7]

In accord with the intuitive perception of motions, Aristotle defined the velocity of a moving object as the path traversed in a certain amount of time. He further defined this path as proportional to the force that caused it, and as inversely proportional to the resistance to be overcome, which is ascribed to the weight of the object and to the medium.

The example of this Aristotelian dynamics shows how a mental model, through the shift from the medium of intuitive thinking to fixed, written language, can become the point of departure and foundation of a scientific theory. Aristotelian dynamics posits general assertions to explicate the structure of intuitive conclusions that are drawn when the motion-implies-force model is applied. Above and beyond this, the example makes clear that the linguistic explication of intuitive conclusions, by constricting the latitude of meaning, can effect a transformation of the knowledge implicit in such conclusions. In the current example, causal and functional relations between the paths traversed, the times elapsed, the forces required and the resistances to be overcome are determined more precisely as proportionalities and thus related to the mathematical theory of proportions, which constituted nearly the only instrument in antiquity that allowed mathematical conclusions to be applied to functional relations.

[7] On the theory of local motion in Aristotle, see Drabkin (1938), Maier (1949, 116ff.) and Maier (1952, 284–286).

1.3.3 The Differentiation of the Mental Model of Aristotelian Dynamics

The shift from the mentally represented model of intuitive thinking to the theory of motion fixed in written language also had consequences other than those directly associated with the change in media. The expanded possibilities for reflecting on conclusions and their results also became the point of departure for a further differentiation of the model. A first form of such a differentiation results from processing experiences that formally fall under the generalized conceptions of the model recorded in writing, but cannot be assimilated into the intuitive model upon which it is based.

Specifically, even in Aristotle's *Physics*, the failure of the model in areas where it was difficult to identify a mover led to the definition and exclusion of particular types of motion. Aristotle excluded from the application area of the concept of motion based on the motion-implies-force model, first, the motion of celestial bodies, and second, the case of heavy objects falling toward the center of the Earth, as well as the ascension of light objects in the opposite direction. The two excluded kinds of motion were demarcated as *natural* motions that require no causal force, as opposed to the *violent* motions, which are the sole subject of his theory of the causation of local motions by forces. According to his theory, the motion of celestial bodies comes about through the natural circular motion of the fifth element, ether, and the natural, rectilinear motion of the heavy and light objects through their aim to reach their natural place.

While these natural motions led to a conceptual differentiation through theoretically excluding them from the area to which the motion-implies-force model could be applied, there is also a motion phenomenon that resulted in an internal differentiation of the model through theoretical reflection, namely the phenomenon of inertial motion. When an object is thrown, after leaving the hand it continues its motion without any detectible cause. However, for the motion-implies-force model to be applicable, it requires the instantiation of the critical variable *mover* through an object that has to be in direct contact with the moved object in order to be able to exert a force upon it.

Philosophers of antiquity, including Aristotle, presented various subtle, and sometimes overly subtle, arguments to try to assign a causal force to such an inertial motion nevertheless. Such arguments generally proceed from the assumption that the throwing hand also induces motions in the medium, which, in turn, can cause the continuation of the movement through the direct contact of the medium with the moved object (Dijksterhuis 1956, I, 30–31). Yet such explanations, especially those of Aristotle himself, were handed down, not so much owing to their explanatory power, but rather due to theoretical dogmatism. The passing on of such attempts to repair the Aristotelian theory of local motion was linked to specific social conditions that guaranteed that dogmas could be established and their validity enforced.

A solution to the problem that was historically more effective emerged not merely as a result of philosophical reflection, but rather, as studies in developmental psychology show (McCloskey 1983; Piaget 1978), was itself grounded in intuitive physics, namely in the everyday experience of inertial motion. An object that is set in motion by a force generally continues to move for a while after the force has ceased to affect it, as if something had been transferred from the causal force to the object. Back in late antiquity this became the foundation of an adaptation of the motion-implies-force model to the phenomena of inertia so difficult to integrate into the model. This adaptation was achieved by transforming the conception of force. The force that causes a motion is no longer conceived as a power of a mover directly affecting the object, but rather as an entity that can be transferred from the mover to the object moved.[8]

This adaptation of the motion-implies-force model to phenomena of inertia by changing the variable *force* was widespread in the Middle Ages. In the early modern period it became one of the points of departure for the theories of motion of the preclassical mechanics of Galileo and his contemporaries.[9] Numerous tracts document particularly the interpretation of projectile motion as a motion that is caused by the transfer of a force to the projectile, whereby this transferred force was generally designated as an *impressed force*—*vis impressa*—or as *impetus*.

1.3.4 The Transformation of Aristotelian into Classical Physics

The differentiation of the motion-implies-force model through the transformation of the concept of force, which it involves as a variable, expanded not only its area of application, but also the complexity of possible conclusions. Depending on assumptions about the moving force, elaborate explanations of different processes of motion resulted, like that of the acceleration of the motion of free fall, of the course of collision processes, or of the shape of a projectile's trajectory.

In the early modern period, such expanded explanatory possibilities constituted the point of departure for overcoming Aristotelian patterns of explanation without having to revoke the basic Aristotelian assumptions in the beginning. For instance, if one assumes that the *impetus* impressed on a projectile is gradually used up, then one obtains an explanation for why its movement slows down. If one assumes, by contrast, that the impressed *impetus* can be used up only by counteracting forces like the resistance of a medium, then the object remains in motion as long as no such forces affect it. In this case the motion-implies-force model thus leads to conclusions that correspond to those of the law of inertia in later classical physics.

[8] The reasons for the emergence of impetus theory were complex and its application was not restricted to physical phenomena. For an overview of the various contexts of the theory, see Wolff (1978).

[9] On the concept of preclassical mechanics as a transitional phase between Ancient and classical mechanics, in which the elements of classical mechanics were elaborated under the prerequisites given by Ancient mechanics, see further Damerow et al. (2004).

In other respects as well, the change to the concept of force evoked by the expanded applications of the model created new explanatory possibilities that made manifest the transformation from Aristotelian to classical physics. Thus, for instance, the concept of a force impressed upon a moving body, which allows it to exert forces on other objects at the expense of its own motion, has the same structure as the concept of kinetic energy in classical physics, which can also be transferred from one object to another through interactions like a collision.

That the transformation of Aristotelian into classical physics did in fact take place historically through the differentiation and change of mental models can be demonstrated using the example of the motion-implies-force model discussed here. Galileo, for instance, is usually considered a pioneer of classical physics because he allegedly refuted the speculations of the Aristotelians through experiments, thereby laying the foundation for classical physics. In fact, however, in Galileo's new science of local motion, the concept of an *impetus* impressed upon a moved object and the Aristotelian differentiation between violent and natural motions continued to play a central role.[10] As for his predecessors, the combination of these two assumptions allowed him to expand the application of the motion-implies-force model from violent motion to include natural motion. In the attempt to use the means at his disposal to determine the dependency of the velocity on the slope of the plane for a ball rolling down an inclined plane by diverting its motion into a horizontal motion, he inevitably encountered an inertial problem, which prompted him to assume a further form of motion, that of *neutral* motion.

The conclusion that led him to this assumption can be summarized briefly as follows: A ball that rolls down an inclined plane is accelerated because ever more *impetus* is impressed upon it during its natural motion. A ball that is pushed to roll up an inclined plane slows down its motion, because something is constantly taken away from the impressed *impetus* by the natural motion. A ball that rolls on a plane parallel to the earth's surface, according to this explanatory pattern, neither gains nor loses *impetus*. Therefore it would have to continue moving at constant velocity along an orbit around the earth if it were not prevented from doing so by friction.[11]

[10] On the relationship between Aristotelianism and classical physics in Galileo's work, see Damerow et al. (2004, ch. 3) and Renn et al. (2000).

[11] An argument that is in striking agreement with Galileo's already appeared in Hero's *Mechanics*, where it was related to an explicit critique of Aristotle's assumption that the force required to move a body is proportional to its weight: "Some people believe that the loads lying on the Earth can be moved only by a force equivalent to them, by rendering homage to incorrect views. So let us prove that loads situated in the described manner can be moved by a force less than any known, and explain the reason for which this phenomenon is not apparent in fact" (*Mechanica* 1, 20, 10–16, translated from the German translation in Heron von Alexandria 1900). Since Hero's *Mechanics* was, before the publication in the late nineteenth century of its Arabic translation, only known in excerpts through quotations in Pappus, it appears impossible that Galileo could have known Hero's argument. In any case Pappus, in contrast to Hero, proceeds from Aristotle's assumption that a force is required for the motion of a body; see Cohen and Drabkin (1948, 194–196). Galileo first presents this argument as early as ca. 1590 in his manuscript *De Motu;* see Galilei (1960, 66–68). Around 10 years later, he repeats the same argument in connection with his derivation of the law of the inclined plane (Galilei 1960, 171–175). Here he connects it with an explicit critique of the

This is the conclusion in the context of the differentiated motion-implies-force model, which forced Galileo to assume a neutral motion, that is, a horizontal motion parallel to the earth's surface, which from the perspective of later classical physics merely represents a special case of inertial motion.

The case of the English mathematician and philosopher Thomas Harriot (1560–1621) shows that the transformation of mental models of motion through the elaboration of theories of projectile motion is a more general characteristic of knowledge development at this time.[12] In particular, Harriot's manuscripts reveal that this development can be conceived of as a process of negotiation between practical knowledge and theoretical models mediated by geometrical constructions. Harriot, just as Galileo and other early modern engineer-scientists, attempted to develop a theory of motion that would capture various of the empirical insights about projectile motion obtained through the contemporary practice of gunnery. To this end, he availed himself of pointwise constructions of projectile trajectories resulting from the combination of a violent motion caused by the propelling force of the powder and a natural motion vertically downward. The geometrical construction served the integration of the theoretical assumptions with the practical knowledge and can be seen as another change in media. In Harriot's case, it led to a weakening of the distinction between natural and forced motion and implied a shift toward a distinction between an eternal uniform motion in a straight line (i.e., an inertial motion) and the motion of fall.

Harriot constructed at least four sets of trajectories, composing them from a motion along the line of the shot and another one vertically downward. He composed them graphically by plotting the points that the motions would reach after equal intervals of time, a method corresponding to the mechanical generation of geometrical curves known from Archimedes' construction of a spiral. To carry out these constructions, Harriot had to specify the laws governing the two linear component motions. From this he obtained insights into the shapes of trajectories following from a specific set of assumptions. These shapes had to conform to the empirical knowledge about projectile trajectories, which particularly concerned the range of a shot in dependence on the angle of projection.

In what must be one of his earliest constructions, Harriot uses arithmetic laws of motion to describe the deceleration of the violent motion as well as the acceleration of the natural motion, that is, the spaces traversed by the two component motions in equal intervals of time decrease and increase according to an arithmetic sequence. The unit of the sequence applied to the motion along the line of the shot depends on the angle of projection in such a way that the resulting motion in vertical direction

Aristotelian assumption of Pappus. Later the observation that no forces are required for horizontal movement becomes an essential prerequisite for the theories developed in his two main works, the *Dialogo* and the *Discorsi*. On the one hand he draws the cosmological consequence that circular motion around a celestial body does not require a causal force; on the other hand, he uses this idea of the deflection of accelerated motion along an inclined plane into uniform motion along the horizontal to develop a measure for the changing instantaneous velocity of the falling motion, a measure that he can master theoretically and thus achieve a proof of the law of fall.

[12] For the following, see Schemmel (2008), in particular ch. 7.

obeys the times squared law. The times squared law states that, in the motion of fall, the distances traversed from the beginning of the motion are to each other as the squares of the times needed to traverse these distances. The resulting trajectories are smooth and their descending branch is steeper than the ascending one. Thus far the trajectories capture the empirical knowledge. Other crucial aspects of the practitioners' knowledge about trajectories are violated by this construction, however. In particular, the construction does not yield a projection angle of maximum range above the horizontal. What is worse, according to this construction, the range of a horizontal shot would be infinite.

In his second construction, Harriot again applied arithmetic laws to the component motions, but now modified the behavior of the range depending on projection angle in such a way that maximum range was attained for a projection angle of 45 degrees and the ranges of shots above and below 45 degrees were symmetrical about that angle. This dependence of range on projection angle was achieved only through the introduction of *ad hoc* assumptions, however. A further drawback of this construction consisted in the fact that the resulting motion in vertical direction no longer obeyed the times squared law.

It was most probably in reaction to this observation that Harriot based his later constructions on quadratic laws, that is, the component motions themselves now obey the times squared law, so that any mixture of them, and in particular the resulting motion in vertical direction, obeys this law. In Harriot's early construction using quadratic laws, the dependence of the range on the angle of projection remained an unsolved problem. He only resolved this problem when he turned to the inclined-plane conception of projectile motion. According to this conception, the dependence of the motion along the line of the shot on the angle of projection obeys the law of the inclined plane, that is, the deceleration of the upward motion is proportional to the sine of the projection angle. Projectile motion is thus conceived as a composition of an upward motion along an inclined plane and the motion of free fall.

This construction rule yields trajectories that not only take into account the smoothness and the observed asymmetry of trajectories but that also display a compelling dependence of range on angle of projection: there is an angle of maximum range between the horizontal and the upright (it lies at about 27 degrees and 55 minutes) and for shots above and below this angle, the range naturally decreases. The inclined-plane conception of projectile motion seems to have presented a satisfying construction to Harriot. In various notes he elaborated on this conception. Thus, he calculated ranges, compared them with empirical ranges given in the contemporary literature on gunnery, and considered the causes for the obvious deviations.

But Harriot's constructions show that with the inclined-plane conception of projectile motion, the clear distinction between violent and natural motions is lost. What at the outset was violent motion, the motion along the line of the shot that followed its own law of deceleration, is now itself composed of two motions: a uniform motion upward along the line of the shot and an accelerated motion downward along the same line. This latter motion is due to the projectile's gravity and may therefore be considered a natural motion. In fact, Harriot starts his later constructions by plotting a uniform motion along the line of the shot from which he then subtracts the natural motion downward along the inclined plane.[13]

Harriot's case therefore shows how the integration of refined practitioners' knowledge on projectile motion into the theoretical model of motion, mediated by geometrical constructions, ultimately changed the theoretical model, a development that, in the long run, brought about the conceptual framework of classical mechanics.

Several decades later, in his *Philosophiae naturalis principia mathematica*, Newton made the law of inertia, as a general principle of all motions, into an axiom of a deductive theory of the motion of bodies. He applied the explanatory thus modified scheme of the causation of motions through forces also to the motions of celestial bodies and was thus able to derive mathematically Kepler's laws of planetary motion.

From the perspective of classical physics, this step appears to be the ultimate victory over the Aristotelian theory of motion. Aristotle's incorrect assumption about the causation of motions was replaced by the correct one. In fact, however, like his predecessors, Newton also defined the motion of a body as caused by the impression of forces.[14] If the historical transformation of this mental model of the causation of motion through forces is reduced to Newton's simple correction of an incorrect theoretical assumption, then Newton's theoretical determination of his basic concepts through definitions taken from the theory of impetus seems an irrational and superfluous insistence on traditional ideas. However, in truth these definitions make clear the constitutive conditions of development given by the framework of the mental model within which the change in knowledge structure from Aristotelian to classical physics took place.

From the assumption that motions are generated and preserved by the impression of forces, Descartes drew conclusions that went even further. If the motion of a body acting on another body is not destroyed, but is instead merely transferred to the other body, then, according to Descartes, the total amount of motion in the world must remain constant.[15] In this basic assumption from his *Principles of Philosophy*, the conceptual structure of the conservation laws of classical physics are preshaped. However, because of its generality, this conclusion could not lead Descartes directly to a more precise determination of the impressed forces in terms of the later differentiation between the conservation of momentum and the conservation of energy.

---

[13] Harriot clearly formulates this idea when he explains: "Now I say because of the bullet's gravity the crooked line is made. If the gravity be abstracted the motion wold be only in the right line [...]; and if the resistence of the air or medium be also abstracted his motion would be infinitely onward." What distinguishes Harriot's solution from that within classical mechanics is the way the effect of gravity is introduced. In Harriot's case gravity decelerates the otherwise uniform motion along the line of the shot and it adds a uniformly accelerated vertical component to the motion. In classical mechanics it only does the latter. Conceptions similar to Harriot's can be discerned in the work of Galileo (Damerow et al. 2004, Schemmel 2006).

[14] See definitions III and IV in Newton (1999, 23f.).

[15] On this, see Damerow et al. (2004, ch. 2, especially 77–82).

Indeed, this differentiation first arose as a result of the *vis-viva* dispute, an argument about the correct measure of the living force, as the force of a moving body was generally called in the subsequent period.[16] The dispute was triggered in 1686 by a critique by Leibniz of Descartes' concept of the force impressed in the motion of a body. For Descartes this force was proportional to the velocity of the body; for Leibniz it was proportional to the square of its velocity. In 1743 d'Alembert attempted to end the long-enduring dispute, which yielded a great variety of interpretations of phenomena of motion that ultimately led to the solution of the problem, by remarking that the dispute was only an argument about words. However, just a few years later this did not prevent Kant from dealing comprehensively with the views represented in this dispute in his first philosophical work, *Thoughts on the True Estimation of Living Forces*. In fact, it were only the interpretations of processes of motion that emerged over the course of the dispute as well as the conceptual differentiations created in their sequel that finally produced the prerequisites for the formulation of the laws of conservation of momentum and of energy. Against this backdrop, in keeping with d'Alembert's dictum, the nineteenth-century dispute could then indeed be regarded as an argument over words.

As an intermediate result, this brief overview of the development of the motion-implies-force model, through which the Aristotelian theory of motion was transformed into classical physics, allows us to illustrate the achievements that can be expected from a theory of mental models for understanding the historical transformation of technical knowledge. To a history of science that takes its categories for describing advances in knowledge from a modern science like classical physics, the transition from the Aristotelian theory of motion to the dynamics of classical physics must appear as a correction to a central theorem: The "erroneous" Aristotelian assumption that the velocity of a moved body is proportional to the moving force is, in classical physics, replaced by the "correct" assumption that it is the *change* in velocity, but not the velocity itself, which is caused by a force and proportional to this force. (Let us mention in passing that, in accordance with the general theory of relativity in modern physics, this assumption, too, is valid only as an approximation.) To such a history of science, the transformation of the theory of motion over the course of history thus appears as a step-by-step correction of theorems in the framework of a fictive deductive system performed on the basis of empirical experiences. However, such a deductive system could not in fact have become explicit and function as a criterion for right and wrong before it matured into the system of classical physics.

If one assumes by contrast that the inferences associated with processes of motion are domain-specific and are drawn in the context of historically developing mental models, then the Aristotelian theory of motion can already be interpreted as a theory which supplies adequate explanations in the domain of application of the central mental model which this theory reflects. Moreover, the applications of this mental model can be interpreted as developmental conditions of the theory of

[16] On the multifarious aspects of this dispute, see Freudenthal (2002), Hankins (1965), Iltis (1970), Laudan (1968), Papineau (1977).

motion, for the inferences drawn in the context of applying the mental model change not only the contents of knowledge, but also the logical structuring of these contents. Transformation processes of knowledge thus no longer appear merely as a correction and accumulation of knowledge on the basis of new experiences. Rather, they appear as the consequence of developmental processes, especially of the external demarcation and the internal differentiation of mental models, which bring forth new structures in the framework of existing structures in order to integrate new experiences.

1.4 Networks of Mental Models

1.4.1 Forms of Networking Mental Models

To illuminate the potential of describing and explaining the historical transformation of knowledge adequately by means of a theory of mental models, we have concentrated on a single model. In fact, however, inferences are drawn within a network of such models, which may be combined in various ways and thereby shape thinking. Due to their interaction, they exhibit a richer spectrum of developmental mechanisms than those considered so far, that is, the external demarcation and the internal differentiation of single mental models.

For purposes of simplification and clarity, we have in our account so far neglected, for example, the fact that the changing ideas of the forces that cause motion were themselves the result of a change in mental models. As a matter of fact, the variable of the force exerted by the mover was historically instantiated not by an isolated item of information, as we have suggested for the sake of simplicity, but rather by a mental model of forces acting as possible causes. This is an example of the most important form of the networking of mental models. The variables of such a model can themselves be determined by mental models, and thus inserted into a network of mental models. One of the developmental mechanisms of mental models consists in the changes that are a consequence of the change in their variables due to the change of the mental models that determine those variables.

Another form of networking is the superposition of various mental models to generate more complex models. For instance, models of material constraints of motions can overlay the motion-implies-force model. This applies, for instance, in case of the rotation of an object that is held together by the constraining forces of material strength, similar to the case of a swinging pendulum, which is held on a spherical surface by the constraining force of a string, and to case of constraint motion of an object along prescribed paths, such as motion on an inclined plane or on rails.

In all of these cases the process of motion can be conceived both as a motion caused by a force, and as a motion that takes place along a certain path because of special material circumstances. The motion can thus be interpreted in terms of two mental models at the same time. One is the motion-implies-force model with the essential variables *mover*, *moved object* and *motion performed*. The second is the model of constraint motion, for instance a rotation with the variables *moved object* and *rotational axis*. Through the superposition of the two models a new mental model emerges, with a restricted area of validity and the essential variables *mover*, *moved object*, *axis of rotation* and *motion performed*.

1.4.2 Sketch of a Theory of the Historical Transformation of Mechanical Knowledge

The complexity of the possibilities for explaining the historical developments of knowledge that emerges from this networking of mental models can be illustrated by the most important line of development, which extends from the beginnings of mechanics in antiquity via its transformation to the classical mechanics of the modern period.

As has been argued in the beginning, mechanical knowledge is much older than the beginnings of theoretical mechanics. The early civilizations in Mesopotamia and Egypt, for instance, already used most of the mechanical devices that later became known as *simple machines*.[17] The inclined plane, for instance, is represented by the enormous ramps needed to produce large-scale buildings. The pulley served to lead and redirect hauling ropes to transport heavy loads, and to realize mechanical devices like the locking mechanisms of the Egyptian royal tombs. The lever existed in the form of tools like the digging stick, as an element in the construction of wine and oil presses, and as an aid in moving and positioning heavy loads. What these tools have in common is that, in contrast to the assumption of the motion-implies-force model that equal forces cause equal motions, they have a force-saving effect. The use of these tools by professionalized craftsmen and engineers was based on mental models specific to these tools, the function of which was to optimize the relation between the exerted force and the effective force.

Mental models of this kind can be regarded as typical for the knowledge of practitioners, acquired through participation in work processes and passed down through the ages, the theory of which is quite inadequately described theoretically by the standard pejorative designation as "empirically acquired rule-based knowledge." Mental models of this kind are distinguished from the motion-implies-force model, which is based on universal conditions for action, in that they are linked to material conditions that change historically, and as such are subjected to a different kind of developmental dynamics than are models that are affected only slightly or not at all by historical changes in framing conditions. The transformation of ancient technical knowledge in the Middle Ages and early modern period can be traced back to a large degree to such changes in historically conditioned mental models of practical knowledge.

[17] See, for instance, Russell (1991) and Arnold (1991).

Also in the case of these mental models, whose emergence is historically conditioned, philosophers of Greek antiquity explicated the inherent conclusions in a meta-language. This reflection yielded the concept of mechanics, which signalized the emergence of a generalized mental model of force-saving tools. At the same time, theoretical mechanics emerged as a discipline of its own besides natural philosophy.

The oldest evidence of this development is Aristotle's *Mechanical Problems*.[18] Since the nineteenth century the traditional attribution of this text to the body of Aristotle's work has been questioned, but even if it might have been written by one of his pupils, there is no doubt that it constitutes an earlier form of the theorization of the mental models of the use of tools than is known from other surviving mechanics treatises of antiquity, especially those by Archimedes and Heron.

The point of departure for the theorization of the use of mechanical tools in the Aristotelian text was the invention of the balance with unequal arms. By equating the lever with the beam of a balance, this device allowed the connection between applied force and effective force to be made comprehensible in the sense of an equivalence of cause and effect. The balance with unequal arms made apparent the precise conditions under which differences in force could be compensated for by differences in length (Damerow et al. 2002).

The persistent theme of Aristotle's text is the question of why a small force is able to overcome a greater one. This question is raised at the beginning of most of the problems, which define sections of the text. In the majority of the mechanical devices in the focus of a problem the answer is given by equating the device with a lever, in which individual elements are identified respectively as a lever beam, as a pivot, or as applied and effective forces. Then the shared principle is applied according to which a greater radius moves faster than a smaller radius in a circular motion caused by the same force. Here the lever equated with the beam of a balance functions as a mental model of any force-saving device. As such, the model thus integrates all of the individual models of mechanical devices into the framework of a generalized structure of knowledge.

In antiquity, in the Middle Ages, and even in the early modern period, this structure of knowledge became the point of departure for all treatises on theoretical mechanics (Clagett 1959; Moody and Clagett 1952; Rose and Drake 1971). The central theorem of all of these treatises is the law of the lever, which results from a differentiation of the Aristotelian principle for the application of forces of different magnitudes. While the Aristotelian principle referred to the different effects of equal forces depending on the pivot of a circular motion, starting from Archimedes' text on the equilibrium of planes (Heath 1953, 189–220), this principle now became the starting point for deriving the conditions under which different forces could be brought into equilibrium.

This further-reaching theorization was accompanied by a further-reaching generalization. Closely linked with the law of the lever is Archimedes' conclusion that every object has a point, known as the center of gravity, relative to which it finds

[18] Aristotle (1980). On the dating of this text, see Krafft (1970). See also Chapter 5 in this book.

equilibrium in any position. The mental model of the lever—more precisely, of the lever equated with the beam of the balance—thus becomes a mental model applicable to every object, the *center-of-gravity model.*

An example for the power of this generalized model is presented by its complex application, passed down through Heron and attributed to Archimedes, to determine the load supported by columns that carry a beam with or without additional, suspended loads. The procedure to calculate such loads is based on a multiple application of the model. In the simplest case, to which the more complicated building construction can be traced back, a beam is supported equally by two columns. The weights of the beam and of the loads located upon it can be thought of as concentrated at the center of gravity, and each of the columns as the fulcrum of a lever through which this weight is transferred from one column to the other.[19]

The actual potential for application and development of the mental model of the lever generalized to a center of gravity was realized, however, only in the context of the technical challenges faced by Renaissance and early modern engineers and the expanded applications they developed. In Galileo's foundation of a theory of the strength of materials, the first of his two new branches of knowledge concerning mechanics and the law of falling bodies,[20] the mental model of the generalized lever played a central role. He derived the lateral load capacity of a beam fixed on one end from its tensile strength by conceiving of the beam as a bent lever whose fulcrum is located at the lower edge of the breakage. According to this model, two forces are exerted perpendicularly on the imagined lever, the load and the tensile strength of the material. Galileo defined the load as the total weight at the beam's center of gravity plus the suspended weight. The tensile strength was defined as an area force exerted on the center of gravity of the fracture plane, the size of which can be determined empirically. If both forces are in equilibrium, the limit is reached beyond which the beam would break under further load.

In addition to innovative applications of traditional mental models, for which Galileo's mechanics of materials serves as an example, new mental models emerged in many areas of knowledge even back in the Middle Ages and increasingly in the early modern period, like, for instance, the model of momentary velocity and the model of an accelerated motion characterized by a changing momentary velocity.

---

[19] The problem of determining the weight posed on two columns had already been discussed by Aristotle in his *Mechanical Problems*. Problem 29 (Aristotle, *Mechanica*, 857b, 9–20) asks about the relative load on the shoulders of two men carrying a load using a carrier bar when the load is not hung from the middle of the bar. The answer is derived from the interpretation of the carrier bar as a lever with its fulcrum at the point from which the load is suspended. This solution has been criticized from a modern perspective, see Heath (1949, 252f.). In his comprehensive discussion of numerous variations of the problem, Heron (*Mechanica* 1, 25, 8–13), invokes a treatise by Archimedes entitled *The Book of Columns*, which did not survive and about which nothing more is known. He emphasizes that he adopts from the treatise only that which is useful for the student to perform the calculations. In fact, because his depiction largely omits the proof of the procedure, it cannot be traced back to the theory of the lever with any certainty.

[20] Galilei (1638). The mechanics of materials is the subject of the first two days of the *Discorsi*. On the historical placement of Galileo's theory of the strength of materials, see Portz (1994).

The mental models of antiquity, from which the emergence of classical physics originated, were confronted with an ever-expanding application horizon in the course of the emergence of classical mechanics in the seventeenth century, the age of Newton, and in the eighteenth century, the age of his successors. But in the confrontation with challenging objects like firearms, the pendulum and elastic springs, they, too, increasingly were networked with new mental models. The temporary end of this process was marked by the emergence of analytical mechanics, the foundation of which was constituted by a novel system of mental models.

The mental models used in analytical mechanics show some marked differences to those that resulted from the reflection of the intuitive knowledge of practitioners in preclassical mechanics. A search for generalization and abstraction was, of course, already inherent in earlier mechanics. When Archimedes reduced mechanical problems to the law of the lever, this implied that the mental model of a lever was used in situations where its application had not been obvious. In this way, it was stripped of specific conditions of applicability (such as the lever being a simple beam with a fixed pivot) and generalized into a more abstract model. Analytical mechanics, however, added an essential new aspect to this process of abstraction and generalization: its defining feature was the translation of mechanical problems into algebraic or differential equations, going beyond geometrical and constructive intuitions. Lagrange, in the introduction to his *Mechanique Analytique*, famously announced that no diagrams were to be found in his book. The universality of the analytical method made the help of intuitive representations unnecessary.

Analytical mechanics subsumed the specific mental models of traditional mechanics under much more abstract models. One of its central models, for instance, allowed natural occurrences to be traced back to minimum principles. In his work on the shape of a suspended chain, Galileo made use of the insight that the chain takes on a form in which its center of gravity is located at the lowest possible point.[21] Fermat formulated the minimum principle that a beam of light refracted by media will always seek the path with the shortest transit time. As different as the subject appears to be in these two cases, they both follow the same model, which Maupertuis called the *principle of least action*. According to this initially quite abstract model, nature always strives to proceed as economically as possible. In the subsequent period, this principle was concretized for ever more application areas and received a precise mathematical form, primarily through the works of Euler and Lagrange. As *Hamilton's principle*, it became a general model of physical systems from which the basic laws of classical physics can be deduced.

However the use of such generalized models required another level of even more abstract mental models embodied in a new conception of mathematics. The most prominent example for this is the mental model of a *function* emerging in Euler's work on the calculus of variations (Euler 1744, 1748). The geometrical model of an orbit or a trajectory of a material body became a single instance of a continuum of

[21] This insight by Galileo is not found in his published works, but from his manuscripts it turned out that Galileo knew the minimum principle and was skilled in its use. See Renn et al. (2000, 374–383).

possible paths, no longer defined by a specific geometric construction or algebraic expression, but merely by an unspecified analytical expression $y = f(x)$. The development of a consistent mathematical theory of functions remained a challenge for mathematics for centuries and only came to a (preliminary) conclusion in the functional analysis of the twentieth century. Characteristically for mental models, however, this did not impede the successful use of the model within analytical mechanics and its eventual extensions to field theory and quantum theory. Rather, the model of a function became the central tool for incorporating more specific and intuitive mental models into a mathematical theory of mechanics.

The mental models of mechanics hence were connected into a network not only horizontally, as discussed above, but also vertically, by subsumption under more abstract mental models. The motion of a projectile can be understood as an instance of a principle of least action, and this principle in turn can be seen as an instance of a functional calculus. The mathematization of mechanics therefore leads to a much more complex networking of models than mere physical considerations would have. It also changed the dynamics of the development of these mental models, since the knowledge embodied in them was no longer limited to empirical knowledge taken from the manipulation of physical objects. Rather, the practice of mathematics, such as methods of calculation and proof, became essential for the development of these models. Not surprisingly, the central actors in the history of analytical mechanics were mathematicians, and its development occurred in parallel with the formation of the modern form of notation in analysis.

In turn, the mathematized mechanics inspired the application of the model of functional dependence on other physical phenomena and so led to new mathematical theories of heat starting in the work of Laplace and Lavoisier, of electricity and magnetism in the work of Coulomb and Ampère, and of ray optics in the work of Hamilton, which in turn led to a new and more powerful version of the principle of least action in mechanics. Similar to the way the mental models of ancient mechanics shaped the emergence of classical mechanics in the early modern period, the mental models of analytical mechanics determined the flourishing analytical sciences of the nineteenth and some of the twentieth century as well. The mental model of the principle of least action even survived the revolution from which the modern physics based on quantum theory and the theory of relativity emerged. Thus, for instance, the structure of stable atoms is explained today by the fact that their ground state is characterized by minimum energy, and the geodesics of the theory of relativity, along which objects move in curved spacetimes, are, like the straight lines of classical physics, the curves of least curvature.

This short sketch of an explanation of the development of mechanical knowledge from the implicit knowledge of practical mechanics up to the replacement of the basic concepts of classical mechanics by the relativistic concepts of space, time and matter of modern physics, shows the importance of relatively simple mental models that span across such developments. This sketch proceeded from the question as to how technical knowledge is transferred historically. The conception of mental models offers a new methodological approach to answer this question, because mental models represent cognitive instruments of the transformation of such knowledge.

A particular strength of this methodological approach is its power for analyzing the interactions among different layers of knowledge, in particular the intuitive, practical, and theoretical layers introduced in the beginning. In this way, questions such as the question of which aspects of knowledge are universal and which may change with culture and history, or the question of the impact of technology on the development of scientific thinking, or that of the conditions of the applicability of theoretical knowledge in practice can be addressed in novel ways.

Another strength is the new perspective that this approach offers on the issue of theory change or rather that of the developmental processes leading from one system of knowledge to another. Here we have seen that classical mechanics did not arise as an alternative to, and rejection of, the Aristotelian system of knowledge. Rather it was only the exploration of the mental models underlying the latter system of knowledge in the context of new domains of application that triggered a transformation process eventually bringing about a new system of knowledge. As we have seen, this transformation process consisted in external demarcations, internal differentiations and networking of mental models. In its course, the developing system of knowledge displayed tensions and ambiguities that created both the possibility and the necessity for separating a newly emerging knowledge structure from the roots of the extant system of knowledge.

References

Aristotle. 1980. Mechanical problems. In *Minor works*, 331–411. Transl. W.S. Hett. Cambridge, MA: The Loeb Classical Library.

Arnold, Dieter. 1991. *Building in Egypt: Pharaonic stone masonry*. New York: Oxford University Press.

Besnard, Philippe. 1989. *An introduction to default logic*. Berlin: Springer-Verlag.

Brewka, Gerhard. 1991. *Nonmonotonic reasoning: Logical foundations of commonsense*. Cambridge: Cambridge University Press.

Charniak, Eugene. 1972. *Toward a model of children's story comprehension*. Technical Report 266. Cambridge, MA: Artificial Intelligence Laboratory, Massachusetts Institute of Technology.

Clagett, Marshall. 1959. *The science of mechanics in the Middle Ages*. Madison: University of Wisconsin Press.

Clement, John. 1983. A conceptual model discussed by Galileo and used intuitively by physics students. In *Mental models*, ed. Dedre Gentner and Albert L. Stevens, 325–339. Hillsdale: Erlbaum.

Cohen, Morris R., and I.E. Drabkin. 1948. *A source book in Greek science*. New York: McGraw-Hill Book Co.

Damerow, Peter, and Wolfgang Lefèvre. 1980. Die wissenschaftshistorische Problemlage für Hegels 'Logik'. In *Hegel-Jahrbuch 1979*, ed. W.R. Beyer, 349–368. Meisenheim am Glan: Anton Hain.

Damerow, Peter, Jürgen Renn, Simone Rieger, and Paul Weinig. 2002. Mechanical knowledge and Pompeian balances. In *Homo faber: Studies on nature, technology, and science at the time of Pompeii*, ed. Jürgen Renn and Giuseppe Castagnetti, 93–108. Rome: L'Erma di Bretschneider.

Damerow, Peter, Gideon Freudenthal, Peter McLaughlin, and Jürgen Renn. 2004. *Exploring the limits of preclassical mechanics – A study of conceptual development in early modern science: Free fall and compounded motion in the work of Descartes, Galileo, and Beeckman*. 2nd ed. New York: Springer.

Davis, Robert B. 1984. *Learning mathematics: The cognitive science approach to mathematics education*. London: Croom Helm.
Dijksterhuis, E.J. 1956. *Die Mechanisierung des Weltbildes*. Berlin: Springer.
———. 1986. *The mechanization of the world picture: Pythagoras to Newton*. Princeton: Princeton University Press.
Drabkin, Israel Edward. 1938. Notes on the laws of motion in Aristotle. *American Journal of Philology* 59: 60–84.
Euler, Leonhard. 1744. *Methodus inveniendi lineas curvas maximi minimive proprietate gaudentes, sive solutio problematis isoperimetrici lattissimo sensu accepti*. Lausanne: Bousquet.
———. 1748. *Introductio in analysin infinitorum*. Vol. 2. Lausanne: Bousquet.
Fillmore, Charles J. 1976. Frame semantics and the nature of language. *Annals of the New York Academy of Sciences: Conference on the Origin and Development of Language and Speech* 280: 20–32.
———. 1985. Frames and the semantics of understanding. *Quaderni di Semantica* 6: 222–254.
Freudenthal, Gideon. 2002. Perpetuum mobile: The Leibniz-Papin controversy. *Studies in History and Philosophy of Science* 33: 573–637.
Galilei, Galileo. 1638. *Discorsi e dimostrazioni matematiche intorno à due nuove scienze*. Leiden: Appresso gli Elsevirii.
———. 1960. *On motion and on mechanics: Comprising De Motu (ca. 1590)*. Translated with introduction and notes by I. E. Drabkin, and *Le Meccaniche* (ca. 1600), translated with introduction and notes by Stillman Drake. Madison: University of Madison
Gentner, Dedre, and Albert L. Stevens, eds. 1983. *Mental models*. Hillsdale: Erlbaum.
Hankins, Thomas L. 1965. Eighteenth-century attempts to resolve the *vis viva* controversy. *Isis* 56: 281–297.
Heath, Thomas Little. 1949. *Mathematics in Aristotle*. Oxford: Clarendon Press.
———, ed. 1953. *The works of Archimedes*. New York: Dover Publications.
Heron von Alexandria. 1900. *Mechanik und Katoptrik*. Transl. L. Nix and W. Schmidt. Leipzig: Teubner.
Iltis, Carolyn. 1970. D'Alembert and the vis viva controversy. *Studies in History and Philosophy of Science* 1: 135–144.
Johnson-Laird, Philip N. 1983. *Mental models*. Cambridge, MA: Harvard University Press.
———. 1989. Mental models. In *Foundations of cognitive science*, ed. M.I. Posner, 469–499. Cambridge, MA: MIT Press.
Krafft, Fritz. 1970. *Dynamische und statische Betrachtungsweise in der antiken Mechanik*. Wiesbaden: F. Steiner.
Kuhn, Thomas S. 1962. *The structure of scientific revolutions*. Chicago: University of Chicago Press.
Laudan, Larry. 1968. A postmortem on the vis viva controversy. *Isis* 59: 130–143.
Lehrer, Adrienne, and Eva Feder Kittay, eds. 1992. *Frames, fields, and contrasts: New essays in semantic and lexical organization*. Hillsdale: L. Erlbaum Associates.
Maier, Anneliese. 1949. *Die Vorläufer Galileis im 14. Jahrhundert*. Rome: Edizioni di Storia e letteratura.
———. 1952. *An der Grenze von Scholastik und Naturwissenschaft*. Rome: Edizioni di Storia e letteratura.
McCloskey, Michael. 1983. Naive theories of motion. In *Mental models*, ed. Dedre Gentner and Albert L. Stevens. Hillsdale: Erlbaum.
Minsky, Marvin. 1975. A framework for representing knowledge. In *The psychology of computer vision*, ed. Patrick Henry Winston, 211–276. New York: McGraw-Hill.
———. 1985. *The society of mind*. New York: Simon and Schuster.
———. 1990. *Mentopolis*. Stuttgart: Klett-Cotta.
Moody, Ernest A., and Marshall Clagett. 1952. *The medieval science of weights: Scientia de ponderibus: Treatises ascribed to Euclid, Archimedes, Thabit Ibn Qurra, Jordanus De Nemore and Blasius of Parma*. Madison: University of Wisconsin Press.

Newton, Isaac. 1999. *Die mathematischen Prinzipien der Physik*. Berlin: De Gruyter.

Papineau, David. 1977. The *vis viva* controversy. *Studies in History and Philosophy of Science* 8: 111–142.

Piaget, Jean. 1978. Die historische Entwicklung und die Psychogenese des Impetusbegriffs. In *Die Psychologie des 20. Jahrhunderts, vol. 7. Piaget und die Folgen*, ed. Gerhard Steiner, 64–73. Zurich: Kindler.

Piaget, Jean, and Rolando Garcia. 1989. *Psychogenesis and the history of science*. New York: Columbia University Press.

Portz, Helga. 1994. *Galilei und der heutige Mathematikunterricht: Ursprüngliche Festigkeitslehre und Ähnlichkeitsmechanik und ihre Bedeutung für die mathematische Bildung*. Mannheim: B.I. Wissenschaftsverlag.

Renn, Jürgen, and Peter Damerow. 2007. Mentale Modelle als kognitive Instrumente der Transformation von technischem Wissen. In *Übersetzung und Transformationen*, ed. Hartmut Böhme, Christof Rapp, and Wolfgang Rösler, 311–331. Berlin: De Gruyter.

Renn, Jürgen, Peter Damerow, and Simone Rieger. 2000. Hunting the white elephant: When and how did Galileo discover the law of fall? *Science in Context* 13: 299–419.

Rose, Paul Lawrence, and Stillman Drake. 1971. The pseudo-Aristotelian *Questions of mechanics* in Renaissance culture. *Studies in the Renaissance* 18: 65–104.

Russell, John Malcolm. 1991. *Sennacherib's palast without rival at Nineveh*. Chicago: University of Chicago Press.

Schemmel, Matthias. 2006. The English Galileo: Thomas Harriot and the force of shared knowledge in early modern mechanics. *Physics in Perspective* 8 (4): 360–380.

———. 2008. *The English Galileo: Thomas Harriot's work on motion as an example of preclassical mechanics, 2 vols. Boston Studies in the Philosophy of Science*. Vol. 268. Dordrecht: Springer.

Ueberweg, Friedrich. 1882. *System der Logik und Geschichte der logischen Lehren*. 5th ed. Bonn: Adolph Marcus.

Wilson, Robert A., and Frank C. Keil, eds. 1999. *The MIT encyclopedia of the cognitive sciences*. Cambridge, MA: MIT Press.

Wolff, Michael. 1978. *Geschichte der Impetustheorie: Untersuchungen zum Ursprung der klassischen Mechanik*. Suhrkamp: Frankfurt am Main.

Chapter 2
Preclassical Mechanics in Context: Practical and Theoretical Knowledge Between Sovereignty, Religion, and Science

Rivka Feldhay

Abstract This chapter delineates a map of mutual dependencies between sovereign rulers, religious establishment, scientists, engineers as well as artists in the early modern Catholic world, circa 1550–1650. The juridical structure of sovereign states is the subject of the first section. The effect of military and economic pressures is discussed in the second section and the disciplining power of religion in the third. The fourth section shortly presents a case study that exemplifies the way traditional mechanical knowledge was transformed and also how it operated in the service of state and church.

Keywords Preclassical mechanics · Sovereignty · Mental models · Jean Bodin · Giovanni Botero · Jesuits

2.1 Introduction

From the point of view of traditional historiography of the Scientific Revolution, the physics engaged in by prominent historical figures—from Galileo's terrestrial mechanics to Kepler's celestial mechanics, from Descartes' mechanical philosophy to Newton's astrophysics—was perceived as the core and model of the "new science." In such historiographical context the heart of the "new science" seemed to be a coherent body of knowledge called "classical mechanics," which represented a radical break with the antecedent Aristotelian physics predominant since antiquity. Historiography of early modern science of the last twenty years, however, gave birth to the investigation of a variety of discourses on mechanics that existed already in

The original version of this chapter was revised. A correction to this chapter is available at https://doi.org/10.1007/978-3-319-90345-3_11

R. Feldhay (✉)
Cohn Institute for the History and Philosophy of Science and Ideas
& Minerva Humanities Center, University of Tel Aviv, Tel Aviv-Yafo, Israel
e-mail: feldhay@post.tau.ac.il

R. Feldhay et al. (eds.), *Emergence and Expansion of Preclassical Mechanics*,
Boston Studies in the Philosophy and History of Science 270,
https://doi.org/10.1007/978-3-319-90345-3_2

antiquity and late antiquity, were partly revived in medieval universities, and rapidly proliferated in early modernity (Laird and Roux 2008; Meli 2006).

This chapter purports to show that a version of mechanical knowledge, referred to here as preclassical mechanics, which is not reducible to Newtonian physics yet was rich in new results, gained a certain degree of hegemony circa 1550–1650. This thesis is anchored in an account spanning a long period of the emergence of theoretical knowledge about mechanical phenomena from everyday experiences, the accumulation of practical know-how concerning tools and machines, and reflection and subsequent conceptualization of intuitive explanations for such knowledge (Chap. 1 in this volume). From this long-term perspective, preclassical mechanics is seen as yet another transformation of old "mental models" enriched by emerging technological innovations, new practices and discourses about them, and a fruitful encounter between ancient intellectual traditions within new sites of knowledge.

A variety of natural and artificial phenomena, such as free fall, percussion and oscillation were investigated in preclassical mechanics using concepts such as center of gravity, local motion, acceleration and early formulations of the concepts of momentum. Preclassical mechanics used heterogeneous resources such as Aristotelian natural philosophy, Archimedean statics as well as sophisticated logical and mathematical techniques developed within the medieval tradition of the *calculatores*. The incorporation of all these elements underscores the role played by ancient and medieval science in preclassical mechanics. At the same time, these investigations reveal how, in the constellation of interest here, a discourse on preclassical mechanics distinguished itself from "the art of mechanics" in the old sense—namely the practices and know-how related to machines—but also from Aristotelian physics, the medieval science of weights and Archimedean statics. It is in this framework that a new theory of motion—including that of machines—started to emerge in a broad context that incorporated cosmological and astronomical issues as well as a theory of matter. This new theory also entailed preoccupation with questions of causality, the role of mathematics and the concept of nature. Finally, an additional feature of preclassical mechanics is revealed in the efforts invested by its carriers to incorporate practical knowledge associated with contemporary mechanical technology and related to such objects as the pendulum, the flywheel and the trajectory of cannonballs, raising questions about the very meaning of weight and force in machines and other constructions.

Preclassical mechanics was first advanced by a new class of engineer-scientists, many of them Italian (Bennett 2011; Valleriani 2010, 2017). It was diffused by way of various knowledge sites in Europe and beyond, including building sites, workshops and arsenals, universities, academies and courts. Some of these, such as the universities, were traditional centers for the production and diffusion of knowledge, which nevertheless underwent transformations in the early modern era (Feingold and Navarro-Brotons 2006). Others, such as the academies, were new types of institutions (Boschiero 2007). Still others, such as the court or the arsenal, were for the most part not conceived as sites of knowledge by traditional historiography of science, yet came to assume importance in more recent forms of such historiography (Biagioli 1993).

Undoubtedly, the configuration of a new map of sites of knowledge in early modernity has drawn historians' attention to the type of political regimes that tended to patronize and make use of this newly transformed knowledge about mechanical phenomena. It has broadened the gaze of historians toward including in their analyses the military and economic needs that enhanced the relevance of mechanical knowledge in its broad sense—namely not simply the sciences of motion and machines but also the manual arts and sciences such as agriculture, geology, architecture, fortification as well as painting and sculpture (Bennett 2006). It has also encouraged awareness of the role of religion in disseminating new types of knowledge while still attempting to control and discipline their use (Feldhay 2011). Such a broad, contextual view is needed if the institutional conditions in which preclassical mechanics was embedded are to be clarified.

Preclassical mechanics emerged in the bosom of the "sovereign state"—a new type of political entity born out of the political and religious upheavals of the long sixteenth century, incorporating Roman and ecclesiastical legal discourses and theological conceptions of temporal and divine power (Kalmo and Skinner 2010). By the second half of that century, Europe was flooded with literature concerning political matters (Birely 1981). This literature—initiated by Machiavelli's radical texts—comprised responses by jurists and humanists engaged in political affairs, like Bodin and Botero, academic neo-stoic philologists like Lipsius, neo-scholastic theologian-philosophers like Suarez and Mariana, Jesuit humanists-historians like Possevino and Ribadeneira and high-ranked statesmen-intellectuals like Cardinals Bellarmine and Richelieu. All of these writers offered a broad vision of the sovereign state. First and foremost, they imagined the role of the ruler as legislator: the maker of law. They then analyzed the ruler's interests and the resources of his force (*potenzia*)—primarily military, but also economic. Finally, they presented the sovereign's power anchored in his virtu` and reputation which deepened his legitimacy. Simultaneously, these texts also testify to the constraints on sovereign power, despite the pretense to omnipotence or absolute sovereignty. The sovereign—a representative of either God, or of his own will, or of "the people"—was bound to depend on his image produced through public representations, as well as on the support of his people. Legitimacy was also thought to be gained through the cultivation of religion and (mainly but not only) through religion-oriented education in the arts and sciences. These sources thus expose the intricate structure through which political and religious ties were interwoven during the period later named "Baroque." These relationships constituted the pattern of command and obedience within the territorial boundaries of a state, but also constituted an international system of sovereigns states, whose rulers were free to do whatever they saw fit within their boundaries but were committed to respect the same rights for other sovereigns as well.

In the following I would like to demonstrate in what sense the new political-institutional conditions allowed for or enhanced the emergence, expansion and success of preclassical mechanics circa 1550–1650, but also bounded science, which lacked an autonomous place of its own, within a set of dependencies and constraints. This analysis, anchored in contemporary as well as modern historical literature, seeks to exemplify the kind of contextualization that takes into account the necessities created by the new political-institutional structures of the sovereign state on what is termed here preclassical mechanics. Each of the following sections will deal with one dimension of this sovereign-political context: the juridical structure of sovereign states (1), the effect of military and economic forces (2), religion and its power of disciplining and educating (3) and finally, the way traditional mechanical knowledge was transformed but also operated in the service of state and church, using the example of the Jesuits in Parma (4).

2.1.1 The Sovereign State as Paradoxical Juridical Structure: Seeking Self-Legitimation Through New Knowledge-Sites

Jean Bodin's *Six Books of the Commonwealth* (Bodin 1967) is one of those rare texts that both reflect the events at a particular moment in history and simultaneously take part in shaping the historical reality in which the text will be read and have continuous impact. Bodin substituted old questions about the powers that rule—temporal or divine, secular or sacred sword, emperor or pope—with new questions about the state: what is it and how does it operate? For Bodin, the state has a telos, which is order and security. Hence he defines the state as "the rightly ordered government of a number of families and of those things which are their common concern by a sovereign power." Order and sovereign power are thus intrinsically and necessarily bounded in Bodin's text. Furthermore, sovereign power appears to be the sole alternative to chaos, anarchy and total lack of security, for "before such things as cities and citizens, or any form of commonwealth whatsoever, were known among men" (Bodin 1967, 10), the strong fight the weak and enslave them. And so the sovereign state must be the guarantee for the existence of human society in conditions of peace and security. Bodin's text is not just formal, as may appear from his definition of the commonwealth. No less legal and scientific, it is also historical and humanistic. His definition of the commonwealth as being composed of a few families, and his subsequent discussion of families as the ground and basis of the state allude to the meaning and goal of his text: to embed the discourse on the sovereign state within a natural kind of authority and to legitimate its power through an analogy to the natural authority of the *pater familiae*, the father to whom all members of the family owe obedience.

But what does sovereign power mean? And who is a sovereign? "Sovereignty," according to Bodin, "is that absolute and perpetual power vested in a commonwealth" (Bodin 1967, Book I, ch. viii). Bodin argues that a definition of real sovereignty does not exist in any civil law (meaning Roman civil law), even though this term is the most fundamental one for any treatment of politics. He thus claims urgency for his discussion and invests tremendous effort in articulating the uniqueness, unshared nature and unlimited power of the sovereign through the terms "absolute" and "perpetuate." The sovereign is the law-maker: "A law proceeds from him who has sovereign power, and by it he binds the subject to obedience, but cannot bind himself." This is the essence of the "absolute." Before Bodin's articulation, God was conceived as the source of all laws pertaining to man and nature (Elshtain 2008). Whether through His reason and intellect, as the Augustinian-Thomistic tradition would have it, or through an act of His will, as fourteenth-century voluntarists believed later on, God was the legislator of the law which human beings—kings, estates, jurists, scholars—had "to find" "to discover," to study and apply. God was also free to abide by his law—his "*potentia ordinata*"—or not, thus applying his "*potentia absoluta*." Late medieval Popes tried to appropriate this Godly power for themselves, developing a notion of the Pope's "*plenitudo potestatis*" in temporal as well as divine matters (Elshtain 2008). But Bodin directly attributed "absolute power" to "the Sovereign, a human being assuming Godly power that is unbounded. For the King—like God—cannot bind himself to his own law" (Bodin 1967, Book I, ch. viii), just as the Pope can never tie his own hands, according to the canonists. Otherwise the sovereign becomes like any other subject owing obedience to another, which contradicts the absoluteness of his sovereign power: "The true sovereign remains *always* seized of his power" (Bodin 1967, Book I, ch. viii). Thus, sovereign power is "perpetual." It persists in time, as well as in space, claiming its right to obedience from subjects without leaving them a right of resistance, for "sovereign power given to a prince charged with conditions is neither properly sovereign nor absolute" (Bodin 1967, Book I, ch. viii).

A historical understanding of the phenomenon of the sovereign state in terms of "all power is conveyed to him and vested to him" [*Ei et in eum omnem potestatem contulit*] is, however, not enough. While the notion of sovereignty indeed entails the ruler being attributed with inhuman omnipotence, it also creates a new field of dependencies between ruler and subject, structured by the need to *mediate* the huge gap opened between them. Bodin continues his vision of the state, arguing that:

> besides sovereign power there must also be something enjoyed in common such as the public domain, a public treasury, the buildings used by the whole community, the roads, walls, squares, churches, and markets, as well as the usages, laws, customs, courts, penalties, and rewards which are either shared in common or of public concern. There is no commonwealth where there is no common interest. (Bodin 1967, Book I, ch. ii–v)

Here we find the new space between sovereign and subject. In principle, it is a space governed according to the same hierarchical pattern of relationship between ruler and ruled, granting the first the right to obedience, and imposing on the second the duty to obey. Thus, "the government of all commonwealths, colleges, corporate bodies, or households whatsoever, rests on the right to command on one side, and the obligation to obey on the other" (Bodin 1967, Book I, ch. ii–v). Yet the sovereign's law does not exhaust this relationship, for "the absolute power of princes and sovereign lords *does not* (author's emphasis) extend to the laws of God and of nature"; and also: "a law and a covenant must …not be confused" (Bodin 1967, Book I, ch. viii). All sorts of covenants regulate public and private life in relation to the ruler, a covenant meaning "a mutual undertaking between a prince and his subjects, *equally binding on both parties*, and neither can contravene it to the prejudice of the other, without his consent. The prince *has no greater privilege* than the subject in this matter" (Bodin 1967, Book I, ch. viii). Furthermore, the sovereign, invested though he is with inhuman power, is in fact more bound to the laws of God and nature. Through these, he is obliged to provide justice and protection for his subjects, and generally support their interests. "The prince is obliged to safeguard the persons, possessions, and families of his subjects, by force of arms, and by force of law" (Bodin 1967, Book I, ch. vi and vii); and moreover: "the bare word of a prince should be as sacred as a divine pronouncement. It loses its force if he is ill-thought of as one who cannot be trusted except under oath, nor relied on to keep a promise unless paid to do so" (Bodin 1967, Book I, ch. viii). Hence a prince cannot go back on his promises and obligations, since he cannot afford to lose his credibility in the eyes of his people.

The political structure delineated in Bodin's text is precariously located between divine omnipotence and human fragility, revealing a desperate need for order and security and the difficulty in attaining it in the here and now. It is thus marked by the paradoxes and ambivalences that haunt the work of historians, political scientists and jurists who find themselves in endless debates about its significance, both in the past and in the present. While considered by some as the only solution available at the time for a Europe torn by civil and religious wars and ravaged by epidemics and famine, this political structure is considered by others as part and parcel of a Europe seized by power fantasies, which brought about more destruction than redemption. None of these interpretations, however, are able to throw further light on the problems that are central to this paper, namely the fact that the paradoxical structure of sovereignty—the pretense to omnipotence as a response to an actual effective weakness—became favorable to the development of a specific culture of the arts and sciences.

The political pragmatism of the time sought to stabilize the ruling structure through two *mitigating channels*: first, by balancing extreme claims to omnipotent, absolute power with sheer physical force in the form of "men and money" (Machiavelli 1988, X, 23)—namely the military and the economy. And secondly, by cultivating religion as a source of legitimization and collective identity, as provider of secondary and higher education for broader circles of the nobility and the upper-middle classes, and as propagator of faith through missions abroad. In order to rule effectively, sovereigns turned to the battlefield, which, among other things, became a site where knowledge-traditions fused and provided solutions to practical problems of "the war machine," thereby propagating further theoretical reflections on these problems, especially in the field of mechanics. Simultaneously, rulers also turned their attention to the economy, encouraging the development of agriculture, investing in the construction and reconstruction of their capital cities, and attempting to attract artists and humanists to their courts. Finally, an effective ruling power meant the reversal of traditional, intermediate loyalties of vassals to their lords in favor of an exclusive and central loyalty of subjects to their sovereign—a process that would be carried out largely by religion, as we shall see in what follows.

2.1.2 Military and Economy: Knowledge-Sites in Transformation and Expansion

The early modern view that the military is the foundation of the state has always been attributed first and foremost to Machiavelli, who wrote that "the chief foundations of all states … are good laws and good arms," and that "there cannot be good laws where there are not good arms" (Machiavelli 2003, XII, 25). Less recognized is the compliance of most anti-Machiavellists with this view, even though they clearly invested much effort in the justification of wars and the necessity of militaries in attempts to reconcile Christian morality with the art of ruling. Thus Bodin, writing in response to the French religious wars, and especially to the St. Bartolomey massacre, agreed that "offensive weapons must also be provided if one would extend one's frontiers and subjugate the enemy…" (Bodin 1967, Book I, ch. i). For him too, armed men, fortifications and military discipline were conceived as complementary to the normative power of the law. The two should be firmly separated, but are both necessary. "I hold that the prince should provide for the thorough fortification of his frontiers, and if he suspects that any enemy contemplates invading his territory, he ought to anticipate him and wage war as far from his own frontiers as possible ..." (Bodin 1967, Book V, ch. 5). But Bodin also admonishes that "…all nations in their turn separated the callings of arms, and of justice and civil administration…" And he continues: "… subjects ought to be practiced in arms, not only for defensive, but also for offensive purposes, in order to shield the innocent and repulse the wicked" (Bodin 1967, Book V, ch. 5). Bodin even admits that there is some truth to the (Machiavellian) argument that, in order to avoid rebellion and civil war within the state, the best way is "to find an enemy against whom they can make common cause" (Bodin 1967, Book V, ch. 5). His conclusion on matters of war and arms, recognized as among the most difficult problems of statehood, is thus:

> It seems to me that the well-ordered commonwealth of any type whatsoever should keep its approaches and frontiers well fortified and should provide itself with an adequate force of trained fighting men … In time of peace these men should be trained in military discipline from their youth up, in garrison duty on the frontiers, after the example of the ancient Romans. For the rest, the carrying of arms should be forbidden to all other subjects in order that laborers and craftsmen should not be tempted to desert the plough and the workshop and take to robbery. (Bodin 1967, Book V, ch. 5)

This last quotation testifies to Bodin's appreciation of the sovereign's need for a strong army, which had to be balanced, however, with the development of the arts and sciences as a basis for a strong economy, as we shall see ahead.

Writing in Italy at the end of the sixteenth century, after the peak of wars on Italian land, Botero's text *The Reason of State* (Botero 1956) concentrates more on the maintenance of the state than on its foundation. He is, perhaps, the first political theorist to interpret the "reason of state" in terms of the prince's "interest" (Bireley 1990), while his focus turns from survival and defense to the expansion that would provide the prince with "glory." To the topic of the force and power of independent states, Botero dedicates four books in his text, opening with the statement that there are two kinds of sovereign states. The first is merely legal, making the ruler the "absolute and supreme lord" (Botero 1956, 169). This kind of sovereignty he calls "external and accidental" (Botero 1956, 169). The second, however, "means that he [the sovereign] is a powerful ruler with strength sufficient to maintain his rule …" (Botero 1956, 169). These general observations are followed by a detailed discussion of weapons, differentiating between defensive and offensive arms, and advising that the rulers' interest should be anchored in good weapons within a technical discourse of experts. Thus, "the ruler should first of all consider what sort of arms his subjects use and then consult men skilled in these matters about possible improvements" (Botero 1956, 176). The power of states, however, is not exhausted by physical force. No less important is a sense of justice in waging wars as well as a feeling of reward among warriors in the form of honors, material gains and care of the injured by the state: "a just cause inspires valor because he who is in the right is always hopeful of victory, and this gives him more spirit …" (Botero 1956, 180). Hence "a ruler or commander should always insure that his men believe that the war they are persecuting [prosecuting sic!] is a just one," thereby defending "the state and his honor" (Botero 1956, 180). Botero dedicates long chapters to issues of rewards for fighting soldiers, but also to the necessity of discipline and punishments. Augmentation of power is attributed not just to physical force, but also to the glorious image of war. Thus, "splendid and beautiful arms" (part of a lengthy discussion of the necessity to decorate weapons) will make soldiers "more daring and courageous," and will "strike fear into the hearts of their enemies." Moreover, the representation *and commemoration* of war is essential for the necessary image of success: "a ruler should also ensure that his wars and campaigns are recorded in writing; thus his own virtues will be celebrated, as well as those of the commanders and soldiers" (Botero 1956, 190).

With all his interest in war and the military, Botero strove to stay within the confines of a Christian reason of state. Offensive war was thus hard for him to justify: "… offensive warfare is only justified by defense" (Botero 1956, 221). Aware of the immense tension between his Christian values and the need for glory and resources that offensive war may bring about, he chooses to confront the problem explicitly by asking: "How then…may I extend my state." His answer is given in a short statement whereby an extension of the state can only be achieved "in the defense of the public good" (Botero 1956, 221). But what was the public good? Just as Bodin's vision indicates, most sovereign rulers in early modern Europe thought that the public good is best articulated through the advice of theologians and canonists to the prince—a point we shall expand on in the following sections.

Not by chance is the long sixteenth century often called in Italian "*il secolo del soldato*" (the century of the soldier) or "*il secolo di ferro*" (the century of iron) (Paci 2005). Indeed, it was a century of wars. Wars that became more and more bloody, fought by means of ever-growing armies, and tolling an ever-growing number of dead soldiers and civilian victims; wars that entailed periods of famine and plague; a cycle of wars that did not die out at the end of the century, but stretched—with some interruptions—into the middle of the next one. The invasion of Charles VIII Valois into Italy in 1494 marked the beginning of this series of wars, which eventually reconstituted the relations between European powers, consolidating a new type of political entity in the form of large sovereign states. Wars also determined Italy's division into middle-sized, more or less independent states like Venice, Florence and the Papal Monarchy, as well as states under Spanish rule, like the kingdom of Naples and the Duchy of Milan. Finally, war also brought about a new cultural splendor to some city states like Ferrara, Mantova or Urbino (Paci 2005, 11).

In the first phase of these wars—between the invasion of 1494 and the peace of the "two ladies" in 1529—battles on Italian territories were mainly fought between France and the Spanish Habsburgs. In the second phase (1535–1559) the theatre of war between France and the Emperor (Charles V) moved to Provence, Northern France and the Low Countries (Alfani 2013). Simultaneously, Charles was fighting in Germany against the reformers and lost. Confirmed by the peace of Augsburg (1555), the confessional division of Germany became a recognized fact and with it the principle that "*cuius regio eius religio*." Following his defeat, however, Charles abdicated and divided his territories between Spain, under Philip II—comprising the low-countries, Burgundy, Milan, Naples as well as the American colonies—and Austria, under his brother Maximilian, who received the Habsburg inheritance, including the kingdoms of Bohemia and Hungary, and the Imperial Crown. Apart from the Duchy of Savoy, which suffered the loss of most of its territories, most of the second phase of these wars did not take place on Italian land. While this stage came to an end with the peace of Cateau-Cambrésis (1559) and the restoration to Savoy of its lost land, many of the Italian states, after 1559, enclosed themselves under Spanish and Papal patronage in the spirit of the Council of Trent (1545). Yet the will for political superiority, for territorial gains and, more generally, dynastic rivalries and religious vehemence continued to fuel war in the low-countries, where Calvinist merchants and noblemen fought against the absolutism of Philip II (Paci 2005). One should add to this picture the English wars with the Scots, Burgundians and the Spanish (De Vries 2006), and the big, continuous battle against the supreme religious foe—the Ottomans, which presented a true threat to Christian Europe in Vienna in 1529, but also brought about a momentary union of powers and triumph in Lepanto in 1571. This religious conflict continued to rage under the leadership of Venice, and later on of the Habsburgs, while the fragile peace of Augsburg was again broken in 1618, continuing until 1648, when the devastating Thirty Years' War reached a new climax of destruction and death (Alfani 2013; Bireley 1990; De Vries 2006; Paci 2005).

No doubt it was a growing body of knowledge about the professionalization of war in the sixteenth century that pushed the historian G. Parker to write about the "military revolution" that took place in that period (Hale 1998; Parker 1996). It involved large, mobile artillery guns loaded at first with stone projectiles, later replaced by the firebombs that came to be used extensively in France by the end of the fifteenth century. Not only was this development connected with metallurgy and the composition of materials for producing gunpowder, it also promoted the invention of machines able to carry and transport such big canons. These developments, in turn, completely transformed the fortifications techniques used for defending city populations against canons (Paci 2005). While the new techniques of fortification endowed military architecture with new functions, Italians also contributed a great deal to reorganizing the military exercises of the great armies of mercenaries (Paci 2005, 18). Finally, writings about the war created a new vocabulary—whether related to techniques of fortifications, military organization and tactics, ballistics or trigonometry—which was necessary for collective communication, reflection and diffusion of the new kinds of highly appreciated varieties of practical and technical pieces of knowledge. The new vocabulary developed not only through the great variety of political-military manuals, but was also enhanced by the invocation of comparisons with the Greek and Roman past by humanist writers like Machiavelli or Francesco Patrizi (Del Negro 2002; Machiavelli 2001; Patrizi 1583).

The early modern battlefield thus turned into a site of knowledge, a huge laboratory requiring trained operators who specialized in loading and firing, as well as smiths, carpenters and masons working together with carriers of knowledge about the composition of substances that could produce effective gunpowder. Most of these workers came from the artisanal and lower classes and remained anonymous. But the proliferation of weapons and warfare also attracted intellectuals capable of reflecting on gunnery, war machines and the construction of fortifications using the mathematical tools developed within the scholarly traditions of antiquity and the Middle Ages. The intellectuals were looking for patrons who would hire them to work in the courts of princes and dukes, where they engaged in writing about gunnery, war machines and the construction of fortifications, developing the conceptual foundations of their field of expertise. Eventually they were joined by painters, sculptors and humanist writers trained to present and record the glory gained by their patrons through warfare.

Weapons, fortifications and the glory of war, however, were not the sole source of power and strength for princes and states. Botero developed a discourse on economy as an additional resource or dimension of princely power. In a chapter entitled "Of Forces" he writes: "…we shall discuss by what means he [the prince] may increase his possessions and these means are his forces, the instruments of prudence and valour" (Botero 1956, 131). His language here, as in many other places, is purposefully mechanical, referring to instruments by which forces are to be augmented. The means, he then adds, are diverse, and include "people, money, victuals, munitions, horses and weapons of attack and defense." Botero continues: "…It is essential that a prince should always have a good sum of money by him, both for the sake of his prestige (since the power of a state is today judged as much by its wealth in money as by its size) and for the requirement of peace and war… it is essential to have money ready for such emergencies" (Botero 1956, 134).

At the same time, Botero tries once again to balance the need for money with the demands of Christian values. Thus, he advises against accumulation without proportion, which may "seem monstrous." He admonishes the boastful display of riches, claiming that "money must be accumulated without ostentation," that a prince should not commit himself to heavy expenditure, that he should always balance his expenditure against his revenues and avoid giving away unwise gifts (Botero 1956, 135). A main source for the ruler's revenues are taxes, used for the public good, and hence lawful and just. Again, Botero is among the first to discuss taxes as a major and steady source of money to be used for the public good. But here too, he does not push aside other moral issues connected with money. Taxes should only be levied, he stresses, on land or wealth, not on movable, personal property: "otherwise the whole burden of taxation will fall upon the poor" (Botero 1956, 136). His concern with the poor stems not only from Christian charity. He is definitely preoccupied with the dangers of striking inequalities in sharing the burden, as well as with the need to use money wisely for the benefit of the public. Another proper use of money is the "greatness and prestige of the king" (Botero 1956, 141), which may be neglected if the pursuit of treasure is excessive. The delicate balance between the powerful ruler with his right to command vis-à-vis the subjects' duty to obey is best seen in his discussion of the interest-free loans that a prince could ask of the estates as a token of their good will toward him. Henry II, he tells us, thus accumulated three million, and with them "centered upon new wars and made important conquests. In this way he was able to win military glory without burdening his people" (Botero 1956, 7:5).

The complicated system of mutual dependencies between the prince and his subjects is even more apparent in the comparison between money and people as the two basic resources of the prince's strength. It is here that an inversion of the traditional hierarchy between nature and culture, science and arts is disclosed, throwing light on the subtle connections of power and knowledge in early modernity. "The true strength of a ruler," Botero opens, "consists in his people: for upon them depend all his other resources. The ruler who has plenty of men will have plenty of everything which the ingenuity and industry of man can provide" (Botero 1956, 143–144). It soon transpires that people are more important than money and treasure since it is through their industry, their skills, their arts and their sciences that art rivals nature and wins. To the question of which is more important to make the state great, gold and silver mines, the fertility of the land or the industry of people, he answers: "…without hesitation I shall say industry. Firstly, the products of manual skill of man are more in number and of greater worth than the products of nature, for nature provides the material and the object but the infinite variations of form are the result of the ingenuity and skill of man" (Botero 1956, 151). Using an Aristotelian idiom of matter and form, he justifies this conclusion by pointing out the active force ingrained in form versus the passivity of matter. The worth of wool, iron, marble or wood in the form of dead matter, he claims, is nothing compared to the worth of man-made products which human skill, industry, art and science are capable of creating out of them. Thus "silk is another simple fruit of nature; but what variety of beautiful clothes are fashioned from it by art! Art contrives that the excrement of a vile worm becomes the admiration of princes and the delight of queens, to be held in honour by all" (Botero 1956, 151). Thus, it is the business of the prince to encourage agriculture—a science already known to the Egyptians—by which "we understand all labor which has to do with the soil and its products…" (Botero 1956, 148) "…it is his business to imitate and direct all works that are for the public benefit, such as draining marshes, cleaning and preparing for cultivation any forests that are useless or superfluous … the prince must also continue to assist nature by leading rivers or lakes through his country…he must keep alive and flourishing whatever serves to make his country fertile and highly productive," and finally: "A Prince…who wishes to make his cities populous must introduce every kind of industry and craft by attracting good workmen from other countries … by encouraging new techniques and singular and rare works, and rewarding perfection and excellence" (Botero 1956, 153).

Two concluding remarks follow from the historical materials presented above. The first concerns the general acknowledgement that mechanical knowledge became a resource for states. A political discourse saturated with mechanical vocabulary, such as Botero's, testifies to this fact. An explicit quotation from Richeleu's *Testament Politique* (published posthumously in 1688) provides yet another strong evidence to the status of such knowledge from the point of view of rulers:

> Si les Lettres étoient profane's à toutes sortes d'Esprits, on verroit plus de Gens capable de former des doutes, que de les resoudre; & beaucoup servient plus propres à s'opposer aux Veritez, qu'à les deffendre. C'est en cette considération que les Politiques veulent, en un Etat bien reglé, plus des Maîtres e's Arts-Mechaniques, que des Maîtres e's Arts-Libe'raux pour ensseigner les Lettres. (Richelieu, *Testament Politique*, 140)

This last quotation from Richelieu's pen is a late reminder of the power and relevance of "the mechanical arts" at a time when the "mechanical philosophy" of Descartes as well as Newton's "classical mechanics" already gained a certain hegemony among natural philosophers. And yet it seems that for all practical purposes, it was still the usefulness of the preclassical "mechanical arts" for a "well ordered state" that prevailed in political thinking about scientific education. Richelieu's words recall both the persistence of preclassical mechanics, as well as its dependence on the material and legitimization needs of princes and kings in the seventeenth century.

2.1.3 Religion Within the Boundaries of Newly Emerging States: The Power of Disciplining and Education

Reflection on the emergence of the sovereign state in early modern Europe invokes thoughts about inherent tensions between church and state implied in the structure of sovereignty. The absolute power of legislation given to the prince meant an expulsion of the church from jurisdiction, not only in earthly matters, but also in spiritual ones. Thus, Bodin claims that "princes are not subject to any jurisdiction which delimits their claims over their subjects" (Bodin 1967, Book I, ch. 6–7), and a sovereign "cannot in any way be subject to the commands of another," meaning he is superior to the jurisdiction of another (Bodin 1967, Book I, ch. 8). And yet by no means did the legal situation under sovereignty diminish the power of religion mobilized by all European rulers for serving their states and aggrandizing their own status within them. Not only was the idea of absolute power—embedded in the theological discourse about God's absolute power (*potential dei absoluta*)—enhanced by fourteenth-century voluntarists (Elshtain 2008; Prodi 1987). No less it was the reality of the papal monarchy—collapsing the distinction between the Pope's temporal and spiritual power and exhibiting the effects of sovereignty through the exercise of "*plenitudo potestas*" combined with military force—that struck Italy with amazement at the conquests that turned the papal monarchy into a major Italian force. In other words, it was by following the sovereign power claimed by the Pope that all other European states crystallized the sovereign structure of their own states.

Traces of the ideology that led to the bondage of state with the power of religion in the early modern era are again to be found in Bodin's text. The first chapter of his

Six Books on the Commonwealth actually opens with a religious vision about the end of the commonwealth, identical in Bodin's eyes with the end of each individual member or subject. As "contemplation is the end and form of the good," he says, so "the moral virtues are … ordained to the intellectual" and therefore "sustenance and defence of the subject … are ordained to moral activities …" (Bodin 1967, Book I, ch. 1). Divine law, he continues, is obligatory in these matters, to which the ruler himself is subject, and so faith and homage are among the most important attributes of sovereignty. However, the bondage of ruler and religion is not only embedded in the moral-spiritual sphere. Rather, its practical implications are well understood and articulated by Bodin: "… nothing so tends to the preservation of commonwealths as religion, since it is the force that at once secures the authority of kings and governors, the execution of the laws, the obedience of subjects, reverence for the magistrates, fear of ill-doing, and knits each and all in the bonds of friendship" (Bodin 1967, Book IV, ch. 7). First and foremost, within religion there is the core of faith in the sense of trust that promises the predominance of fairness among members of the commonwealth, so that "faith is the sole foundation and prop of that justice on which all commonwealths, alliances, and associations of men whatsoever, is founded" (Bodin 1967, Book V, ch. 5). At the same time, religion has the means to control the people who in absence of such control tend to fail to do justice, to oppress the poor and humble, to be "unfair in the distribution of punishments and honors," to bring about "excessive riches in a few and excessive poverty in the rest, idleness in the subject, and impunity in ill doing" (Bodin 1967, Book IV, ch. 7). Thus, "priests, bishops, and popes" concludes Bodin, "have always claimed the censorship of morals and religion" (Bodin 1967, Book VI, ch. 1). Their role is to cultivate the exclusive loyalty to the sovereign—at the expense of local loyalties to the feudal lords—and thus to provide the commonwealth with unity and identity: "Those who wish to diminish the estate of bishops, ministers, and overseers, and deprive them of their powers of ecclesiastical censure, their possessions and their privileges, to trample them underfoot, dishonour God, and destroy all religion" (Bodin 1967, Book VI, ch. 1).

Botero's arguments about the role of religion in a sovereign state are embedded in prudence and raison d'état rather than in any contemplative ideal or moral vision. Following the Romans who "regarded religion as one of the chief bases of government," Botero reminds the prince of his humble position in the face of "Divine Majesty and knowledge," urging him to recognize that "from Him proceeds the power of a ruler and the obedience of his subjects" (Botero 1956, 63). Like Bodin, Botero too believed that religion is the source of unity and collective identity of the commonwealth. As religion is "the foundation of kinship," the prince has "to win the favour of God by piety" and "do all he can to introduce and foster religion and piety in his country" (Botero 1956, 65), in particular the Christian law "according to which not merely the bodies and possessions but even the souls and consciences of his people are subject to him." The Catholic faith, he continues "makes men obedient to their rulers, binding their consciences and making them lovers of peace and enemies of uproar and scandal." Thus the role of the church in disciplining and controlling people on the one hand while fostering their virtues and providing them with moral education on the other transpires in Botero's text, leading him to suggest some pragmatic means toward attaining that goal. "A prince should never bring a matter before a Council of State without first submitting it to a spiritual council containing doctors of theology and of canon law...," he writes, predicting the immense involvement of Catholic confessors of kings and emperors in matters of state, not only in the Italian courts but in France, Spain and Austria as well. Finally, Botero's "reason of state" recommends fostering a mutual empowerment of sovereigns and priests through the enhancement of their social bonds, hence:

> a ruler should select clergy outstanding in learning and virtue, and do all he can to win them the esteem of his people by hearing them often, if they are preachers, by profiting from their wisdom if they are men of experience... attending divine services... inviting them occasionally to his table, asking them advice... giving them every facility and opportunity for using their talents in the public interest. (Botero 1956, 68)

Aware of the political dimensions of religion in early modernity, contemporary historians prefer to speak about a process of confessionalization undergone by European societies, rather than focus their attention on the narrow prism of the protestant reformation versus the catholic counterreformation. By confessionalization they mean the process whereby European societies were politically and culturally structured along contrasting religious lines, a process involving the crystallization of religious identities through a wide variety of practices, including discipline by force, censure, persuasion and ritual, propaganda through catechism and the arts, and education in schools, colleges and universities (Feldhay 2006, 29). In many respects, educational reforms in both Protestant and Catholic areas responded to similar concerns and followed similar patterns. In the largely Protestant German world, the key institution that channeled and controlled religious, social and political energy was the university, and the sixteenth century saw a series of wide-ranging reforms in university education that were modeled on the changes instituted at the University of Wittenberg by Luther's collaborator and successor, Philip Melanchthon (1497–1560). Melanchthon strengthened the control of teaching masters over students and charged the university's rector with enforcing Lutheran orthodoxy. He aimed to secure simultaneously the intellectual and the moral character of graduates by imposing on them the Augsburg Confession (the Lutheran profession of faith), emphasizing the study of scripture, the works of Augustine and the history of church councils, and by mandating the practice of annual disputations over theses censured by the rector (Feldhay 2006, 29). Other Protestant universities implemented educational reform along the Wittenberg model, and began to produce a new elite of professors, priests and counselors to princes, who became the principal administrators in Protestant lands.

In Catholic Europe, the most ambitious innovators in the area of education were the Jesuits, the new religious order founded by Ignatius Loyola (1491–1556) in 1540. They aimed to reform the Catholic world through what they called "studies and moral formation," and developed an educational program intended not only to train—or "form"—priests but also to educate the entire Catholic population, not least generations of future rulers (Feldhay 2006, 29). The Jesuits administered a network of hundreds of colleges, centered in Rome, unified by common curricular goals and pedagogical practices (Feldhay 2006, 32). Their success in the field of education brought under their direction not only teaching colleges for novices and lay students, but also special colleges for the nobility and seminaries for the priesthood, such as the German College in Rome (Feldhay 2006, 33). These schools were the Catholic answer to the problem of discipline and control that preoccupied Melanchthon, and, like the reformed universities he envisaged, were to be responsible not only for the intellectual formation of their students but also for their religious observance, social conduct and moral education.

In neither the Protestant nor the Catholic world, however, was educational reform confined to issues of discipline and administration, but rather involved profound curricular changes as well. A major theme in the new Protestant curricula was a deep interest in studying the heavens, which were considered, even more than other parts of creation, to be imprinted with God's signs (Feldhay 2006, 36). One result of this emphasis on astronomy was what historian Robert Westman dubs the "Wittenberg interpretation" of Copernicanism (Feldhay 2006, 37). The latter enabled astronomers to use Copernican models to determine and predict the angular positions of planets without changing the picture of the universe organized around a central, stationary earth. Members of the circle of brilliant students around Melanchthon—including Joachim Camerarius, Jacob Heerbrand and Samuel Eisenmenger (Johannes Kepler's teacher)—carried this interpretation from Wittenberg to Leipzig, Tübingen and Heidelberg, turning Germany into what the French pedagogical reformer Petrus Ramus called "the nursery of mathematics" (Feldhay 2006, 38).

At the same time, the Jesuit colleges that rivaled Protestant universities stressed useful knowledge, which included rhetoric—also prized by Protestant educators—as well as "mixed mathematics," namely mathematics applied to the physical world—a field of knowledge that at best had been marginal in the context of medieval natural philosophy and theology (Feldhay 2006, 39). In this area, the Jesuits were building on new humanist editions and translations of mathematical texts, as well as new genres of writing about practical geometry and the mechanical arts. Scholars such as Niccolò Tartaglia (1505–1557), Federico Commandino (1509–1575) and Francesco Maurolico (1494–1575) had begun the work of recovering the lost works of Greek mathematicians (Andersen and Bos 2006, ch. 28). At the same time, mathematicians from academic backgrounds began to take an increasing interest in the work of practitioners of the mechanical arts in areas such as measuring and surveying. In that context, they applied geometrical theorems and techniques—triangulation, projection, squaring, ratio and proportion—to physical objects in the real world (Bennett 2006; Renn and Damerow 2012). Jesuit mathematicians also combined mathematics with experimental techniques, arguing that the resulting natural knowledge glorified God through the study of his creation. Inspired by this vision, many Jesuits worked to develop not only astronomy, but also optics and mechanics in physico-mathematical directions. These included Christoph Clavius (1537–1612), the architect of the general enterprise of "Jesuit science," Christoph Scheiner (1573–1650), who did original work on sunspots, Josephus Blancanus (1566–1624), who wrote a treatise on mechanics, Paul Guldin (1577–1643), who investigated centers of gravity, and many others who are not as well known.

The complex role and function of religion within premodern sovereign states throws light on the need to avoid simplistic, dichotomous historical narratives, such as those that speak either of an inherent conflict, or alternatively, an inherent complementarity between science and religion in early modernity. Rather, a nuanced reading of the texts discussed here draws attention to the non-dichotomous and subtle system of relationships between state-religion and science, in an attempt to reconstruct the more complex dynamics that characterized them. Religion, as I have tried to argue, played an active, even prominent role in education at large, and scientific education in particular (Dear 1995). Whereas the old universities were loosely under Papal control and cared a lot about philosophy as a handmaid to religion, in early modernity many of these universities became state universities, open to the needs of sovereign rulers. Moreover, universities of the Protestant world, such as Wittenberg, and even more so the large network of Jesuit colleges and universities, provided solid scientific education, even though they were not free of pressures exercised through internal and external mechanisms of supervision and control. The next section provides an example that shows, on the one hand, the intricate relationships of patronage and mutual support between rulers and religious educational institutions, and on the other, the way in which preclassical mechanics became a viable option for the dissemination of relevant scientific education which could still be presented in traditional terms as *orthodox* education.

2.1.4 Mechanical Knowledge Between State and Church: The Case of Parma[1]

On November 5, 1614 the provincial Congregation of the Jesuits convened in Parma, Italy. Parma was then a flourishing university town and capital of the rich and successful state of Rannucio I Farnese. In 1600, the Jesuits had been invited by the Farnese family to assist in building a system of higher education in their state. Eventually, they controlled the most prestigious, interrelated institutions in the city of Parma: the old university, the Jesuit college of St. Rocco, and a "*collegio dei nobili*," a kind of boarding school where the sons of noble families from all over Europe used to spend a few years while their formation as the future serving elite of many Catholic states was meticulously planned (Baldini 2002; D'Alessandro 1980; Turrini 2006).

The opening day of the congregation—November 5, 1614—was carefully chosen by the Jesuits after consultation with the duke. All of them believed that the city would be full of scholars and citizens by that time, the beginning of the academic year.[2] In preparation, the Jesuit church was decorated from top to bottom with rich

[1] This chapter is based on an earlier paper published in *Il Caso Galileo: una rilettura storica, filosofica, teologica*, eds. M. Bucciantini et al., 47–74. Florence: Olschki 2011.

[2] Author's pagination. All translations from the *Relazione* have been done by the author: "…fu intimate la Congregazione per gli 5 Novembre, tempo quando La Citta', e le Cittadine, e le Scolari suole essere ripiena" Relazione Parma (1614).

drapery. The duke ordered to send over from his own *guardarobba* beds, tables, chairs and other paraphernalia necessary for hosting a large number of guests from the province.[3] He endowed the Jesuits with a large sum of money–*buona soma di danaro*—for buying generous quantities of food and useful equipment. On the opening day, his highness, *Sua altezza Serenissma*, came to the Jesuit church to take part in mass. When it was over he called the Fathers and delivered his speech, speaking with much affection, showing how much respect he held for the Society, and the love with which he cared for them.[4] As the Father Provincial expressed his wish to pay tribute to the duke before the congregation began, the duke sent two coaches and received a small group of Jesuits in his palace. He welcomed them with the same kind of majesty that he extended to great personalities, with torches and candles in the halls and rooms to enhance the clarity of day.[5] The congregation started in the evening, in the presence of the duke, who took part in almost all of the sessions. It lasted for ten days, during which some of the Jesuits were invited to a ceremonious meal in the palace.

These details are relevant as part of a thick description of the cultural practices that enveloped the transmission of knowledge in Jesuit colleges, in this case, the "*collegio dei nobili*" in Parma. For now, let us concentrate on the afternoon routine of the congregation:

> After lunch, they attended the disputations that were four in Theology and one in Philosophy..., frequented by "literati," scholars and religious people, many of them coming from other cities in the surroundings in order to participate and argue. Also, two most "curious" problems in mathematics were demonstrated. One: Why, when a ray of sun passes through a hole of whatever figure, it nevertheless shows itself with a circular shape on the place [surface] where it arrives [it is projected]; and the other: how—or in what way—someone [concerned with] a certain product made of various metals: gold, silver, bronze or others, can know how much of each metal is inside, though the object [the product] is kept together [namely is not made to disintegrate]. And they were successful, with the grace of God, to the great satisfaction of all those who were present.[6]

---

[3] "coltelli, forcine, cuchiari, boccali, bichieri, Lucerne, cattini per Le mani, vasi per l'acqua benedetta, insino la carta, penne, calamari, temperini, forbicette, et altro. Segno evidente dell'isquisitissima sua providenza," Relazione Parma (1614).

[4] [Author's pagination]: "...e con molto affetto si parlo', monstrando quanta sia la stima che egli fa della Compagnia, quanto L'amore che le porta, quanto il contento di vederli congregate," Relazione Parma (1614).

[5] "Li riceve' con quella grandezza; e maiesta; colla quale suole ricevere gran Personaggi. Ardevano nelle Scalle, nelle Sale, e nelle Camere torchie, e candelle, che uguagliavano il chiaro del giorno" Relazione Parma (1614).

[6] Author's pagination: "Il doppo pranzo, si attendeva alle dispense, che fuorono quarto di Teologia e una di Filosofia (per che un'altra si differi' per la causa che abbasso si dira') con grande frequenza di Letteratti, scolari, e Religiosi, molti de quail venero da alter Citta' circonvicine, per intervenirvi, et argomentare. Si dimonsstrarono, anche due Problemi curiosissimi nella matematica/ L'uno per che quando il raggio del Sole, per buco di qualsivoglia figura, nel Luogo non dimeno ove termina, si mostra circolare; l'altro come qualunque opera fatta di metallie varij, oro, argento, bronzo, o' altri restando intiera, si sappia quanto di ciaschedun metallo vi sia dentro," Relazione Parma (1614).

The first problem was an attempt to solve mathematically and explain physically the problem of light passing through a pinhole, which dates back to ancient Greek times (Lindberg 1968). Originally presented in pseudo-Aristotle's *Problemata*,[7] the problem was formulated as a question about the image of the sun reflected through a hole of quadrilateral form with some finite size. It was then extensively discussed by all the great figures of the medieval optical tradition—including Alhazen, Witelo, Bacon, Pecham, as well as by Leonardo da Vinci, and the great Renaissance mathematician Francesco Maurolyco, and finally by Kepler.

Galileo was certainly quick to take advantage of the combination between Kepler's new optics and the know-how of linear perspective shared by both mathematicians and artists, as has been so well shown by Filippo Camerota (Camerota 2004; Malet 2003, 2005). But the idea was immediately taken up by many Jesuit mathematicians as well. Thus, Christoph Scheiner used real and model eyes as *camera obscura* to demonstrate the principle of inversion of images (Gorman 2007; Scheiner 1619 and 1626–1630). And he used the helioscope and the pantograph for observing and mechanically reproducing celestial images that were essential for the development of visual astronomy (Camerota 2004). Moreover, the Jesuits soon developed an insatiable appetite for optical instruments, which they used not only for research and teaching purposes, but also in their theatres as well as for their missionary propaganda outside Europe, and more generally for buttressing their status as an intellectual elite.[8] The decision to publically perform, in front of the audience of political elite, the problem of the sunray passing through a pinhole could function as an emblem, smacking of traditional respectability but also invoking all the connotations that accompanied the most recent scientific and epistemological debates and controversies.

The second problem dates back to the legend of Archimedes and the crown of Hieron, King of Syracuse. Thus, it enjoyed the same aura of tradition and respectability as the *camera obscura* problem. It also had a long history in the Society of Jesus. According to the historian Ugo Baldini, the Archimedean treatise *De his quae vehuntur in aqua* was translated sometime during the mid-sixteenth century by B. Torres, who taught mathematics at the Collegio Romano between 1553–1560. Likewise, Baldini quotes from the fifteenth-century book of the *Bibliotheca selecta* (*De mathematica*), published by A. Possevino in 1593, the following judgment on Archimedes' treatise: "qui sane libellus, si aetatis nostrae Mathematici sibi familiariorem adhibuissent, reprissent sententias multas, quas ipsi

[7] "Why is it that when the sun passes through quadrilaterals, as for instance in wickerwork, it does not produce a figure rectangular in shape but circular?" Aristotle (1936, 1:333).

[8] On the complex uses of the magic lantern, for example, see Vermeir (2005); on the use of projection in the theatre, see Gorman (2003). Particularly interesting is the following passage relying on the description of Jean Baptiste du Halde's *Description geographique, historique, chronologique, et physique de l'empire de la chine et de la Tartarie Chinoisie* published in Paris in 1735, where he narrates how: "Claudio Filippo Grimaldi (1639–1712) entertained them [namely the emperor and his entourage] in the gardens of the summer residence, using convex lenses, *camera obscura* and cylindrical and pyramidal mirrors to cast shadows and project images from the outside world," Gorman (2003, 14).

firmas et ratas esse docent, subtilissime atque verissime convulses, et labefactas" (Baldini 1992, 66, n. 54). In 1603, the young mathematician Marino Ghetaldi (1568–1626)—a frequent visitor to Clavius's academy, but no Jesuit—published a treatise entitled *Promotus Archimedes* in which the Crown problem was discussed (Ghetaldi 1603).[9] Ghetaldi was concerned with comparisons between the gravity and magnitude of various kinds of bodies, as testified by the subtitle of his treatise, *De variis corporum generibus, gravitate & magnitudine comparatis*. The treatise included numerical tables representing the specific weights of a long list of metals and other materials discovered through experimentation.[10] It also contained an explanation of buoyancy in terms of the difference between the specific weights of the body and the medium, which had become a matter of contention between Aristotelian philosophers and mathematicians in the tradition of Archimedes.[11] Hence, Ghetaldi's book acquired new relevance as the controversy on floating bodies between Galileo and a number of Aristotelian philosophers in Florence heated up in the years 1611–1614.

In his discourse on floating bodies, *Discorso intorno alle cose che stanno in su l'acqua o che in quella si muovono*, published in April 1612 at the explicit request of the Duke of Tuscany Cosimo II de Medici, Galileo chose to sharpen his disagreements with his Aristotelian interlocutors by emphasizing his Archimedean solution to the problem of buoyancy and by presenting it as anti-Aristotelian. In fact, however, right at the beginning of his text, Galileo declared that he would present an Archimedean solution by other means. These means consisted of principles which he borrowed from the mechanical tradition of the pseudo-Aristotelian *Problemata* (Shea 1971). It is within this context that one should understand the inclination of many Jesuit mathematicians to support Galileo's position against his Aristotelian opponents. The Jesuit mathematicians were committed to follow Aristotle in physics, but also took active part in the revival of Archimedes' work of the sixteenth century (Baldini 1989, 246). No wonder they attempted to reconcile between the two traditions. This was not too difficult to do. Aristotle in *On the Heavens* had maintained that "...a body which is regarded as light in one place is regarded as heavy in another. In air, for instance, a talent's weight of wood is heavier than a mina of lead, but in water the wood is lighter." (Aristotle 1941). Since no vacuum is admitted within the Aristotelian cosmos, the heaviness and lightness of all bodies can only been discussed in medium. Hence, in principle, there seemed to be no contradiction in presenting Archimedes' mathematical theorems concerning floating bodies as continuous with Aristotle's more general and less mathematical discussion in his *On the Heavens*.

---

[9] The letter to the reader mentions Clavius and the good will he showed toward Ghetaldi: "Etenim cum Clauium, quod iam diu cupiebam, vidissem, nec minorem tanta scientia, & fama viri benignitatem comperissem ..." Ghetaldi (1603).

[10] For a thorough discussion of Ghetaldi's work, see Napolitani (1988, 139–237).

[11] Sixteenth proposition of the eighth theorem, Ghetaldi (1603, 28); for a critique of Archimedes' principle of buoancy from an Aristotelian perspective, see Helbing (1989, 64–74; 203–225).

Thus, when Johannes Bardi, a former student of Galileo who was studying mathematics with Christoph Grienberger at the Collegio Romano in 1614 expressed his wish to recite in public a problem concerning bodies that float in water, his teacher Grienberger, an old correspondent of Galileo, probably wrote the problem for him.[12] Indeed, Bardi opened his performance by declaring: "I shall tell you briefly the things contended at length by Galileo Galilei, my mentor in the old days, concerning things that stay in water" (Bardi 1614, 5). Basically, Bardi's contentions justified Galileo's two main conclusions against his Aristotelian interlocutors. First, that the form of the body was irrelevant for its position in water, whether under the surface, on it, or protruding above it. And second, that the behavior of bodies in liquids should be explained mainly by means of the difference between their densities—namely their specific "gravity," as Archimedes had shown (Bardi 1614, 8). In addition, Bardi took upon himself to account for such exceptional cases, which were brought up by Galileo's opponents. These, he had rejected through the identification of the previously unknown phenomenon of a small dip formed around the floating body which, combined with the body's weight, made it lighter, and hence float (Galluzzi 1979, 227–246). In fact, Bardi dedicated much space to a meticulous description of that phenomenon on the basis of experiments, declaring that:

> Material will be abundantly supplied by experience which, just as it has existed for the most serious fight, so is equal firstly to take position in this skirmish, to urge the soldier, to equip with arms, and to encourage with rewards; and then, having encouraged them to fight bravely, (is equal) to be the first to take a stand, and to be the first in line.[13]

There is also strong evidence that those experiments were designed, and perhaps even performed, *in situ* by Grienberger himself (Gorman 2003, 25). The problem was performed in Rome on 23 June 1614, in the presence of "il detto Sig. Principe [Cesi], con Mons. Suo fratello [Bartolemeto Cesi] et altri Prelati et signori letterati, con il Sig. Valerii [Luca Valerio] et Sig. Fabri [Giovanni Faber]" (Favaro 1968, 78). These well-known persons, together with other prelates of high status in the church hierarchy and several other Roman intellectuals, inspired the event with their authority and brilliance. Stelluti, a Roman correspondent who reported to Galileo on the event from Rome, emphasized in his letter just how much Galileo's views were promoted and defended during that evening (Favaro 1968, 78). One last detail should not be omitted: a few days later the problem saw light in the form of a text published by Zanetti, the printing shop the Jesuits favored and made their own.

Recitation of problems was a regular event not only in the Jesuit's public events, with an eye to securing their relations with political power, but also in their colleges. The text of the *Ratio Studiorum* confirms the testimony emerging from the Parma *Relazione*. Similar events all over Europe and beyond found their pedagogical justification in stimulating not only intellectual capacity but also the capacity for

---

[12] On Bardi's problem, see Celia and Paolo (1997); On the identification of Grienberger as the probable author of Bardi's problem, see Gorman (2003, 25).

[13] "Materiam abunde suppeditabit experientia; quae, ut gravissimae pugnae causa extitit, ita par est., primum ut ipsa locum in hac velitatione obtineat, militem ipsa cogat, armis instruat, ac praemijs invitet; quaeve ad fortiter pugnandum adhortata est., prima in acie, prima ipsa constituat" Bardi (1614, 6).

presentation and interaction. Jesuit education placed enormous emphasis on the development of rhetorical skills, which accompanied many of the common activities in the college such as repetitions, disputations, public defense of theses and theatrical productions. Ignatius's preliminary justification, according to which "the purpose is that the intellectual powers may be exercised more, and that difficult matters occurring in these branches may be clarified, unto the glory of God our Lord," was indeed developed by the Jesuits into a complete educational philosophy—the full extent of which cannot be fully surveyed here.

What should be further emphasized, however, is that the Society also hoped to gain some advantages in the public sphere from such activities. No less than training students in presenting their work to their fellow students as well as to wider circles of dignitaries and intellectuals, the recitation of problems was also meant to attract visitors to the college and to contribute to the cultural life of the city. The fusion of educational purposes with more general cultural-political goals found its ultimate expression in the *Ratio*'s rules for the professor of rhetoric which captures the dramatic, baroque spirit of Jesuit education in the following words:

> Nothing, in fact, so develops resourcefulness of talent as frequent individual practice in speaking from the platform in the lecture hall, in church, and in school ... as well as in the refectory. (Farrel 1970, 79)

The lecture hall, the church, the school tribune and the spaces allocated for public disputations, mathematical problems and defenses of theses all emerge in the text of the *Ratio* itself as meeting places between novices and externs, used in a process of mediation between the controlled spaces of the Jesuits enclosed behind the gates of the college and the less regulated, sometimes chaotic space of the city bursting in so many different rhythms and a plurality of cultural forms.

The *Relazione* of Parma testify to the interpenetration of the "refectory" and the city, even though they do not specify the location of the after-lunch disputations and problems conducted during the Provincial Congregation of 1614. However, later-dated descriptions of the concrete spaces built by the Jesuits of Parma for their public intellectual activities exist and demonstrate a growing tendency toward refinement and grandeur. One example is a description accompanied by a series of aquarelles of the great hall, first called *sala d'armi* at the Collegio dei Nobili. A series of allegorical representations of music, geography, dance and other fields of intellectual fascination for the Jesuits, situated in historical scenes derived from classical and contemporary sources, decorated the hall,[14] which came to be known as *il salone*. Another, smaller hall built for weekly philosophical disputations, called *la saletta,* was also used by the fathers to convene around the fire in the evenings. In addition, the complex of the college also included two theatres, *grande* and *piccolo*. In fact, the plan of the ground level of the complex[15] that survived the buildings to the present day testifies to the ways in which the complex of the college was divided

[14] A description of the hall with reproductions of the representations is in BBP, Ms. Parm. 1250, [Pietro Mazza] *Gran sala d'armi e delle academie che esisteva nella parte dell'acntico edifizio del collegio de' nobili di Parma...* . See in Turrini (2006, 43).

[15] Pianta del collegio dei nobili di Parma. Il primo piano, Turrini (2006, 310).

between public spaces designed for the interaction of insiders and outsiders on special occasions, and the internal spaces where everyday scholarly routine was taking place. Thus, the whole southern part of the complex seems to have been built especially for the special events. It contained the great *salone*, the *saletta*, the two theatres and a *foresteria* with a special kitchen and other services for visitors. This part had its own gate and entrance at the south and was separated from the refectory in the north by a large court. Such an architectonic structure testifies to the complex cultural messages embodied in the material culture within which Jesuits scholars and students were living their life. The college was supposed to provide a closed, tightly regulated space, which protected the students and boarders from the influences and noise of the general culture around. At the same time, it contained a well-demarcated arena for monitored exchanges, which pointed out the desire for cultural hegemony but also betrayed the impact left by the surrounding culture on the norms that guided Jesuit activities.

The performative dimension of the "mathematical problems" is not exhausted, however, by pointing out the material location where such public events were situated. The academic preparation for reciting a problem was deeply rooted in the pedagogic traditions of the Jesuits. Starting from the lower classrooms, contests with allocated "roles" were frequently practiced in order to intensify the appetite for learning:

> Class contests are to be highly valued and are to be held whenever time permits, so that honorable rivalry which is a powerful incentive to studies may be fostered. It is customary in these contests to have the teacher ask the questions and the rivals correct the errors or to have the rivals question one another. Individuals or groups from opposite camps, particularly from among the officers, may be pitted against each other, or one pupil may engage several opponents. (Farrel 1970, 68)

Second, practices of public disputations were constitutive for the philosophical and theological curriculum:

> On Saturday or any other day dictated by local custom, disputations should be held in the classes for two hours, or longer where there are many extern students … Other professors of the Society, though they belong to different faculties, should attend the disputations whenever possible. To enliven the discussions, they should press the objections that are proposed … The same privilege may be extended to extern doctors, who may even be invited expressly to take part in the argument … . (Farrel 1970, 27–28)

Thus, discussion and exchange of ideas between students and professors, and among professors of the different disciplines were regularly celebrated in special events and provided an opportunity for outsiders to take part and express their criticisms and opinions. These well-imprinted habits culminated, however, in the various "academies" that grew up in Jesuit colleges already in the sixteenth century: literary academies, as well as academies of philosophers and theologians, and finally also the mathematical academy mentioned above:

> By the word 'academy' we mean a group of students chosen from the entire student body on the basis of their devotion to learning who will meet under a Jesuit moderator to take part in special exercises connected with their studies. (Farrel 1970, 103)

Moreover, these advanced seminars called "academies" were always partly open to scholars who did not necessarily belong to the student body or professorial staff, but were somehow closely associated with the local college:

> Besides, where custom sanctions it, the rector may approve the admission of others who are not members of the sodality or even students in our school. (Farrel 1970, 103)

It was especially in the context of the academies that private teaching took place, and it was precisely at the academies—to a large extent structured around disputations, defense of theses or the solving of difficult problems—that interpenetration between the inside and the outside of the college took place.

The Parma case study well exemplifies the network of dependencies in which sovereign rulers, religious institutions, scholars, artists and professionals were entangled in the early modern Catholic context. It points out the two facets of the institutionalization, success and expansion of preclassical mechanics in early modernity. The first has to do with the relevance of "mechanical knowledge" to the ruling power, enhanced by the special, paradoxical nature of sovereign regimes (discussed in Sect. 1). It was a type of knowledge that was useful in many senses, for practical purposes but also for enhancing the glory and legitimization of rulers. The case-study shows how mechanical knowledge was not only produced in the courts by engineer-scientists patronized by the prince, but was also institutionalized, often within the religious establishment, and widely diffused through the dense network of the Jesuits (Leach et al. 2006). At the same time, the case study directs our attention to the advantages of Jesuits as transmitters interested in incorporating new types of knowledge into their schools, albeit on the condition that such knowledge would still "appear" under the guise of tradition and orthodoxy.

References

Alfani, Guido. 2013. *Calamities and the economy in Renaissance Italy: The grand tour of the horsemen of the apocalypse*. Hampshire: Palgrave Macmillan.

Andersen, Kirsti, and Henk J.M. Bos. 2006. Pure mathematics. In *The Cambridge history of science: Early modern science*, ed. Katharine Park and Lorraine Daston, 697–723. Cambridge: Cambridge University Press.

Aristotle. 1936. *Aristotle in twenty-three volumes. Vol. 1, Aristotle problems*. Cambridge, MA: Harvard University Press, The Loeb Classical Library.

———. 1941. In *On the heavens*, ed. Richard McKeon, 395–466. New York: Random House.

Baldini, Ugo. 1989. Archimede nel seicento italiano. In *Archimede: mito, tradizione, scienza*, ed. Corrado Dollo, 237–289. Florence: Leo S. Olschki.

———. 1992. *Legem impone subactis: Studi su filisofia e scienza dei gesuiti in Italia 1540–1632*. Rome: Bulzoni.

———. 2002. S. Rocco e la scuola scientific della provincial veneta: il quadro storico (1600–1772). In *Gesuiti e universita in Europa (secoli XVI– XVIII)*, ed. Gian Paolo Brizzi and Roberto Greci, 283–323. Bologna: Clueb.

Bardi, Giovanni. 1614. *Eorum quae vehuntur in aquis experimenta*. Rome: Bartolomaei Zannetti.

Bennett, Jim. 2006. The mechanical arts. In *The Cambridge history of science: Early modern science*, ed. Katharine Park and Lorraine Daston, 677–695. Cambridge: Cambridge University Press.

———. 2011. Early modern mathematical instruments. *Isis* 102 (4): 697–705.

Biagioli, Mario. 1993. *Galileo, courtier: The practice of science in the culture of absolutism*. Chicago: University of Chicago Press.

Bireley, Robert. 1981. *Religion and politics in the age of the counter reformation*. Chapel Hill: University of North Carolina Press.

———. 1990. *The counter-reformation prince anti-Machiavellianism or Catholic statecraft in early modern Europe*. Chapel Hill: University of North Carolina Press.

Bodin, Jean. 1967. *Six books of the Commonwealth, 1576*. Trans. M.J. Tooley. Oxford: Blackwell.

Boschiero, Luciano. 2007. *Experiment and natural philosophy in seventeenth–century Tuscany: The history of the Accademia del Cimento*, Studies in History and Philosophy of Science. Vol. 21. Dordrecht: Springer Science & Business Media.

Botero, Giovanni. 1956. *The reason of state*. Trans. P.J. and D.P. Waley. New Haven: Yale University Press.

Camerota, Filippo. 2004. Galileo's eye: Linear perspective and visual astronomy. *Galileana. Istituto e museo di storia della scienza* 1: 143–170.

D'Alessandro, Alessandro. 1980. Materiali per la storia dello studium di Parma 1545–1622. In *Università, Principe, Gesuiti. La politica farnesiana dell'istruziane a Parma e Piacenza 1545–1622*, ed. Gian Paolo Brizzi Alessandro and Alessandra del Fante. Rome: Bulzoni.

De Celia, Francesco Paolo. 1997. *Relazzioni romane: l'idraulica galileiana negli scritti di Giovanni Bardi e Giuseppe Biancani*. Vol. 3. Bari: G. Laterza.

De Vries, Kelly. 2006. Sites of military science and technology. In *The Cambridge history of science: Early modern science*, ed. Katharine Park and Lorraine Daston, 306–319. Cambridge: Cambridge University Press.

Dear, Peter. 1995. *Discipline and experience: The mathematical way in the Scientific Revolution*. Chicago: University of Chicago Press.

Del Negro, Piero. 2002. Una lingua per la guerra: il Rinascimento militare italiano. In *Storia d'Italia 18. Guerra e pace*, ed. Walter Barberis, 299–336. Torino: Einaudi.

Elshtain, Jean Bethke. 2008. *Sovereignty: God, state, and self*. New York: Basic Books.

Farrell, Allan P., ed. 1970. *The Jesuit Ratio Studiorum of 1599*. Washington, DC: Conference of Major Superiors of Jesuits. http://www.bc.edu/sites/libraries/ratio/ratio1599.pdf.

Favaro, Antonio. 1968. (Reprint). *Le opere di Galileo Galilei*. Florence: G. Barbèra.

Feingold, Mordechai, and Navarro–Brotons Victor, eds. 2006. *Universities and science in the early modern period*, Archimedes. Vol. 12. Dordrecht: Springer Science & Business Media.

Feldhay, Rivka. 2006. Religion. In *The Cambridge history of science: Early modern science*, ed. Katharine Park and Lorraine Daston, 727–755. Cambridge: Cambridge University Press.

———. 2011. The Jesuits: Transmitters of the new science. In *Il caso Galileo: Una rilettura storica, filosofica, teologica*, ed. Massimo Bucciantini, Michele Camerota, and Franco Giudice, 47–74. Florence: Olschki.

Galluzzi, Paolo. 1979. *Momento*. Rome: Edizioni dell'Ateneo.

Ghetaldi, Marino. 1603. *Promotus Archimedis*. Rome: Aloysium Zannettum.

Gorman, Michael John. 2003. Mathematics and modesty in the Society of Jesus: The problems of Christoph Grienberger. In *The new science and Jesuit science*, Archimedes, ed. Mordechai Feingold, vol. 6, 1–120. Dordrecht: Springer.

Gorman, Michael J. 2007. Projecting nature in early modern Europe. In *Inside the camera obscura: Optics and art under the spell of the projected image*. Preprint 333, ed. Wolfgang Lefèvre, 31–50. Berlin: Max Planck Institute for the History of Science,. https://www.mpiwg-berlin.mpg.de/Preprints/P333.PDF.

Hale, John Rigby. 1998. *War and society in Renaissance Europe, 1450–1620*. Montreal: McGill-Queen's Press.

Helbing, Mario Otto. 1989. *La filosofia di Francesco Buonamici: professore di Galileo a Pisa*. Pisa: Nistri-Lischi.

Kalmo, Hent, and Quentin Skinner, eds. 2010. *Sovereignty in fragments: The past, present and future of a contested concept*. Cambridge: Cambridge University Press.

Laird, Walter Roy, and S. Sophie Roux, eds. 2008. *Mechanics and natural philosophy before the Scientific Revolution*. Dordrecht: Springer.

Leach, J., J.W. O'Malley, G.A. Bailey, and S.J. Harris, eds. 2006. *The Jesuits II: Cultures, sciences and the arts 1540–1773*. Toronto: University of Toronto Press.

Lindberg, David C. 1968. The theory of pinhole images from antiquity to the thirteenth century. *Archive for History of Exact Sciences* 5 (2): 154–176.

Machiavelli, Niccolo. 1988. In *The prince*, ed. Q. Skinner and R. Price. Lancaster: Lancaster University Press.

———. 2001. *Dell'arte della Guerra*. Rome: Salerno.

———. 2003. *The prince and other writings*. New York: Barnes and Noble Classics.

Malet, Antoni. 2003. Kepler and the telescope. *Annals of Science* 60: 107–136.

———. 2005. Early conceptualizations of the telescope as an optical instrument. *Early Science and Medicine* 10 (2): 237–262.

Meli, Domenico Bertoloni. 2006. *Thinking with objects: The transformation of mechanics in the seventeenth century*. Baltimore: Johns Hopkins Press.

Napolitani, Pier Daniele. 1988. La geometrizzazione della realtà fisica: il peso specifico in Ghetaldi e in Galileo. *Bollettino di storia delle scienze matematiche* 8: 139–237.

Paci, Renzo. 2005. *La guerra nell'Europa del cinquecento e il generale Achille Tarducci da Corinaldo*. Ancona: Proposte e ricerche.

Parker, Geoffrey. 1996. *The military revolution: Military innovation and the rise of the west, 1500–1800*. Cambridge: Cambridge University Press.

Patrizi, Francesco. 1583. *La Militia Romana di Polibio, di Tito Livio, e di Dionigi Alicarnaseo. Da Francesco Patricii dichiarata, e con varie figure illustrata*. Ferrara: Domenico Mamarelli.

Prodi, Paolo. 1987. *The papal prince, one body and two souls: The papal monarchy in early modern Europe*. Trans. S. Haskins. Cambridge: Cambridge University Press.

Relazione Parma. 1614. *Relazione dell'accoglimento fatto dal Serenissimo Duca Rannuzio primo, alla Congregazione Provinciale tenutasi in Parma l'anno 1614*. ASPr, Casa e Corte farnesiane, b. 56. fasc. 15., second page.

Renn, Jürgen, and Peter Damerow. 2012. *The equilibrium controversy: Guidobaldo Del Monte's critical notes on the mechanics of Jordanus and Benedetti and their historical and conceptual backgrounds*. Berlin: Edition Open Access. http://www.edition-open-sources.org/sources/2/index.html.

Richelieu, Armand Jean du Plessis. 1688. *Testament politique d'Armand du Plessis, cardinal duc de Richelieu, ... Partie 1*. Amsterdam: H. Desbordes. http://gallica.bnf.fr/ark:/12148/bpt6k5427451c.

Shea, William. 1971. Galileo's discourse on floating bodies: Archimedean and Aristotelian elements. In *Actes du XIIe Congres international d'Histoire des Sciences, Paris 1968*, vol. IV, 149–153. Paris: A. Blanchard.

Scheiner, Christoph. 1619. *Oculus hoc est: Fundamentum opticum*. Innsbruck: apud Danielem Agricolam.

———. 1626–1630. *Rosa Ursina sive Sol*. Braccioni: Apud Andream Phaeum. Typographum Ducalem.

Turrini, Miriam. 2006. *Il giovin signore in collegio: i gesuiti e l'educazione della nobiltà nelle consuetudini del collegio ducale di Parma*. Bologna: Clueb.

Valleriani, Matteo. 2010. Was Galileo an engineer? In *Galileo Engineer*, 193–211. Dordrecht: Springer.

———, ed. 2017. *The structures of practical knowledge*. Cham: Springer Nature.

Vermeir, Koen. 2005. The magic of the magic lantern (1660–1700): On analogical demonstration and the visualization of the invisible. *The British Journal for the History of Science* 38 (2): 127–159.

Chapter 3
Socio-Political Coordinates of Early-Modern Mechanics: A Preliminary Discussion

Pietro D. Omodeo

Abstract How does a cultural-political understanding of science integrate socio-economic treatments? How can a historiography that takes subjectivity into account avoid the pitfall of post-modern relativism? The history of mechanics is a paradigmatic field to use in answering these questions and, in fact, it has always been at the center of much political-epistemological skirmish. This chapter first recounts the main motives and features of early twentieth-century social accounts of science. Further, it deals with the issue of how the need for a non-reductionist treatment of intellectual history (neither economicist nor monocausal) calls for an integration of the economic context and the political element for a more appropriate understanding of scientific development.

Keywords Political epistemology · Science and cultural hegemony · Socio-political history of mechanics · Marxist historiography of science · Boris Hessen · Henryk Grossman · Edgar Zilsel · Antonio Gramsci

In the history of science of the last century, in an age of irreconcilable dichotomies mirroring the geo-political divide between Capitalism and Real Socialism, Marxist historians undertook a socio-economic investigation of science fostering an approach that came to be labeled as "externalist."[1] On the opposite front, that of the so-called "internalists," an idealistic conception of science prevailed, one that was seen as *politically correct*. Its rejection of *everything* "material" was meant to immunize the discipline against socialist contamination.[2] By the Nineties, the turn away from *modernist* concerns about all-encompassing interpretations generated a plethora of *post-modern narratives* in which historians' subjectivity came to the

[1] See Porter (1990) and Young (1990). Also see Long (2011).

[2] See Omodeo (2016b).

P. D. Omodeo (✉)
Ca' Foscari University of Venice, ERC EarlyModernCosmology, Venice, Italy
e-mail: pietrodaniel.omodeo@unive.it

R. Feldhay et al. (eds.), *Emergence and Expansion of Preclassical Mechanics*, Boston Studies in the Philosophy and History of Science 270,
https://doi.org/10.1007/978-3-319-90345-3_3

foreground—often at the expense of careful source analysis and theoretical clarity.[3] Broadly speaking, the new mainstream shows a *culturalist* bias whose ostensible (and celebrated) advantage is to substitute the opposition between socio-economic accounts and theoretical/technical investigations for a more circumscribed treatment of local environments that takes into account (and often overemphasizes) the *actors' perspectives*.[4] To be sure, this outlook (often micro-historical, ethnological/anthropological) has served to remedy rough anachronisms and generalization, and the post-modernist emphasis on rhetoric and narrativity gave a worthy caution against the preposterous assumption/prescription of objectivity and detachment on the historians' part. However, it has also produced increasing historical fragmentation and theoretical vagueness about epistemology and the foundations of science. The postmodern-cum-culturalist conjuncture raises two interrelated questions: How does a cultural understanding of science integrate earlier economic and technical treatments? Furthermore, how can a subjectivity aware historiography avoid the pitfall of relativism, well captured by the epistemological slogan "anything goes"? The legacy of earlier *paradigms* has remained quite obscure. In particular, the extent of their validity has not been fully reflected.

It seems appropriate to deal with these issues beginning with a historical/historiographical reflection on early modern mechanics. In fact, this field has been at the center of much ideological skirmish since the 1930s. The most notable contrast was that which opposed Edgar Zilsel's economic explanation of the rise of modern science to Alexandre Koyré's defense of its purely speculative origins (Lefèvre 2001). In the following discussion I will recount the main motives and features of early twentieth-century social accounts of science. Further, I will point to how the need of a non-reductionist (neither economicist nor monocausal) treatment of intellectual history calls for an integration of the economic context and the political element for a more appropriate understanding of scientific development. In order to articulate my proposal, I will eventually delve into a case study in Chap. 7 of this volume: a discussion of the sociopolitical coordinates of a *typical* Renaissance scholar in mechanics of the generation between Girolamo Cardano and Galileo Galilei, Giovanni Battista Benedetti.

3.1 The Marxist Paradigm: Mechanics Between *Basis* and *Überbau*

One of the most quoted and controversial passages of Karl Marx's *œuvre*, his preface to *Zur Kritik der politischen Ökonomie* (*A Contribution to the Critique of Political Economy*) (1859), offers a significant sketch of his conception of culture in

[3] I discuss this issue with Roger Cooter in Omodeo (2015).

[4] The reference work that is mostly mentioned as a watershed in the history of science is Schaffer and Shapin (1985). See their "Introduction to the 2011 Edition: Up for Air: *Leviathan and the Air-Pump a Generation On*," in Schaffer and Shapin (2011, xi–l).

relation to history and society. In this text he introduces the distinction between economic structure (*Struktur* or *reale Basis*) and juridical and political superstructures (*Überbau*), which correspond to determinate historical forms of consciousness. He famously asserted that: "It is not the consciousness of men that determines their existence, but their social existence that determines their consciousness" (Marx 1987, 263). Thus he affirmed the dependence of consciousness on socio-historical settings. Similarly, Marx saw the *ideological formations*—juridical, political, religious, artistic, and philosophical—as rooted in the terrain of the economic conditions of production. It should be remarked that he did not enlist *science* among the *ideologische Formen*. Rather, he ascribed to *Naturwissenschaft* a special status.[5] By arguing that the evolution of the conditions of production (*Produktionsbedingungen*) can be investigated with the accuracy of the natural sciences, he implied that the latter have a degree of objectivity that is superior to that of the other disciplines. He did not discredit the aforementioned intellectual (ideological) realms, as they are the means through which people become aware of their positioning in society. And he compared science, in this case economy, to Dante's descent to the hell in the third Canto of the *Inferno*...

> Qui si convien lasciare ogni sospettoOgni viltà convien che qui sia morta.[6]

For Marxist historians of science reflecting on these pages, science constituted an epistemological puzzle: Does it belong to the structural sphere or to the superstructural? On the one hand, according to the preface to the *Kritik der politischen Ökonomie*, natural science appears to escape cultural determination; this would make the analysis of its relationship to philosophy, art, and religion as well as to the political context superfluous. On the other hand, science cannot be reduced to economy *tout-court*. Rather, it occupies an intermediate epistemological position between *Basis* and *Überbau*, as its history clearly shows (see Fig. 3.1). From the 1930s to the 1950s, Marxist inquiries in the history of science, from Boris Hessen to John Bernal, Henryk Grossman and Zilsel, especially stressed the dependency of science on economics—and this inquiry was made more difficult by controversies over the relation between economics and technology. Moreover, within the Marxist tradition, debates about the interrelation of structure and superstructure, politics and consciousness, and science and ideology determined fluctuations in the understandings of science.[7] Later social accounts, often departing from Marxism, explored the connection of science with cultural realms such as religion (Merton's Weberian school), philosophy (e.g. Paolo Rossi), art (e.g. Pamela Smith) and politics (e.g. Mario Biagioli).

---

[5] Cf. Stachel (1994).

[6] Here all misgiving must thy mind reject./ Here cowardice must die and be no more.

[7] On the oscillating assessment of the relationship between structure and superstructure and its relevance in cultural studies, see Hall (1980). The problem of consciousness is typically Lukácsian, whereas that of ideology and science is generally connected to the *Frankfurter Schule*. See for instance Habermas (1969). More recently, the Foucauldian strand has pointed out the biopolitical dimension of science, which goes beyond the "mental" emphasized by the concept of *ideology*. For an insightful discussion and case study, see Bruskell-Evans (2015).

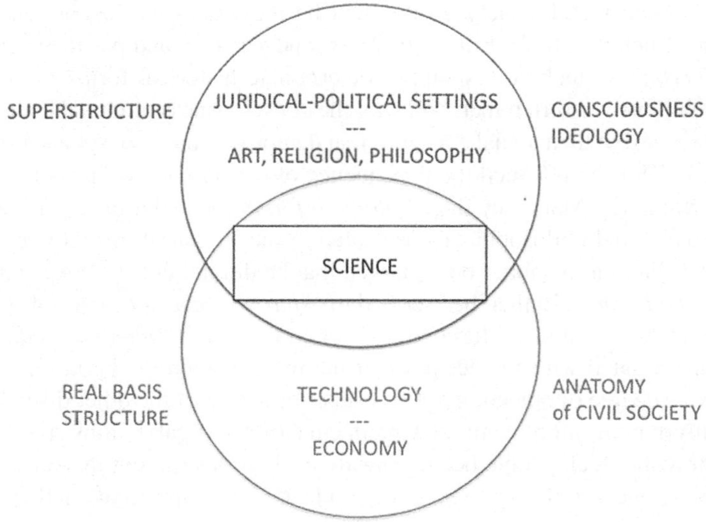

Fig. 3.1 A Marxist problematic: Science between basis and superstructure

I will begin my assessment with the earlier trend pointing out the economic roots of science. Boris Hessen's "The Social and Economic Roots of Newton's *Principia*"—a *classic* of Marxist historiography of science, as Gideon Freudenthal and Peter McLaughlin have called it—was a talk delivered at the second *International Congress of the History of Science and Technology* in London in 1931.[8] Hessen was one of the Soviet delegates deputed to present and promote *the Marxist way* of the history of science. His talk on Newton was one of the most articulated among those delivered by the Soviet delegates, and had a lasting impact on subsequent studies on the Scientific Revolution. The aim of his talk was to reject the semi-mythical treatment of Newton's physics as the product of a lonely genius detached from mundane interests by demonstrating that the *Principia* was the historical product of a scientific process induced by economy-driven technological advance.

In order to set the theoretical stage of his analysis, Hessen began with Marx's preface to the *Kritik der Politischen Ökonomie* whose treatment of the structure/superstructure distinction he summarized as follows: "The mode of production of material life *conditions* the social, political and intellectual life process of society" (Freudenthal and McLaughlin 2009, 42, emphasis added). Although he stressed the dependency of the superstructure on the economic basis, in my opinion he left open for debate the question of whether this *conditioning* had to be seen as monocausal and deterministic or not. By contrast, the leader of the Soviet delegation, Nikolai

[8] Hessen's essay first appeared in *Science at the Cross Roads* (London: Kniga, 1931), reprinted in 1971 (London: Frank). I will cite from the most recent edition in Freudenthal and McLaughlin (2009).

Bukharin, had a clear-cut opinion on this matter. In his popularizing introduction to Marxist philosophy, *The Theory of Historical Materialism: A Popular Manual of Marxist Sociology* (1921), he had argued, on a materialist basis, for the naturalization of the laws of social development as dealt with by Marxist sociology (or "historical materialism," according to his understanding of it). He also embraced a strictly deterministic conception of both natural and social processes. He wrote that "in nature and society there is a *definite* regularity, a *fixed* natural law. The definition of this natural law is the first task of science. This causality in nature and society is objective" (Bukharin 1934 [1921], 20). Social developments, according to him, are as predictable as eclipses or the boiling of water (Bukharin 1934, 51).

Hessen refrained from such a crude naturalization of history. Instead, he reflected on the dynamics of science in relation to economy and technology, on the one hand, and politics, philosophy and religion, on the other. However, according to the assumption of a fundamental dependency of the superstructural upon the structural, Hessen considered the socioeconomic linkage of science to be more fundamental than its cultural embedment. To sum up his model, "economics [...] *present[s]* demands, which *pose* technical problems, which *generate* scientific problems" (Freudenthal and McLaughlin 2009, 4).

The economy of Newton's time was that of the rise of early capitalism, marked by the expansion of merchant capital and manufacture. A new class of entrepreneurs emerged in late medieval and early modern Europe, and their interests were closely linked to the expansion of communication (and transport), industry and war. Hessen detailed the technical problems raised by the expansion of these three areas and the corresponding scientific fields that flourished during the sixteenth and seventeenth centuries in Europe. First, the needs of communication and transport, mainly maritime, required the improvement of vessels, the development of new navigation techniques, especially in the Oceans, and the building of canals and locks. In order to tackle the technical problems of navigation, improved hydrostatics and hydrodynamics were needed as well as astronomical, geographical, mathematical and optical knowledge. Second, the most important realms of industry were mining and war, and the technical problems of these realms could be solved by chemical and mechanical experts of a variety of topics ranging from simple machines to hydro- and aerostatics and the science of materials. Third, the military requirements of a time marked by the introduction of firearms led to the perfection of ballistics and fortification techniques, the study of dynamics and of architecture.

After his overview of the economic, technical and scientific characteristics of Newton's age, Hessen offered a summary of the central problems faced by early modern mechanics. Here the main issues follow: (1). simple machines, inclined planes and statics in general; (2). free fall and projectile trajectories; (3). hydrostatics and aerostatics, atmospheric pressure and the motion of bodies through a medium; and (4). celestial mechanics and the theory of tides. As collateral subjects, Hessen mentioned optics and magnetism. Additionally, he produced a list of the scientific protagonists—it comprised Leonardo da Vinci, Cardano, Del Monte, Stevin, Galileo, Tartaglia, Riccioli, Guericke, Pascal, Boyle, Kepler and a few other *big names* (Hessen in Freudenthal and McLaughlin 2009, 52). He also mentioned *en passant* Agricola, for mining, and Gilbert, for magnetism, a discipline whose economic-technological *raison d'être* resided in navigation. According to Hessen, Newton brought all of these branches of physics to a *theoretical* synthesis but this *intellectual* achievement should not obscure its deep economic and technical roots.

After considering the *structure* underlying early modern mechanics, Hessen dealt with its *superstructure*. As he cautioned his listeners,

> It would [...] be a gross oversimplification to derive *every problem* studied by various physicists, and *every task* they solved, directly from economics and technology. [...] The economic situation is the basis. But the development of theories and the individual work of a scientist are also affected by various superstructures, such as political forms of class struggle and its results, the reflection of these battles in the minds of the participants—in political, juridical, and philosophical theories, religious beliefs and their subsequent development into dogmatic systems.

The political context of Newton was that of "reformism" sanctioned by the Glorious Revolution of 1688, that is, a compromise between the interests of the Monarchy and those of the bourgeoisie that had heralded the English Revolution of 1648. According to Hessen, this climate of moderation and settlement (the "class compromise of 1688" as he called it) informed Newton's piety which, in turn, affected the *philosophical views* underlying the *Principia*. Thus, God and idealistic assumptions entered his conception of nature. Although radical materialistic and mechanistic views were available in his time (such as those by Richard Overton, Thomas Hobbes, René Descartes and John Toland), Newton refrained himself from these excesses. He assumed that a teleological principle of divine origin preserves the universe and its order; he considered motion to be a *mode* superimposed on essentially inert matter and viewed space as a *sensorium Dei* separable from matter. To sum up Hessen's considerations on superstructures and science, the cultural environment Newton operated within—in one word, his "ideology"—*only* accounted for the perceived shortcomings of his system, essentially its distance from an accomplished material and physical (as well as evolutionary) account of nature like the one that would be produced by Laplace in the time of the French Revolution.[9]

It should be remarked that Hessen did not posit a unidirectional dependency of science on technology. Rather, he pointed out a possible feedback mechanism, that is, a dialectical reinforcement of technology through science and vice versa: "The immense development of technology was a powerful stimulus to the development of science, and the rapidly developing science in turn fertilized the new technology" (Hessen 2009, 84). However, he did not go so far as to expand this idea on a more general explanatory level: He did not explicitly admit a similar dialectical loop between technology and economy and, even less so, between economy and politics, or politics and philosophy. In other words, his conception of the relation of structure and superstructure did not overtly challenge economic reductionism. However, he

[9] A similar idea, that ideology only accounts for the shortcomings of science, has been defended by the influential exponent of the French *épistémologie historique* George Canguilhem (2009).

acknowledged that science is affected by the cultural element alongside the economic-technological. As he did not survive the Stalinist purges, he could never develop and circulate the details of his research program.[10] It was Robert Merton's achievement to develop Hessen's insight by considering in more detail the incidence of technology, on the one hand, and puritan *ethics* (distinct from theology), on the other, in the natural debates of Newton's time in his classic of Weberian sociology of science, *Science, Technology and Society in Seventeenth Century England* (1938).[11]

With the benefit of hindsight some limitations of Hessen's approach become apparent. Although he embraced a view of technology as the product of collective and largely anonymous human activity and, on this basis, discarded Romantic views about scientific genius, he nonetheless focused on the *explanation of an individual work* through its context and did not put too much emphasis on the collective production of knowledge. Ludwig Fleck's concept of *Denkkollektiv*, referring to the collective dimension of science, and his emphasis on education as a central moment in the formation of a scientific culture, constitute a valid complement to Hessen's conception. Regrettably, even after the publication of Fleck's *Entstehung und Entwicklung einer wissenschaftlichen Tatsache* (The Genesis and Development of a Scientific Fact, 1938), the reception of his ideas was arrested by the outbreak of WWII, Fleck's deportation, and the dispersion of the German-speaking community of *Mitteleuropean* philosophers of science to whom his work was directed.[12] Eventually, Thomas Kuhn's *The Structure of Scientific Revolutions* (1962) disseminated and popularized Fleck's collectivist view of science, style of thought, and pedagogy by introducing categories such as "paradigm" and "normal science." Hessen did not deal with the epistemological problem of shared knowledge and paid but little attention to its collective character, although he incidentally referred to seventeenth-century academies, the *Accademia del Cimento* and the *Royal Society*, as new places of scientific experimentation and knowledge production. As for teaching institutions, characterized by the long-lived scholasticism, he saw the early modern universities as a hindrance to the advance of natural and technical knowledge. As I will discuss, Zilsel's position on the role of traditional education in the rise of modern science proved more detailed and comprehensive than that.

In recent years, our understanding of the shared character of theoretical knowledge has been improved through consideration of its long-lasting cognitive models, or "mental models." These have been defined as "instruments for drawing conclusions in the context of given knowledge" which "link present with past

[10] For the intellectual context of Hessen's work, see Winkler (2013). At his death, Hessen had an anthology of sources of early modern science ready for print. I am very thankful to Rose Luise Winkler and Peter McLaughlin for making it available at: http://www.philosophie.uni-hd.de/md/philsem/personal/hessen_textbook.pdf (accessed September 2, 2016). Cf. Winkler (2007).

[11] Merton (1938, b). Merton openly acknowledged his intellectual debt toward Hessen but limited this to the issue of technology. See 501–502.

[12] Cf. Schäfer (2012). On the split of rationality, *Spaltung der Rationalität*, resulting from the violent interruption of a virtuous synergy of natural science and philosophy from the mid-eighteenth century to the 1930s, see Engler and Renn (2010).

experiences by embedding new experiences in a cognitive network of previous experiences."[13] Unlike the platonically conceived scientific concepts investigated by Koyré in his idealistic history of science,[14] cognitive models bypass the sharp separation between experience and theory by identifying durable conceptual-experiential instruments, for instance the basic model of *motion-implies-force* that was crucial to a millenary tradition of Aristotelian and post-Aristotelian dynamics.[15] Attention to such elementary components of scientific thought permits us to bring into focus deep theoretical continuities in spite of apparent discontinuity and vice versa. For instance, one can mention the enduring presence and transformation of scholastic concepts in the physics of eminently anti-Scholastic scientists such as Galileo or Descartes (Damerow et al. 2004). Moreover, the inner constraints of shared models account for concomitant discoveries in the history of science, for instance the convergence of the methods and results in scholars working independently from each other, as in the case of Galileo and Thomas Harriot's dynamics (Schemmel 2008). In the light of shared knowledge, issues of priority become less relevant. In the case of planetary theory after Copernicus, it can be argued that the acceptance of his epicyclical models and parameters combined with the rejection of terrestrial motion and eccentricity *necessarily* led to the independent "discovery" of the geo-heliocentric system by several astronomers agreeing on common premises. Consideration of such theoretical constraints sheds a very different light on the past polemics over priority of discovery. In this case, the heated and well-known polemics over the paternity of geo-heliocentrism that burst out between the Danish astronomer Tycho Brahe and the Imperial mathematician in Prague, Nicolaus Reimers Ursus, should not obscure the fact that it is possible, even likely, that scholars working on the same *research program* which rests on the same premises can come to similar solutions to given problems.[16] This instance also shows that the inquiry into the conceptual dimension and the inner developments of science is not unrelated to its social context. Rather, consideration of shared knowledge offers an additional tool for the comprehension of the dynamics of knowledge at both a social level and a cognitive level.

Shortly after Hessen and independently from him, Henryk Grossman, a Marxist scholar linked to the Frankfurt Institute for Social Research, dealt with questions concerning science and society in the early phases of modernity in *Die gesellschaftlichen Grundlagen der mechanistischen Philosophie und die Manufaktur* (The Social Foundation of Mechanistic Philosophy and Menufacture, Paris, 1935). It was a critical review commissioned by Max Horkheimer, directed against the

[13] See Renn and Damerow (2007). See Chap. 1 of this volume.

[14] Cf. Koyré (1943). This essay can be seen as the author's manifesto of a disembodied history of science, as developed in his major works. The most important for the history of mechanics are *Études galiléennes* (Paris, 1939) and *Newtonian Studies* (Cambridge, Mass., 1965). In "Galileo Engineer" Lefèvre criticized Koyré's speculative attitude and his neglect of the social context of early mechanics.

[15] Cf. Renn and Damerow (2007).

[16] See Omodeo (2014).

Marxist historian Franz Borkenau's simplistic treatment of early mechanics. In *Der Übergang von feudalen zum bürgerlichen Weltbild* (The Transition from the Feudal to the Bourgeois World Picture, Paris, 1934), Borkenau argued that modern mechanics rose from a direct abstraction from the labor conditions of rising capitalism. Grossman reprimanded Borkenau for his lack of historical accuracy, which he saw as particularly due to a literal and rigid application of Marx's periodization, as well as an insufficient understanding of early-capitalist social developments. Most importantly for our present concern, Grossman argued that mechanics originated from reflection on machine technology in the late Middle Ages and the Renaissance. According to a thesis, which is also known as the "Hessen-Grossman-Thesis," "technology was developed *in order to* facilitate economic development, and science developed *by means of* the study of the technology that was being applied or developed" (Freudenthal and McLaughlin 2009, 4).

Grossman mentioned Da Vinci as the *Idealtyp* of modern mechanical thought. Moreover, he took Descartes into consideration, alongside Boyle, Galileo and Hobbes, to make the point that, for these early scientists, mechanics was *first of all a theory of machines*. As Grossman observed:

> the machines mentioned by Descartes [...] can be divided into four categories: artillery, clock, water and lifting machines, [they] also represent the most important areas of practical mechanics, by which the basic concepts and laws of theoretical mechanics could be developed. Mechanics was only slowly created by the struggle of human ratio with the empirical material.[17]

Hessen, who shared this view, also brought forward the reverse argument: The fields of physics that did not come to the forefront of the discipline in the seventeenth century are those that could not rely on a sufficiently developed technology (e.g. the science of heat, due to rudimentary steam engine technology) (Freudenthal and McLaughlin 2009, 20).

This insight by Hessen and Grossman has proved fruitful in the history of mechanics. In recent years, the study of the interrelation of technology and science has received new impetus. Accurate studies on Renaissance mechanics have deepened the link of this discipline with machine technologies. As has been argued, theoretical knowledge was gained through abstraction from experience and the reflection on "challenging objects" such as balances, turning wheels or projectiles.[18] Practice-oriented mathematicians like Cardano, Tartaglia, Del Monte, Benedetti, Stevin and Galileo built on craftsmen's experience and based their theories on technological objects from architecture, ballistic and navigation, to mention only the most visible developing fields of the early modern period. Grossman's idea that seventeenth-century "mechanistic philosophy and scientific mechanics derived their basic mechanical concepts from the observation of mechanisms and machines" is still valid (Grossman 2009, 107). Today we can rely on detailed accounts of the

[17] Cf. Grossman in Freudenthal and McLaughlin (2009, 141).

[18] On statics, cf. Renn and Damerow (2012). On turning objects, cf. Büttner (2008). On Ballistics, cf. Valleriani (2013). For studies in the history of mechanics making use of the concept of "challenging object," see Renn (2001) and Bertoloni Meli (2006).

interconnection between modern mechanics and machine technology, at the intersection of theoretical and practical activity. Indeed, the Renaissance was a time in which the rediscovery of Archimedes and the pseudo-Aristotelian *Mechanical Problems* merged with the scholastic science of weights, paving the way for the physics of Galileo and Newton and seventeenth-century mechanistic ontologies.[19]

3.2 From Economic Structures to Organic Intellectuals: The Gramscian Moment

Working in the same years as Hessen and Grossman, the political leader and thinker Antonio Gramsci, who was a prisoner of the Fascist regime, reflected on epistemology, culture and the status of science in a direction that was specifically aimed at avoiding economic reductionism. His *Quaderni del Carcere* (Prison Notebooks) bear witness to his indirect knowledge of and interest in Grossman's approach to economy.[20] As far as science is specifically concerned, Gramsci built up his positions in direct and explicit opposition to Nikolai Bukharin's *scientist* Marxism. He accused the Soviet intellectual of *vulgar* materialism, rejected his assumption that the study of human society and history should be founded on positive science (historical materialism seen as Marxist sociology) and refused a monocausal and deterministic outlook on society. In order to reject Bukharin's positions, as expounded in *Theory of Historical Materialism* and in the intervention at the London congress of 1931, Gramsci offered a different reading of Marx's preface to the *Kritik der politischen Ökonomie*, one that emphasized the superstructural element. As one reads in his Notebook XI (64), Gramsci (2007a)

> The issue of 'objectivity' of knowledge according to the philosophy of praxis [i.e. Marxism] can be developed beginning with the statement (entailed in the *Critique of Political Economy*) that 'men become conscious (of the conflict between material productive forces) in the ideological terrain' of juridical, political, religious, artistic and philosophical forms.[21]

These words clearly show that Gramsci—who even undertook an Italian translation of Marx's preface in his *translation notebooks* (Gramsci 2007b vol. 2, Notebook 7[a], 745–737)—did not see the architectonic metaphor of *Basis* and *Überbau* as implying economic necessity and the unidirectional dependency of the ideological forms on the structure.[22] Rather, he set the accent on Marx's observation, in the

---

[19] For a recent treatment of this trajectory, see Renn and Damerow (2010).

[20] Gramsci (2007a, 890). The title of this section is inspired by one of the most updated introductions to Gramsci, Thomas (2009).

[21] Gramsci (1975, 1492) (author's translation): "La questione della 'obiettività' della conoscenza secondo la filosofia della prassi può essere elaborata partendo dalla proposizione (contenuta nella prefazione alla *Critica dell'economia politica*) che 'gli uomini diventano consapevoli (del conflitto tra le forze materiali di produzione) nel terreno ideologico' delle forme giuridiche, politiche, religiose, artistiche, filosofiche."

[22] For an insightful treatment of diverging perspectives on structure and superstructure and their interconnection in Marxist thought, see Williams (1973).

same preface, regarding the epistemological relevance of superstructures as the means by which people grasp and conceive of their positioning in the world. Thus, Gramsci pointed out the *superstructural* conditioning of science but added a significant specification:

> One can say, however, that in the study of superstructures, science occupies a special place [*un posto privilegiato*], for its relation to the structure has a particular character, wider in extension and closer as far as the continuity of its development is concerned.[23]

Gramsci resorted to another metaphor, a *biological* one, taken from the same pages of the *Kritik der politischen Ökonomie* in which the *Basis-Überbau* idea is introduced. Marx stated there that economy deals with the *anatomy* of society.[24] Gramsci used the image of a living organism to *translate* the architectonic relation of structure and superstructure into the interdependency of skeleton and skin in a living body, in which the parts concur with the constitution of the whole:

> Structure and superstructure have a necessary and vital connection. [...] One cannot say that, in the human body, the skin and the type of physical beauty that prevails historically are mere illusions and that the skeleton and anatomy are the only reality, although something similar has been said for a long time. By enhancing the value of anatomy and the function of the skeleton nobody intended to assert that man, and even less so woman, can live without the latter.[25]

Gramsci's perspective aimed at undoing the allegations of "economicism" and "fatalistic mechanicism" often raised against Marxist thought. He took very seriously the historicist and idealistic objections raised by the philosopher Benedetto Croce, and partly accepted his criticism. Gramsci derived from Croce the idea that, complementary to economic analysis, a historical comprehension of society has to take into account culture, thought and *hegemony*, that is, intellectual and moral leadership and the construction of consensus, as essential elements of politics, besides domination and coercion.[26]

Such considerations motivated Gramsci's inquiries into the history of *intellectuals* in the thirteenth prison notebook. He stressed the social relevance of those intellectuals who embody the political aspirations of large social groups and are able to reinforce their interest by means of cultural and educational programs. Using a biological metaphor, Gramsci called them "organic intellectuals." Their realm of activ-

[23] Gramsci (2007, 1457): "Si può dire, tuttavia, che nello studio delle superstrutture la scienza occupi un posto privilegiato, per il fatto che la sua relazione sulla struttura ha un carattere particolare, di maggiore estensione e continuità di sviluppo [...]."

[24] Marx (1987, 100) "[...] daß aber die Anatomie der bürgerlichen Gesellschaft in der politischen Ökonomie zu suchen sei."

[25] Notebook X, 41. See Gramsci (2007, 1321) (author's translation): "Tra struttura e superstruttura esiste un nesso necessario e vitale. [...] Nel corpo umano non si può certo dire che la pelle (e anche il tipo di bellezza fisica storicamente prevalente) siano mere illusioni e che lo scheletro e l'anatomia siano la sola realtà, tuttavia per molto tempo si è detto qualcosa di simile. Mettendo in valore l'anatomia e la funzione dello scheletro nessuno ha voluto affermare che l'uomo (e tanto meno la donna) possano vivere senza di essa." Also, see Notebook X, pt. 1, 12, note 5: vol. 2, 1237–1238.

[26] Gramsci (2007, vol. 2, 1211).

ity is civil society: They can be seen indeed as *superstructural functionaries* (Gramsci 2007a, Notebook XIII 1, 1518–1519). Their *action* is as important as the government of political leaders and the economic activity of entrepreneurs, since culture is indispensable for the *living body* of human society. It should be remarked that organic intellectuals are not only lay people, as was the case with the public figure of the liberal philosopher Croce, but also exponents of the clergy. Gramsci did not neglect the hegemonic force of the Church in the organization and direction of society by means of education, culture and religion. Gramsci—who was sympathetic with Max Weber's inquiry into the ethical-religious side of capitalism—looked at Catholic institutions, such as the Jesuit Order, as historical examples of cultural hegemony. In such a perspective, the activity of a Renaissance Inquisitor such as the Jesuit Cardinal Robert Bellarmine proved much more effective in shaping modern Italy than the *rather individualistic* intellectual work of a "philosopher of immanence" like Giordano Bruno or a *modern* scientist like Galileo.[27]

Gramsci also addressed the delicate issue of the relation between politics, culture and religion. He especially reflected on the Jesuit cultural endeavor, which he treated from a cultural-political perspective as part of a wide struggle for hegemony in the field of ideas. In this light, he regarded a theological champion of the Counter-Reformation such as Bellarmine as an organic intellectual who effectively contributed to forge and develop a vision and project of social order. Not only was Bellarmine's epistemology embedded in a deeply political discourse, but his rehabilitation and his posthumous career as a Saint and as a Doctor of the Church was also part of a political project. This happened in particularly obscure years of European history. Bellarmine was in fact beatified in 1923, sanctified in 1931 and eventually elevated to the dignity of a Doctor of the Church in 1931. As Gramsci remarked in his *Prison Notebooks*, this apotheosis can be understood only against the background of the evolving relations between the Church and the State, of the expansion of the Jesuit influence within the Catholic Church and within society, and of the eventual collaboration between the Fascist State and the Vatican (Omodeo 2011). In those years, through the *Patti lateranensi* (*Lateran Accords*, 1929), Benito Mussolini obtained the support of Pius XI at the cost of a series of economic and civil concessions to the Church, including the teaching of religion in public schools. These new political liaisons between Italy and the Catholic Church facilitated the transfer from Madrid to Rome of the editorial project of the *Monumenta Historica Societatis Jesu*, a vast operation aimed at presenting archival documents relative to the early years of the Jesuit Society to learned scholars and thus to induce a reappraisal of its history (Koch 1934). As for the reasons for Bellarmine's sanctification, they shall not be searched far from these events. According to the Jesuit Xavier-Marie Le Bachelet in the *Dictionaire de théologie catholique* (1932), the major theological merits of Bellarmine were genuinely "political" (Le Bachelet 1932). The significance of this strenuous opponent of the Reformation and of *all heresies* should not be restricted to his polemics against the Lutheran *servum arbitrium*. Rather, Bellarmine's doctrine included issues such as

[27] Cf. Omodeo (2011, 41–48).

the affirmation of the primacy of the Roman pontiff, his *indirect* power over the worldly sphere, and the superiority of his *divine* monarchy over *human* civil powers. Gramsci's reflections on such a prominent early modern intellectual caution us against the pitfalls of studies on so-called "Jesuit science" if the political dimension is not taken into due account.

As important as it is in general terms, Gramsci's treatment of politics, culture and religion (and of intellectuals as cultural-political agents) does not include much discussion of *scientists*, that class of intellectuals of particular concern here.[28] In the following, I will recount the most perceptive analyses about early modern scientists and then consider the perspectives that can be opened by integrating them within a cultural-political framework à la Gramsci.

From the viewpoint of the study of scientists as intellectuals, Edgar Zilsel's seminal work on the social context of science in the time of the Scientific Revolution is particularly relevant. In his best-known essay "The Sociological Roots of Science," which appeared in *The American Journal of Sociology* in 1942, he investigated the origins of the figure of the *modern scientist* tracing his[29] legacy back to three types of "knowledge bearers" corresponding to three "strata of intellectual activity:"[30] university teaching, humanistic literacy and craftsmanship.

Zilsel's explanatory framework is economic: The rise of modern science is directly connected with the shift, in European history, from feudalism to capitalism, in the period between 1300 and 1600. He faced the question—similar to that famously investigated by the Marxist historian of science Joseph Needham in his monumental *Science and Civilization in China* (1954, 1st volume)—of why modern science emerged at a certain point in space and time, that is, early modern Europe. Zilsel (who actually identified *science* and *modern science*) sought the answer in the socio-economic conditions of the rising capitalist society, in which urban classes of artisans and merchants augmented their power at the expense of the traditional aristocracy and clergy. Their material interests favored a *worldly* enterprise such as natural science, opposed to transcendent theology and aristocratic military art (i.e. the cultural fields of the ruling classes of feudalism). Bourgeois individualism produced a shift of values. It fueled competition and criticism at the expense of established authority. Criticism—Zilsel argued—is an ingredient of modern science. Moreover, a society emerged based on a rational division of labor. In it, technology and technological inventions had a recognizable economic value, unlike ancient societies, in which slave exploitation did not make the rationalization of labor and the improvement of the means of production appear necessary. By contrast, in the age of early capitalism, technology (e.g. machinery) and rationalization (e.g. bookkeeping) became crucial and furthered the development of quantitative methods (Zilsel 2000, 936–938). At an intellectual level, technological advance, geographical discoveries, and economic transformation created a sense of superiority of

[28] Cf. Garin (1958, 1). "Gramsci risente di tutto un clima culturale […] nella limitata attenzione rivolta […] agli "scienziati." Also, see Geymonat, (1958, 148).

[29] It was only men in the period considered by Zilsel.

[30] Both expressions stem from Zilsel. See Zilsel [1942] (2000).

modernity over tradition producing a profound crisis of both university scholasticism and humanism. At once, a new alliance between superior craftsmen and "respectable" scholars was established which constituted the necessary premise for the new science (Zilsel 1941, 27). Under these conditions a new type of intellectual emerged, as a sort of Gramscian "intellectual" concurrent with the rising bourgeoisie: the modern scientist, dealing with "worldly phenomena," endowed with a critical mindset, and interested in practical matters and technology.

The thesis of a causal dependency of *science* (in the singular) on the emergence of capitalism is also known in the history of science as the "Zilsel-Thesis." Particularly relevant to the discussion of the *intellectuals* involved in scientific advancement is what we could call a "corollary" of the Zilsel-Thesis: The societal transformations induced by economy made age-old prejudices and social barriers fall down, in particular the radical separation between *free* intellectual activity and manual labor. This permitted the new figure of the *modern scientist* to take shape, as somebody capable of connecting the experimental method of the lower craftsmen and the rational and systematic methods of upper-class elites, university-trained or humanistic:

> On the whole, the rise of the methods of the manual workers to the ranks of academically trained scholars at the end of the sixteenth century is the decisive event in the genesis of science. (Zilsel 2000, 945)

This merging of lower and higher cultures was no simple task. It was eased, in Zilsel's eyes, by the establishment of a class of *superior craftsmen* who occupied an intermediary rank between university scholars and the humanistic literati, above them, and the mass of "artisans, carpenters, shipbuilders, carpenters, foundrymen, and miners," beneath them. Superior craftsmen were acknowledged professionals: surgeons, makers of nautical, astronomical and musical instruments, surveyors, navigators and, most importantly for the history of mechanics, those whom Zilsel labeled the "artist-engineers." The latter category—which Zilsel freely derived from earlier scholarship, in particular from Leonardo Olschki—refers to the likes of Filippo Brunelleschi, Leonardo da Vinci, Leon Battista Alberti and Albrecht Dürer, people who "constructed lifting engines, canals and sluices, guns and fortresses [...] invented new pigments, detected the geometrical laws of perspective, and constructed new measuring tools for engineering and gunnery" (Zilsel 2000, 941). These were the "immediate predecessors" of the modern scientist.

The question arises: When did the modern scientist appear on the stage of history? *Real science*, as Zilsel stated, was born with Gilbert, Galileo and Francis Bacon. He extolled Gilbert's work on magnetism, *De magnete* (1600), as "the first printed book, written by an academically trained scholar and dealing with a topic of natural science, which is based almost entirely on actual observation and experiment" (Zilsel 1941, 1). Additionally, he observed that Gilbert's critical spirit was akin to that of radically anti-Aristotelian natural philosophers such as Bernardino Telesio, Tommaso Campanella, Giordano Bruno and Francesco Patrizi. However, Zilsel saw Renaissance naturalism as the "older brother, not the father" of experimental science. Gilbert's science stemmed from the practical knowledge of miners, foundrymen, navigators and instrument-makers (Zilsel 1941, 24). Thus, Zilsel regarded Gilbert as the *first modern experimentalist*.

The second prototype of the *modern scientist*, Bacon, who was seen by Zilsel as the champion of a science-oriented philosophy aiming at the control of nature and the advancement of learning for the common good. Bacon's technological utopia, *Nova Atlantis* (1627), inspired the founders of the first modern scientific societies in London and Paris (Zilsel 2000, 943–945). "Bacon, however, did not make any important discovery in the field of natural science, and his writings abound with humanistic rhetoric, scholastic survivals, and scientific mistakes" (Zilsel 2000, 944). From this viewpoint, Bacon did not "advance" much further than the aforementioned Italian naturalist philosophers.

Thirdly, Zilsel pointed to Galileo as the first intellectual capable of merging and harmonizing "craftsmen-like experimentation and measurement with learned mathematical analysis" (Zilsel 1941, 943). This is a reference to the two poles of Galilean epistemology, *sensate esperienze* and *necessarie dimostrazioni*. Zilsel believed that the different origin of the two constituents could be detected in Galileo's stylistic choices in the *Discorsi e dimostrazioni matematiche intorno a due nuove scienze* (Discourses and Mathematical Demonstrations Relating to Two New Sciences, 1638): Galileo presented the theorems in Latin, the language of erudition, and used Italian, the spoken language, for the discussion and assessment of the theories. This linguistic divide mirrored the social and epistemological separation that Galileo was successfully overcoming between the empiricism of the lower classes and the systematic reasoning of educated scholastics and humanists. What Zilsel neglected to mention, though, is that the elegant Italian spoken in the courts was not the same as the language of the folk. However, although this rectification might shed a different light on the *persona* of Galileo, who was a *courtier* rather than a *popular*, the idea of a double origin of his scientific method does not have to be abandoned.

It seems to me that two of Zilsel's insights are particularly fruitful when considering the history of science: first, his stress on the role of superior artisans and artist-engineers in the development of natural knowledge and mechanics; second, his identification of scholasticism and humanism as a rational-rhetoric culture that was at once rejected and absorbed by scientists and scientifically-minded philosophers of the seventeenth century.

The relationship between artists and early scientists has proved a precious heuristic hypothesis in studies on artisanal and practical knowledge, often connecting science and art history.[31] In recent years, Pamela Smith has convincingly argued that the artistic/artisanal experience of the late Middle Ages and the Renaissance was an important constituent of early modern epistemology. Artistic production, seen as an activity capable of revealing the works of nature from inside, paved the way for experimental science; while naturalism in art corroborated the shift from a principle of authority, that is, the reliance on the scholarly tradition, to the observation of

[31] On artisanal knowledge and its codification see Smith (2004) and Long (2001). On practical knowledge, see Valleriani (2017). On art and science in the Renaissance, one can look at, among others, Bredekamp (2001).

nature as the proper source of worldly knowledge (Smith 2004). As to the social necessity of "hybrid experts" capable of bringing together the technical and the theoretical dimensions of knowledge, their relevance would never diminish from the late Middle Ages to the Industrial Revolution and beyond.[32] During the Renaissance, this mediation was secured by a new group of intellectuals, already singled out by Zilsel, and later labeled as "scientist-engineers." Arguably, the most eminent among them was Galileo, the latecomer of a series of court mathematicians like his protector Del Monte and Benedetti.[33] In *Galileo Engineer* (2010), Matteo Valleriani has offered an accurate description of such scientist-engineers. Galileo and his peers were well versed both in the technical and in the intellectual dimensions of knowledge. They underwent a period of apprenticeship of practical mathematics, in some field of application like architecture or the art of war, but later distanced themselves from artist-practitioners as they aspired to gain higher social recognition and prestige, especially at a court. They had a high degree of education as they mastered theoretical mathematics, the language of the learned, Latin, and the elegant literary Italian that was highly appreciated in the Renaissance courts. Scientist-engineers thus acted as mediators connecting the centers of power and decision, on the one side, and the workshops and building sites, on the other. As experts, they supervised artisanal work or gave advice on technical issues (Valleriani 2010, 208). As courtiers, they were additionally required to participate in the refined dialogical and literary culture of the elites, to serve as educators and to use their expertise to satisfy the curiosity of their rulers, for example, by casting horoscopes. I will come back to the sociopolitical issue of the courtly embedment of early modern scientists in Chap. 7.

A second important aspect highlighted by Zilsel relates to the institutional settings of knowledge: the towns of merchants and artisans, the universities of scholastic professors and the municipalities and courts of the humanists. In this respect it has to be remarked that university history has incredibly expanded as a field of inquiry in recent years. Historians of science today tend to see education as crucial to the understanding of the reception, implementation, systematization and transfer of scientific theories.[34] I will limit myself to mentioning only a few significant contributions to our understanding of the role of institutions of higher education in shaping modern science. In his studies on Renaissance universities and in particular the Studium of Padua in Galileo's times, Charles Schmitt has stressed the centrality of university Aristotelianism as a model of rational strength, epistemological reflection and systematicity. The clearest example of scientific advance fostered by Aristotelian training is William Harvey's functional anatomy. His explanation of the heartbeat by means of a scholastic *vis pulsiva* proved superior to Descartes'

[32] Ursula Klein has made this point most forcefully in Klein (2015).

[33] The figure of the "scientist engineer" was widely discussed in the history of science by Renn (2001), particularly in the contributions by Lefèvre (2001) and by Renn et al. (2001). Valleriani discusses it in detail in *Galileo Engineer* (2010, Chap. 6).

[34] For an assessment of the relevance of university history for the study of knowledge transfer, see my introductory chapter to Omodeo with Friedrich (2016, 3–21).

mechanical substitution of such an "occult quality" for a chemical-hydraulic model of the circulation of blood.[35] Further studies worth mentioning are those by Mordechai Feingold pointing to the decisive role played by English universities in the development of a mathematical culture in Newton's century and those by Antonella Romano on the teaching of mathematics at Jesuit colleges in France (Feingold 1984; Romano 1999). In the history of science, an interesting case is the dissemination of Copernican astronomy in the network of reformed universities gravitating around Wittenberg, in spite of the aversion of the two leading figures of the Reformation, Martin Luther and Philipp Melanchthon, toward the heliocentric and geokinetic theory.[36] This happened as a side effect of the reinforcement of the teaching of mathematics and astronomy for pedagogical and religious reasons—a heterogony of ends that Merton already pointed out with irony (Merton 1970 [1938]).

Regarding the connection of science with humanism, Zilsel rightly affirmed the importance of this cultural trend in the early modern period but his assessment was prejudiced by an exclusive consideration of the literary and rhetorical strands of humanism. Later scholarship has considered many instances of humanists devoted to mathematics and the natural sciences.[37] One of the most significant was Federico Commandino whose contribution to the *Archimedes Renaissance* cannot be neglected in the history of mechanics.[38] Arguably, new editions of classics from antiquity and the Middle Ages—including Latinized sources of Islamic origin—challenged modern readers, fueled debates and polemics, and disseminated intellectual tools that proved fundamental for the advance of science. The humanists Georg Peuerbach and Johannes Regiomontanus contributed to the restoration in Europe of Ptolemy's mathematical astronomy while Erasmus of Rotterdam disseminated the *other Ptolemy* by publishing the *editio princeps* of the *Geography* (1533).[39] Among the innovators of cosmography, Copernicus and Amerigo Vespucci were both imbued with humanistic culture.[40] To be sure, this list could be extended further.

Thus, in the light of more recent studies on institutions of higher education and humanism Zilsel's positions have to be integrated and improved; however, his suggestion to take them into account remains apt. Today, most historians of science agree upon the importance of scholasticism and humanism as cultural strands underlying early modern science alongside practical knowledge, but, unlike in the

[35] See Schmitt (1981). Regarding Descartes's views on the heartbeat see, among others, Grene (2005). As an insightful case study about the connection between mechanics and medicine in the seventeenth century via mechanicism, see Bertoloni Meli (2011).

[36] See, among many publications on the subject, Westman (1975).

[37] The classic reference is Rose (1975), although the emphasis on humanism shows clear bias toward idealistic history and Eurocentrism.

[38] See, among others, d'Alessandro and Napolitani (2013).

[39] The best study on Peurbach and Regiomontanus is Zinner (1968), Engl. transl., *Regiomontanus: His Life and Work* (1990). On the wider humanistic context, also see Omodeo and Pasini (2014).

[40] On Copernicus' humanism, see Hallyn (2000), on Vespucci, Vogel (2006).

seminal debates of the first half of the twentieth century, a clarification of the general historical and epistemological coordinates that permits us to deal with the complex interdependency of science and sociocultural phenomena is often lacking. Hence, Zilsel's program—reflecting a more general Marxist program—of an interconnected understanding of the economic, institutional, and sociological settings, on the one hand, and the individuation of the cultural strands informing science, on the other, is still valuable.

Some limitations of Zilsel's perspective should be mentioned as well. First and most evidently, he conceived of science as a uniquely modern phenomenon. This identification highlights his marked Eurocentric perspective: "Fully developed science appears once only, in modern Western civilization" (Zilsel 2000, 396). In this respect, Hessen's approach was more nuanced. He investigated the early capitalist and social context of Newton's *Principia* without excluding that science has developed in the past, and can be furthered in the future, in different socioeconomic settings—to be sure, with different characteristics than Newton's science.

Moreover, the priority Zilsel allotted to socio-economic factors and his predilection for the popular classes whom he saw as the main makers of modern experimental science produced blind spots in his historical analyses. Take his interpretation of Gilbert. As has been contended by critics, Zilsel's *exclusive* interest in Gilbert's *experimental* method made him neglect his participation in some of the most lively debates of his time, specifically, the insertion of his magnetic theory in the Copernican cosmological debates (Henry 2001). Notably, the controversy over the *worldly* or *heavenly* origin of early modern physics received an ideological connotation in the years of the Cold War. Just as Marxist scholars such as Zilsel emphasized experimentation, intellectual historians such as Koyré overemphasized the "Copernican Revolution" as the starting point of a process culminating with the Newtonian synthesis of terrestrial and celestial physics.[41] Koyré exploited the proximity of astronomy to general views about the world and the divine since antiquity to argue for the *spirituality* of science and its speculative origins.[42]

The last point I would like to raise concerns the insufficient analysis of the cultural dimension of modern science, in particular its relation to politics, by early social historians of science such as Hessen and Zilsel. To be sure, Hessen stressed the *ideological* dimension of the theological conceptions permeating Newtonian science and rightly pointed to the *political* significance of religious positions, but he did not further this line of thought. Moreover, he generally denied a positive function to philosophical and theological conceptions in the advance of science. Instead, he saw them as either accessories or hindrances to scientific and social progress. Zilsel looked closer at the social and cultural settings of modern science. On the one hand, he indicated specific places and institutions (town, university, court); on the

[41] Koyré (1961); Kuhn (1959); cf. Omodeo (2016b).

[42] Ernst Cassirer's understanding of the interconnection of astronomy and general worldviews in the Renaissance was led by a very different cultural agenda; his treatment was informed by the idea that the modern outlook coincided with a secularization of philosophy and of nature. See Cassirer (2002).

other hand, he specified the main intellectual strands (empiricism, scholasticism, humanism). He also named the different categories of knowledge bearers, namely the artist-engineer, the scholastic professor, and the humanist virtuoso who merged into the figure of the modern scientist. However, a comparison with Gramsci's "organic intellectuals" can show some limitations of Zilsel's concept of "knowledge bearers." While Zilsel's scientists and intellectuals are described in terms of the *function* descending from their *position*, Gramscian intellectuals are considered in relation to the *action* related to their *identity*. This difference helps to highlight what was excluded from Zilsel's consideration: the political dimension, that is, a reflection on agency and an understanding of institutional history in terms of hegemony and struggle. Both Hessen and Zilsel used arguments resting upon a rather simplistic treatment of culture, mechanically derived from class interests. My claim is that their lack of in-depth cultural and political analysis was a consequence of their philosophical premises. In Marxist terminology, they focused on the structural at the expenses of the superstructural or, in more plain terms, they almost exclusively treated the socio-economic aspects at the expense of the cultural-political.

This deficit was distinctly perceived by the scholars of the *cultural turn* of the late Eighties. As Steven Shapin and Simon Schaffer observed in a retrospective on their most influential work (speaking of themselves in the third person):

> The authors of *Leviathan and the Airpump* [1985] were [...] dissatisfied with the 'extenalism-internalism debate' [...] The problems, they thought, lay with the identity and coherence of the categories themselves. One incoherence concerned the placement of the boundary between what was deemed internal and what external to science. On what grounds were social and political things accounted not 'intellectual'? And how was it that the making and warranting of scientific knowledge was judged not 'social'? Did other *intellectual* practices—say religion and magic—count as external (since they were not considered to be 'scientific')? (Schaffer and Shapin 2011, XV)

Leviathan and the Airpump sanctioned a historical approach that, since the Eighties, has devoted great attention to actors' perspectives, intentions and interactions:

> Intellectual historians increasingly identified their goal as something like recreating past action in past actors' terms, and, from that point of view, the only pertinent categories and boundaries for interpreting past scientific action were said to be those recognized by those acting in the past. (Schaffer and Shapin 2011, XV)

To be sure, the *culturalization* of science has opened up entirely new fields of research—for instance the investigation of astrology and alchemy as integral parts of the natural discourse of the Renaissance. Moreover, after the cultural turn, the exploration of cultural, philosophical, and religious themes informing upon science has become common.[43] Topics such as the theological and ethical dimensions of science have increasingly attracted the attention of scholars interested in the *ideological* significance of the scientific debates of the past.[44]

---

[43] As an instance of culturalist revision (and revisionism) of earlier views about early modern science that emphasizes constituents such as religion, see Osler (2000).

[44] See for instance Steven Harris's treatment of "Jesuit spirituality" as a science-driving ideology in the context of early modern Jesuit engagement with scientific research and teaching, along a line of inquiry that has been opened up by Rivka Feldhay: Harris (1989).

In spite of the positive fact that the field of inquiry has been expanded, I claim that this move has produced severe losses in our overall comprehension. It seems to me that the culturalist reaction to Cold-War externalism (based on the equation of externalism and Marxist economicism) often undergoes the opposite mistake when compared with economic reductionism. I would call it "*culturalist* relativism" (an epistemological category not to be confused with *cultural* relativism, pertaining to ethics). It mostly rests on shaky ground, due to insufficient clarification of the concept of culture and its relation to knowledge. It reinforces *de facto* the postmodern relativization and subjectivization of historical categories. Apart from epistemological difficulties—and this is not the place to expand on the philosophical definitions of culture from the viewpoint of historical epistemology[45]—the new trend in science studies has abandoned socio-economic considerations as ideologically flawed, thus throwing out the baby with the dirty bathwater. Explicit political concerns and engaged historiography have also become rare, as they have been substituted by "narratives."

In light of these developments, an integrated treatment of economy and politics is desirable, as it can avoid the Scylla of economic reductionism and the Charybdis of culturalist relativism. Since the opening up of such a third way is the problematic with which Gramsci specifically struggled, in particular in Notebooks X, the so-called *Anti-Croce*, and XI, the so-called *Anti-Bukharin*, I deem it useful to retake his insights and some of his crucial categories, in particular his emphasis on hegemony as a means to integrate the social-economic analysis with the cultural-political. For that purpose, it is necessary to readdress fundamental questions such as the interrelation of structures and identities and, on the basis of a *theoretical translation* of Gramscian categories, integrate the social with the political.[46] Back in the Eighties, Stuart Hall saw Gramsci's philosophy of praxis as an alternative to structuralist approaches to cultural studies. Personally, I do not see the humanistic stress on agency, typical of Gramscian Marxism, as irreconcilable with structural analysis. As I have argued in this essay, it is rather the *nesso vitale*, the vital connection between socio-economic constraints and cultural-political action, that permits an integrated socio-political approach (Fig. 3.2) such a perspective will permit

[45] Let me stress the relevance of the philosophical discussions at the beginning of the twentieth century, ranging from neo-Kantianism and empiriocriticism to historical materialism, phenomenology and the philosophy of symbolic forms, as an extremely rich repository of perspectives and unfulfilled potentialities. On the divorce between science and philosophy in the turn of the 1930s, see Engler and Renn (2010). Moreover, for a critical assessment of the epistemological limitations of Cold-War philosophy of science, see Reisch (2005).

[46] Among the assessments on the vitality of Gramsci's thought today, especially see Anderson (2016). The presence and absence of Gramsci in science studies, in particular the concept of hegemony, is discussed in Nieto-Galan (2011), and Omodeo (2016a and 2016c). Nieto-Galan has particularly shown the usefulness of this appropriation for issues such as popularization and the circulation of knowledge in the public sphere: Nieto-Galan (2016). The theoretical reassessment of the Gramscian perspective, aimed at by this essay, is at the basis of the volume project edited by Badino and Omodeo (2019).

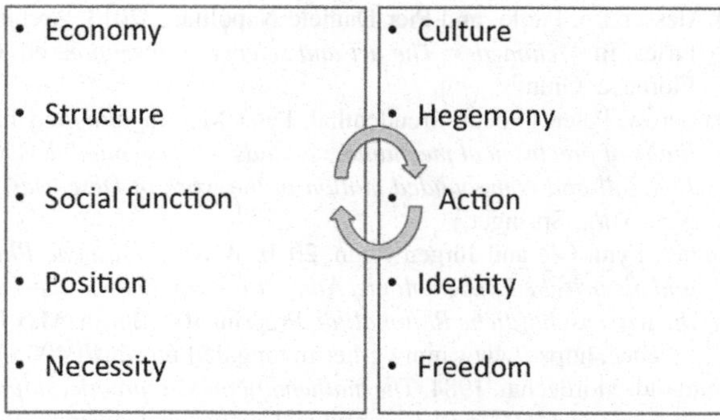

Fig. 3.2 Cultural hegemony as an integrative perspective bringing together structural analysis and the political understanding of cultural history

investigation into both poles of intellectual history without incurring economic monocausal determinism, on the one hand, and the excesses of culturalist relativism, on the other.[47]

References

Anderson, Perry. 2016. The heirs of Gramsci. *New Left Review* 100: 71–97.

Badino, Massimiliano, and Pietro D. Omodeo, eds. 2019. *Cultural hegemony in a scientific world: Gramscian concepts for the history of science*. Leiden: Brill.

Bertoloni Meli, Domenico. 2006. *Thinking with objects: The transformation of mechanics in the seventeenth century*. Baltimore: Johns Hopkins University Press.

———. 2011. *Mechanism, experiment , Disease: Marcello Malpighi and seventeenth-century anatomy*. Baltimore: The Johns Hopkins University Press.

Bredekamp, Horst. 2001. Gazing hands and blind spots: Galileo as a draftman. In *Galileo in context*, ed. Jürgen Renn, 153–192. Cambridge: Cambridge University Press.

Bruskell-Evans, Heather. 2015. The hegemony of psychology. In *Gramsci and Foucault: A reassessment*, ed. David Kreps. Farnham/Burlington: Ashgate.

Bukharin, Nikolai. [1921] 1934. *Historical materialism: A system of sociology*. New York: International Publishers.

Büttner, Jochen. 2008. Big wheel keep on turning. *Galilaeana* 5: 33–62.

Canguilhem, George. 2009. Qu'est-ce qu'une idéologie scientifique? In *Canguilhem. Idéologie et rationalité dans l'histoire des sciences de la vie*, 39–55. Paris: Vrin.

Cassirer, Ernst. [1927] 2002. *Individuum und Kosmos in der Philosophie der Renaissance*. Hamburg: Felix Meiner.

[47] In this respect, I deem post-structuralist readings of Gramsci, such as those of Ernesto Laclau and Chantal Mouffe, to sterilize rather than reinforce a crucial category such as that of hegemony by reducing it to identity-constitutive discursive struggles (Laclau and Mouffe 1985). By contrast, I would emphasize the fruitfulness of an approach to socio-cultural phenomena looking at the interrelation and tension between position and identity, as has been wonderfully done in the framework of Subaltern Studies. Gayatri Chakravorty Spivak, for one, has defined subalternity along a Gramscian line as "a position without identity" thus *appropriating* for subaltern studies a crucial issue of Marxist thought, traditionally addressed as the problem of the relation between class and consciousness. Cf. Spivak (2005, 476).

D'Alessandro, Paolo, and Pier Daniele Napolitani. 2013. Archimedes in the 12th and 16th centuries. In *Archimedes: The art and science of invention*, ed. Giovanni Di Pasquale, 138–143. Florence: Giunti.

Damerow, Peter, Gideon Freudenthal, Peter McLaughlin, and Jürgen Renn. 2004. *Exploring the limits of preclassical mechanics: A study of conceptual development in early modern science: Free fall and compounded motion in the work of Descartes, Galileo and Beeckman*. 2nd ed. New York: Springer.

Engler, Fynn Ole and Jürgen Renn. 2010. *Wissenschaftliche Philosophie, moderne Wissenschaft und historische Epistemologie. Albert Einstein, Ludwik Fleck und Moritz Schlick im Ringen um die wissenschaftliche Rationalität*. Preprint 400. Berlin: Max Planck Institute for the History of Science, https://www.mpiwg-berlin.mpg.de/Preprints/P400.PDF.

Feingold, Mordechai. 1984. *The mathematicians' apprenticeship: Science, universities and society in England 1560–1640*. Cambridge: Cambridge University Press.

Freudenthal, Gideon, and Peter McLaughlin. 2009. *The social and economic roots of the Scientific Revolution. Texts by Boris Hessen and Henryk Grossmann*. Dordrecht: Springer.

Garin, Eugenio. 1958. Antonio Gramsci nella cultura italiana. In *Studi gramsciani: Atti del convegno tenuto a Roma nei giorni 11–12 gennaio 1958*, 3–14. Rome: Editori Riuniti.

Geymonat, Ludovico. 1958. Per un intervento al convegno di studi gramsciani. In *Studi gramsciani: Atti del convegno tenuto a Roma nei giorni 11–12 gennaio 1958*, 147–148. Rome: Editori Riuniti.

Gramsci. [1975] 2007a. *Antonio Gramsci, Quaderni del Carcere*, ed. V. Gerratana. Turin: Einaudi.

———. 2007b. In *Antonio Gramsci, Quaderni di traduzioni (1929–1932)*, ed. Giuseppe Cospito. Rome: Istituto della Enciclopedia italiana.

Grene, Marjorie. 2005. Descartes and the heart beat: A conservative innovation. In *Wrong for the right reason*, ed. Jed Z. Buchwald and Allan Franklin. Dordrecht: Springer.

Grossman, Henryk. 2009. The social foundation of mechanistic philosophy and manufacture. In *The social and economic roots of the Scientific Revolution. Texts by Boris Hessen and Henryk Grossmann*, ed. Gideon Freudenthal and Peter McLaughlin. Dordrecht: Springer.

Habermas, Jürgen. 1969. *Technik und Wissenschaft als 'Ideologie'*. Frankfurt am Main: Suhrkamp.

Hall, Stuart. 1980. Cultural studies: Two paradigms. *Media, Culture and Society* 2 (1): 57–72.

Hallyn, Fernand. 2000. Copernic et Erasme. *Humanistica Lovaniensia: Journal of Neo-Latin Studies* 49: 89–100.

Harris, Steven J. 1989. Transposing the Merton thesis: Apostolic spirituality and the establishment of the Jesuit scientific tradition. *Science in Context* 3 (1): 29–65.

Henry, John. 2001. Animism and empiricism: Copernican physics and the origins of William Gilbert's experimental method. *Journal of the History of Ideas* 62 (1): 99–119.

Klein, Ursula. 2015. *Humboldts Preußen: Wissenschaft und Technik im Aufbruch*. Darmstadt: Wissenschaftliche Buchgesellschaft.

Koch, Ludwig. 1934. *Jesuiten-Lexikon: Die Gesellschaft Jesu einst und jetzt*. Paderborn: Bonifacius Druckerei.

Koyré, Alexandre. 1943. Galileo and Plato. *Journal of the History of Ideas* 4: 400–428.

———. 1961. *La révolution astronomique: Copernic, Kepler, Borelli*. Paris: Hermann.

Kuhn, Thomas S. 1959. *The Copernican revolution: Planetary astronomy in the development of western thought*. New York: Random House.

Laclau, Ernesto, and Chantal Mouffe. 1985. *Hegemony and socialist strategy: Towards a radical democratic politics*. London: Verso.

Le Bachelet, Xavier-Marie. 1932. Bellarmin. In *Dictionnaire de théologie catholique*. Paris: Letouzey.

Lefèvre, Wolfgang. 2001. Galileo engineer: Art and modern science. In *Galileo in context*, ed. Jürgen Renn, 11–27. Cambridge: Cambridge University Press.

Long, Pamela O. 2001. *Artisan/practitioners and the rise of the new science, 1400–1600*. Corvallis: Oregon State University Press.

Marx, Karl. 1987. A contribution to the critique of political economy. In *Collected works*, ed. Karl Marx and Friedrich Engels, vol. 29. London: Lawrence & Wishart.

Merton, Robert K. [1938] 1970. *Science, technology and society in seventeenth-century England*. New York/Evanston/London: HarperTorchbooks

Nieto-Galan, Augustí. 2011. Antonio Gramsci revisited: Historians of science, intellectuals, and the struggle for hegemony. *History of Science* 49: 453–478.

———. 2016. *Science in the public sphere: A history of lay knowledge and expertise*. London/New York: Routledge.

Omodeo, Pietro D. 2011. Roberto Bellarmino: il grande inquisitore. In *Il nostro Gramsci: Antonio Gramsci a colloquio con i protagonisti della storia d'Italia*, ed. Angelo d'Orsi. Rome: Viella.

———. 2014. *Copernicus in the cultural debates of the Renaissance: Reception, legacy, transformation*. Leiden: Brill.

———. 2015. Review-interview with Roger Cooter: The critical intellectual in the age of Neoliberal hegemony. *Journal for the Interdisciplinary History of Ideas* 4 (7): 5:1–5:20.

———. 2016a. After Nikolai Bukharin: History of science and cultural hegemony at the threshold of the cold war era. *History of the Human Sciences* 29 (4–5): 13–34.

———. 2016b. Copernicus as Kuhn's paradigm of paradigms: The epistemological dimension of The Copernican Revolution. In *Shifting paradigms: Thomas S. Kuhn and the history of science*, ed. Alexander Blum, Kostas Gavroglu, Christian Joas, and Jürgen Renn, 61–86. Berlin: Edition Open Access. http://edition-open-access.de/proceedings/8/7/index.html.

———. 2016c. Egemonia e scienza: Temi gramsciani in epistemologia e storia della scienza. *Gramsciana: Rivista internazionale di studi su Antonio Gramsci* 2: 57–84.

Omodeo, Pietro D., and Enrico Pasini. 2014. Erasmian science: the influence of Erasmus of Rotterdam on early modern science. *Journal of Interdisciplinary History of Ideas* 6 (2): 2:1–2:19.

Omodeo, Pietro D., and Karin Friedrich, eds. 2016. *Duncan Liddel (1561–1613), Networks of Polymathy and the Northern European Renaissance*. Leiden: Brill.

Osler, Margaret J., ed. 2000. *Rethinking the Scientific Revolution*. Cambridge: Cambridge University Press.

Porter, Roy. 1990. The history of science and the history of society. In *Companion to the history of modern science*, ed. R.C. Olby et al. London/-New York: Routledge.

Reisch, George A. 2005. *How the cold war transformed philosophy of science*. Cambridge: Cambridge University Press.

Renn, Jürgen, ed. 2001. *Galileo in context*. Cambridge: Cambridge University Press.

Renn, Jürgen, and Peter Damerow. 2007. Mentale Modelle als cognitive Instrumente der Transformation von technischem Wissen. In *Übersetzung und Transformation*, ed. Hartmut Böhme, Christoph Rapp, and Wolfgang Rösler, 311–331. Berlin: de Gruyter.

———. 2010. The transformation of ancient mechanics into a mechanistic world view. In *Transformationen antiker Wissenschaften*, ed. Georg Toepfer and Hartmut Böhme, 243–268. Berlin: de Gruyter.

———. 2012. *The equilibrium controversy: Guidobaldo del Monte's critical notes on the mechanics of Jordanus and Benedetti and their historical and conceptual background*. Berlin: Edition Open Access. http://www.edition-open-sources.org/sources/2/index.html.

Renn, Jürgen, Peter Damerow, and Simone Rieger. 2001. Hunting the white elephant: When and how did Galileo discover the law of fall? In *Galileo in context*, ed. Jürgen Renn. Cambridge: Cambridge University Press.

Romano, Antonella. 1999. *La contre-réforme mathématique: Constitution et diffusion d'une culure mathématique jésuite à la Renaissance*. Rome: École Française de Rome.

Rose, Paul L. 1975. *The Italian renaissance of mathematics: Studies on humanists and mathematicians from Petrarch to Galileo*. Geneva: Librairie Droz.

Schäfer, Lothar. 2012. Einleitung. In *Entstehung und Entwicklung einer wissenschaftlichen Tatsache: Einführung in the Lehre vom Denkstil und Denkkollektiv*, ed. Ludwik Fleck. Frankfurt a.M.: Suhrkamp.

Schaffer, Simon, and Steven Shapin. [1985] 2011. *Leviathan and the air-pump: Hobbes, Boyle and experimental life*. Princeton: Princeton University Press

Schemmel, Matthias. 2008. *The English Galileo: Thomas Harriot's work on motion as an example of preclassical mechanics*. Dodrecht: Springer.
Schmitt, Charles B. 1981. Towards a reassessment of renaissance Aristotelianism. In *Studies in renaissance philosophy and science*, ed. Charles B. Schmitt. London: Variorum Reprints.
Smith, Pamela. 2004. *The body of the artisan: Art and experience in the Scientific Revolution*. Chicago: The University of Chicago Press.
Spivak, Gayari C. 2005. Scattered speculations of the subaltern and the popular. *Subaltern Studies* 8 (4): 475–486.
Stachel, John. 1994. *Marx's critical concept of science*. Preprint 10. Berlin: Max Planck Institute for the History of Science.
Thomas, Peter. 2009. *The Gramscian moment: Philosophy, hegemony and Marxism*. Leiden/ Boston: Brill.
Valleriani, Matteo. 2010. *Galileo engineer*. Dordrecht: Springer.
———. 2013. *Metallurgy, ballistics and epistemic instruments: The Nova scientia of Nicolò Tartaglia*. Berlin: Edition Open Access. http://www.edition-open-sources.org/sources/6/index.html.
———, ed. 2017. *The structures of practical knowledge*. Cham: Springer Nature.
Vogel, Klaus. 2006. Cosmography. In *The Cambridge history of science*, ed. Katharine Park and Lorraine Daston, vol. 3. Cambridge: Cambridge University Press.
Westman, Robert S. 1975. The Melanchthon circle, Rheticus and the Wittenberg interpretation of the Copernican theory. *Isis* 66: 163–193.
Williams, Raymond. 1973. Base and superstructure in Marxist cultural theory. *New Left Review* 82: 3–16.
Winkler, Rose-Luise. 2007. Ein unveröffentlichtes Manuskript von Boris M. Hessen: 'Materialien und Dokumente zur Geschichte der Physik'. *Sitzungsberichte der Leibniz-Sozietät* 92: 133–152.
———. 2013. *An den Urspüngen wissenschaftssoziologischen Denkens. Erstes Drittel des XX. Jahrhunderts. Russland/Sowjetunion*. Berlin: trafo Wissenschaftsverlag.
Young, Robert M. 1990. Marxism and the history of science. In *Companion to the history of modern science*, ed. R.C. Olby et al., 23–31. London/New York: Routledge.
Zilsel, Edgar. 1941. The origins of William Gilbert's scientific method. *Journal for the History of Ideas* 2 (1): 1–32.
Zilsel, E. [1942] 2000. The sociological roots of science. Reprinted in *Social Studies of Science* 30/6: 935–939.
Zinner, Ernst. [1968] 1990. *Leben und Wirken des Joh. Müller von Königsberg genannt Regiomontanus*. English trans: *Regiomontanus: His life and work*. Amsterdam: North-Holland.

Part II
The Ancient Roots and Emergence of Preclassical Mechanics

Chapter 4
The Early History of Weighing Technology from the Perspective of a Theory of Innovation

Jochen Büttner, Jürgen Renn, and Matthias Schemmel

Abstract An extended model of cultural evolution is brought to bear on the development of practical and theoretical knowledge related to early weighing. We argue that this development can be characterized as an iterative process in which the exploration of the inherent potential of external representations of cognitive structures leads to the establishment of new cognitive structures, whereby this exploration is canalized at each step by historically specific contexts that constrain the actors. We argue that this scheme applies to technological innovation processes as well, and, based on the concrete example of the balance with variable arm length, implications are developed. Thus, the first theoretical writings on mechanics in the western tradition were indeed the result of a reflection on the external representations of weighing techniques. This is contrasted to the case of China. Comparing the historical developments of the two major types of balance with variable arm length—the Bismar and the Roman steelyard—we show how earlier developmental stages function as a scaffold for later techniques and, in particular, how the Roman steelyard required a rather elaborate societal and cognitive infrastructure as the basis for its standardized production. Based on an example drawn from Hero, we indicate how the development of weighing techniques and technical knowledge in turn influenced theoretical knowledge. (An earlier version of this chapter was published in 2016 in *eTopoi. Journal for Ancient Studies*.)

Keywords Evolution of knowledge · Mechanics · Weighing technology · Innovation · Practical knowledge · Unequal-arm balance · China · Greece · Roman Empire · Islamicate societies

J. Büttner (✉) · J. Renn · M. Schemmel
Max Planck Institute for the History of Science, Berlin, Germany
e-mail: jochen.buettner@topoi.org

R. Feldhay et al. (eds.), *Emergence and Expansion of Preclassical Mechanics*, Boston Studies in the Philosophy and History of Science 270,
https://doi.org/10.1007/978-3-319-90345-3_4

4.1 The Origin of Weighing Technology and its Conceptual Consequences

The technology of weighing emerged when administrative and economic developments of early urban societies began to involve standards for exchange values. In Mesopotamia standardized weights used for this purpose have been preserved since the ED IIIa (Fara) period (mid-third millennium BCE).[1] In the context of the political and economic globalization processes of the first millennium BCE, the role played by these crucial standards even increased.[2] By the middle of the first millennium, coined money was widespread in Lydia, Greece and India, and somewhat later also in China. In Egypt and probably slightly later in Mesopotamia, the lever balance with equal arms of fixed length was introduced at around the turn from the fourth to the third millennium. Balances evolved as well, but their basic principle remained the same for millennia: the weight of the item to be weighed on one arm of the balance was compensated (or literally "balanced") by the identical weight of one or more standardized balance weights placed on the other arm of equal length. This only changed when a new type of balance obeying a different principle emerged: the balance with variable arm length, more commonly referred to as the unequal-arm balance. This type of balance is recorded in the late fifth century BCE in Greece and may have been in use at the same time or somewhat later in India.[3] The spread and transformation of weighing technology was thus closely associated with economic evolution.

These economic and technological developments went hand in hand with conceptual transformations. The introduction of standards for exchange values together with an emerging practice of weighing gave rise to an abstract and quantitative concept of weight, distinguished from other bodily characteristics such as bulk or material quality. The spread of the unequal-arm balance led to a further differentiation of this concept, also taking the positional effect of a weight into account. In the course of the globalization processes of the first millennium, writing was simplified and spread. In particular, alphabetic writing was fully developed and various literary cultures formed in different parts of Western Eurasia and Northern Africa.[4] In Greek culture, characterized by its marginality to some of the major contemporary empires, its widespread connectivity and exchange with other cultures, plus the emergence of discursive practices beyond political and religious realms, these globalization processes formed the backdrop for the creation of a scientific literature in which the new abstract concepts were taken up and further developed.[5] The first Greek texts dealing with "mechanics" focused in one way or the other on the properties of the balance and the concepts that had been abstracted from weighing technology. These

[1] See Sommer (2013, Chap. 6).

[2] See Geller (2014).

[3] See Renn and Damerow (2012).

[4] See "Survey 1" in Renn (2012).

[5] See Malkin (2011).

concepts, rooted in globalized economic and technological developments, thus became part of a long-lasting literary tradition that persists in science even today.[6]

Various technological traditions evolved parallel to the emergence, transmission and transformation of this literary tradition. In the Mediterranean region during the period under consideration here, machines were circulating in increasing quantities; if not the actual machines, then at least the operative knowledge of how to build, use, maintain and repair them. Engineers and mechanics were rarely able to remain in the place where they accomplished a technological feat, and were obliged to travel to wherever a new machine was required. Thus, the need to communicate and disseminate the know-how associated with the new technology also emerged. It was technology, therefore, that represented the primary vehicle for the transfer of mechanical knowledge. Technologies changed over time and differentiated regionally, and such changes were not limited to the optimization of function. Indeed, a broad range of factors can be identified that triggered or even necessitated changes in certain technologies, such as the availability of new raw materials, new methods of fabrication or the widening of the range of application of a given technology.

The factors regulating the diffusion and development of the knowledge underlying the production, adaptation and use of technology are different from those governing the transmission and development of theoretical knowledge predominantly encoded in texts. For a very long time, the innovation and diffusion processes of these textual and technological traditions followed different pathways. This was due to the knowledge economies generating technological and intellectual novelties not being closely coupled until early modern times when technical artifacts became objects that challenged theoretical traditions.[7]

Nevertheless, the transmission of theoretical texts on mechanics could not be independent from the transmission of a material culture constituting key points of reference for concepts contained in these texts and inducing, at several junctures, important theoretical insights. Thus, it is hardly conceivable that the science of weights in the Arabic and Latin Middle Ages could have flourished without the material basis of widespread weighing practices. Intellectual novelties such as the elaboration of concise concepts for the positional qualities of weight depended, however, not only on technical developments, but also on changing discursive contexts such as those offered by the appropriation of Greek texts by Islamicate scholars and of Arabic texts by Latin scholars.[8] Technological innovations, on the other hand, such as the Roman steelyard, could hardly profit from theoretical knowledge that only dealt with fundamental principles, such as the law of the lever, but not with the intricacies of their material implementation. The extent to which genuinely theoretical insights may have nevertheless affected the course of the development of technology and technological knowledge in antiquity is still largely an open question.

[6] See Damerow and Renn (2010).

[7] See Büttner (2008a, 2008b); Renn et al. (2001); Valleriani (2009, 2010, 2012, 2013, 2014).

[8] See Brentjes and Renn (2016).

4.2 The Cultural Evolution of Practical and Theoretical Knowledge

In light of this—for the most part—independence of the technological and theoretical developments that interacted over a period of more than two millennia, the comparative study of the dynamics of the innovation characteristic of these two strands becomes relevant, as does posing questions about their commonalities and differences. So far, the focus of studies has been primarily on the scientific, theoretical side, mostly neglecting the distinct innovation dynamics of technology.

Scientific innovations are often contingent on technological developments. The law of the lever for instance was formulated on the basis of theoretically motivated reflections on a technical device: the unequal-arm balance. However, the transmission and accumulation of scientific knowledge largely depends on texts such as the early mechanical writings of Greek culture. Given the limited feedback of scientific knowledge on the technological development itself, its textual transmission in turn depended on historical contingencies other than those relevant to the development and spread of technology and technological knowledge.

Until modern times, text transmission in literate societies focused on administrative, practical, legal and religious texts, as well as on other literary texts constituting cultural identity, while philosophical and scientific literature with its limited practical value constituted at best a secondary phenomenon. Nevertheless, the knowledge economy dealing with these esoteric matters participated in general societal processes of incorporating new experiences. Reflecting on its own production, science was able to generate new abstractions—cognitive as well as institutional.

This dynamic, familiar from studies of the evolution of theoretical knowledge,[9] warrants closer inspection to assess its relation to technological innovation. As mentioned, weighing technology was originally introduced for regulating social and cognitive processes dealing with the exchange of goods. In weighing, these regulative processes find "external representations."[10] These external representations comprised among others standard weights, the balance and a specialized technical terminology. The reflection on these external representations gave rise to an abstract concept of weight, as part of a particular conception or mental model of equilibrium. This model turned out to be applicable not just to weights and weighing but also to other abstract values such as justice.[11] In fact, the embedding of the concept of weight in a broader linguistic usage increasingly connected it with other concepts or suggested metaphorical generalizations. Thus, the reflection on the external representations associated with weighing technology eventually led to a transformation and extension of social and cognitive structures, not all of which related directly to weighing.

Based on the concrete example of the balance with variable arm length, a brief explanation will now be given of how a more differentiated picture of the innovation dynamics of technology and the underlying technological knowledge can inform our general understanding of the relation between technology and science in antiquity.

---

[9] See Damerow (1996).

[10] See Laubichler and Renn (2015).

[11] See Renn and Damerow (2012).

4.3 Unequal-Arm Balances as a Technical Weighing Innovation

Balances with variable arm length, or unequal-arm balances as they are more commonly referred to, belong to the more general class of lever balances, characterized by a rigid beam that can turn around a point of suspension: the fulcrum. Such balances are in equilibrium when the sum of the moments of force with respect to the fulcrum is zero, a general condition that, under certain constraints, coincides with the law of the lever. Depending on how equilibrium is established in such balances, one can distinguish balances with a fixed arm length from balances with variable arm length.[12] In the case of balances with fixed arm length, equilibrium is produced by adding or taking away weights, that is, by altering the acting forces. The most common realization of this is the equal-arm balance which, as mentioned earlier, was first introduced around 3000 BCE. In balances with variable arm length, the counterweight remains unchanged. Instead, the distances at which the forces act from the fulcrum are varied to bring about equilibrium. Somewhat misleadingly, this type of balance is referred to as the unequal-arm balance (Fig. 4.1).

Unequal-arm balances can be further subdivided according to the way in which the relevant distances are varied. In the bismar, equilibrium is produced by altering the position of the fulcrum with respect to the beam, that is, by varying the distance at which the weight as well as the distance at which the load acts. In the more familiar Roman balance, also referred to as the steelyard, equilibrium is reached by varying the distance at which a movable counter-weight acts from the fulcrum.[13]

The earliest evidence of the introduction of balances with variable arm length comes from a play by Aristophanes, *The Peace,* which was first staged in Athens in 421 BCE. In the play, a maker of war trumpets is ridiculed because he cannot figure out what to do with his surplus trumpets. Trygaeus, the central character of the play, suggests pouring lead into the bell and to add "a dish hung on strings, and you will have a balance for weighing the figs." Despite being rather abridged, this description of the transformation of a trumpet into a balance makes it rather unambiguously clear that the trumpet is turned into a specific unequal-arm balance: the bismar.[14]

---

[12] An alteration of both the counteracting force and the length of the arms on which the forces act is conceivable and was in fact realized historically in form of the equal-arm balance with an additional counterpoise. A number of finds suggest that this type of balance, which was fairly common in the Roman imperial period, may be of earlier origin than the steelyard and the bismar types discussed in somewhat more detail in this article. The equal-arm balance with additional counterpoise has not yet received due attention in the literature. For an albeit cursory description, see Corti and Giordani (2001). At the current stage of research, we must assume that the steelyard evolved from this type of balance.

[13] The third perceivable variation, a balance in which the distance at which the load acts is altered to achieve equilibrium, was occasionally realized historically but never really established. See Jenemann (1989).

[14] See Büttner (2013).

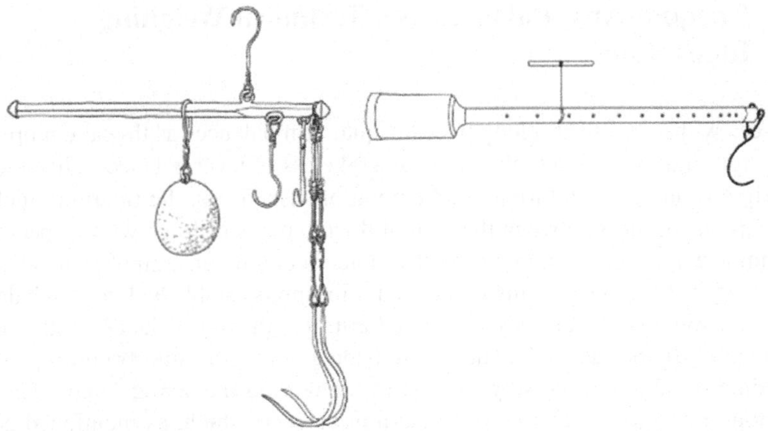

Fig. 4.1 Schematic representation of the two different major types of balances with unequal arms, the Roman steelyard (*left*, subtype Osterburken; redrawn from Franken (1993)) and the bismar or Danish balance (*right*)

4.4 The First Writings on Mechanics as a Theoretical Weighing Innovation

A subsequent step in the cultural evolution of weighing technology occurred when some of the extended cognitive processes it entailed (such as the introduction of an abstract concept of weight or the realization that weights can be compensated by distances in the new balances with variable arm length) were externalized by a new level of external representation: the documentation in written language. This step was, of course, not taken because of an intrinsic logic in the development of weighing technologies but for reasons completely external to it. In particular, the specific context of Greek culture gave rise to a tradition of philosophical writings dealing with natural processes and the astonishing power of human devices to modify their properties. The first documented example of a sustained theoretical reflection on mechanical knowledge is the peripatetic *Mechanical Problems,* written, at least in part, as early as 330 BCE[15] and passed down as authentically Aristotelian. The knowledge presented in the *Mechanical Problems* was the point of departure for later, more advanced work on mechanics that informed the writings of Archimedes, Hero of Alexandria or Vitruvius (Fig. 4.2).[16]

The theoretical knowledge represented by these texts was structured by mental models such as the equilibrium model, a model of causality relating force and effect, or the lever model, according to which a lever can be used to save force. Such mod-

[15] The work is presumably pre-Euclidean and may have been initiated during Aristotle's lifetime. Euclid's *Elements* is generally taken to have been compiled shortly after 300 BCE; Aristotle died in 322 BCE. See Renn and McLaughlin in this volume.

[16] Vitruvius' work on mechanics is contained in chapter X of his *De Architectura,* Vitruvius Pollio (1999). For a recent analysis of the *Mechanical Problems*, see Chap. 5 in this volume.

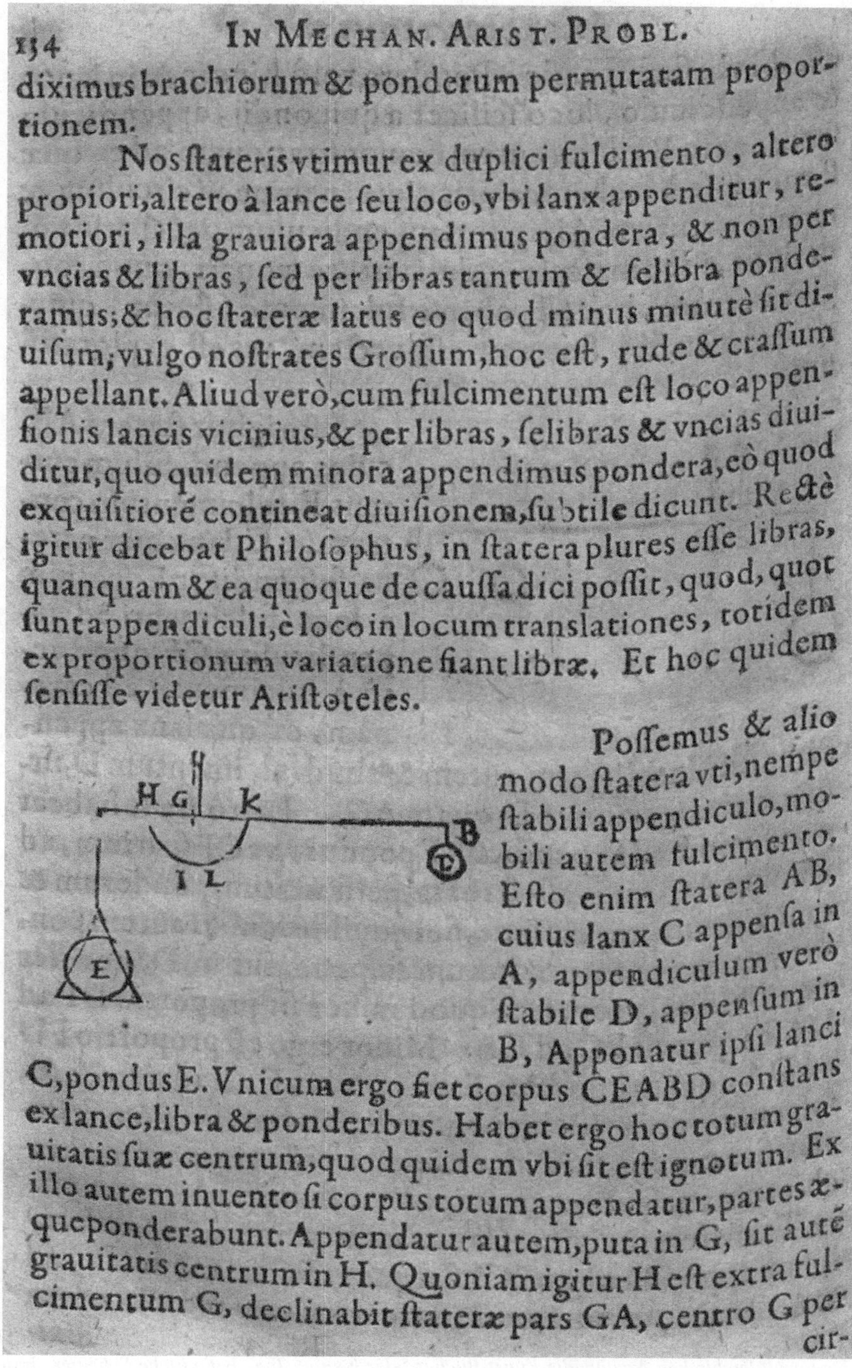

134 IN MECHAN. ARIST. PROBL.

diximus brachiorum & ponderum permutatam proportionem.

Nos ſtateris vtimur ex duplici fulcimento, altero propiori, altero à lance ſeu loco, vbi lanx appenditur, remotiori, illa grauiora appendimus pondera, & non per vncias & libras, ſed per libras tantum & ſelibra ponderamus; & hoc ſtateræ latus eo quod minus minutè ſit diuiſum, vulgo noſtrates Groſſum, hoc eſt, rude & craſſum appellant. Aliud verò, cum fulcimentum eſt loco appenſionis lancis vicinius, & per libras, ſelibras & vncias diuiditur, quo quidem minora appendimus pondera, eò quod exquiſitioré contineat diuiſionem, ſubtile dicunt. Rectè igitur dicebat Philoſophus, in ſtatera plures eſſe libras, quanquam & ea quoque de cauſſa dici poſſit, quod, quot ſunt appendiculi, è loco in locum translationes, totidem ex proportionum variatione fiant libræ. Et hoc quidem ſenſiſſe videtur Ariſtoteles.

Poſſemus & alio modo ſtatera vti, nempe ſtabili appendiculo, mobili autem fulcimento. Eſto enim ſtatera AB, cuius lanx C appenſa in A, appendiculum verò ſtabile D, appenſum in B, Apponatur ipſi lanci C, pondus E. Vnicum ergo fiet corpus CEABD conſtans ex lance, libra & ponderibus. Habet ergo hoc totum grauitatis ſuæ centrum, quod quidem vbi ſit eſt ignotum. Ex illo autem inuento ſi corpus totum appendatur, partes æqueponderabunt. Appendatur autem, puta in G, ſit autē grauitatis centrum in H. Quoniam igitur H eſt extra fulcimentum G, declinabit ſtateræ pars GA, centro G per cir-

Fig. 4.2 Page from Bernardino Baldi's *In Mechanica Aristotelis problemata exercitationes: adiecta succinta narratione de autoris vita et scriptis* of 1621. Problem 20 of the *Mechanical Problems* refers to a bismar. Baldi, as many other sixteenth-century authors studying and discussing the work, however, believed the problem to pertain to the familiar Roman steelyard and thus had considerable problems with its interpretation (Nenci 2011)

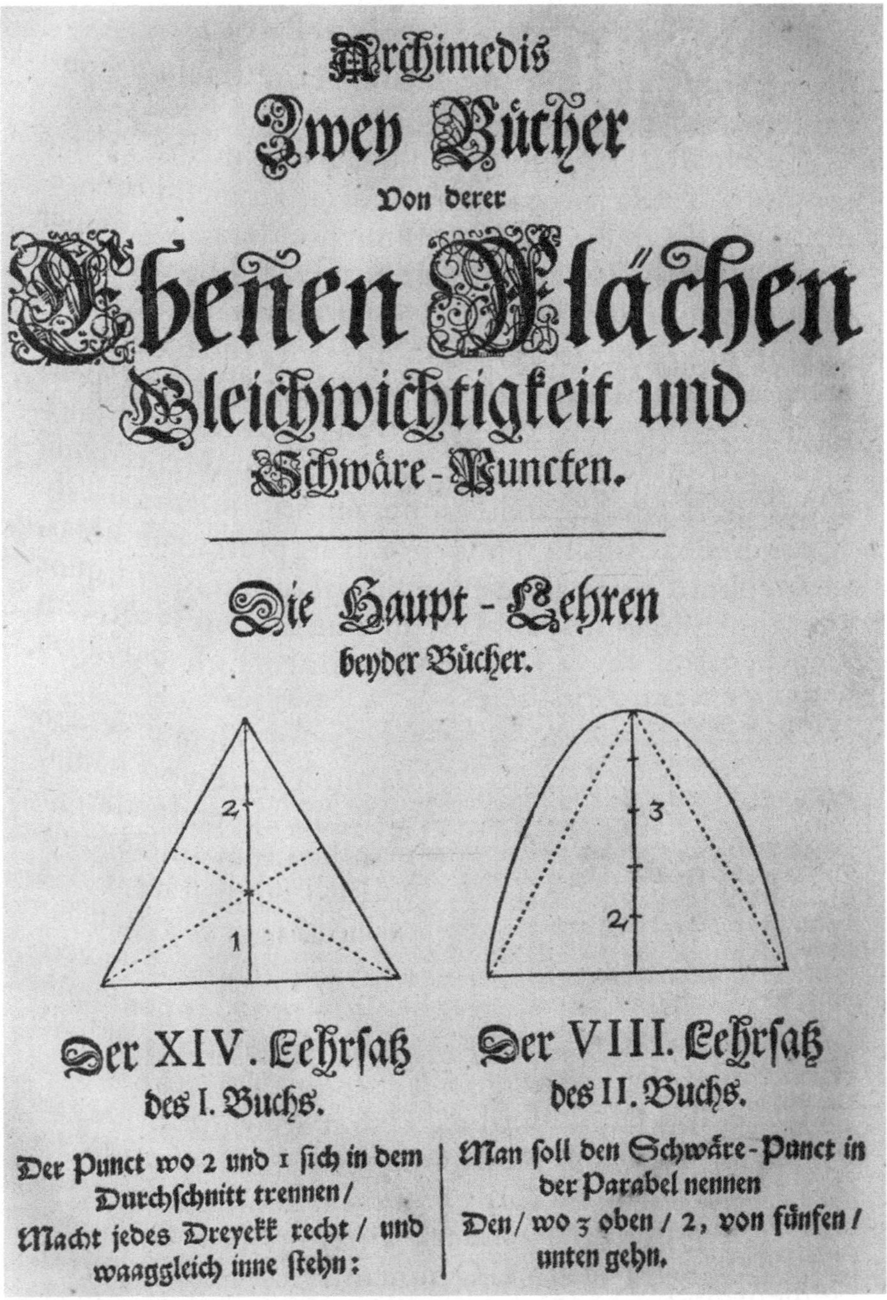

Archimedis
Zwey Bücher
Von derer
Ebenen Flächen
Gleichwichtigkeit und
Schwäre-Puncten.

Die Haupt-Lehren
beyder Bücher.

Der XIV. Lehrsatz
des I. Buchs.

Der Punct wo 2 und 1 sich in dem
Durchschnitt trennen/
Macht jedes Dreyekk recht / und
waaggleich inne stehn:

Der VIII. Lehrsatz
des II. Buchs.

Man soll den Schwäre-Punct in
der Parabel nennen
Den / wo 3 oben / 2, von fünfen /
unten gehn.

Fig. 4.3 Title page of Archimedes' *On the Equilibrium of Planes* in a German translation of 1670 (*Des unvergleichlichen Archimedis Kunst-Bücher oder heutigs Tags befindliche Schriften*, translated and commented by Johann Christoph Sturm, Nürnberg). In this work, Archimedes first introduced the concept of a center of gravity, which can be understood as generalizing the notion of the fulcrum of a balance

els were based on and encoded intuitive and practical physical experiences, among them the experiences gained in weighing. With the medium of writing, it became possible to reflect on the application of these models and relate them to each other. Thus, the joint application of both the equilibrium and the lever model to balances with unequal arms led to a new mental model: the balance-lever model. This balance-lever model provided the means to interpret various force-saving mechanical devices as working due to a compensation relationship between force and lever arm: a precursor of the law of the lever. This allowed an explanation of the apparent conflict between their force-saving power and the proportionality of force and effect suggested by the causality model.

Similarly, the reflection on the application of the equilibrium model, not just to balances but also to other devices, led to a further abstraction of this model by generalizing the fulcrum of a balance to the notion of a center of gravity in principle applicable to arbitrary bodies. The concept of 'center of gravity' played a crucial role, particularly in the work of Archimedes.[17] In his writings, these novel theoretical structures and their implications were represented with recourse to Greek mathematics, in particular, the theory of proportions so that the law of the lever could be quantitatively formulated. Thus, the foundation of a mathematical theory of mechanics was laid (Fig. 4.3).[18]

4.5 A Unique Development or a Universal Pattern: An Excursus to China

So far we have discussed the emergence and spread of weighing technologies and the rise of a theoretical tradition of reflection on its representations, focusing mainly on the western half of the Eurasian continent.[19] The question naturally emerges of how the above-described general scheme of an iterative process, in which cognitive structures are externally represented by artifacts, language or writing, thus enabling new possibilities for a reflective abstraction, works out in other historical contexts. It is especially in view of the path-dependence of the development of knowledge, the constraining force of earlier, historically specific contexts on later developments, that the exploration of alternative, historically given scenarios is of interest. While the developments of theoretical mechanical knowledge in European and Islamicate societies all refer back to Greek antiquity and thus cannot be seen as being independent, there is a text documenting what appears to be the independent emergence of theoretical reflection on the functioning of mechanical devices in

[17] See Di Pasquale and Parisi Presicce (2013).

[18] See Knorr (1982).

[19] For the case of China, we will discuss the evidence, some of which is presented here for the first time, in somewhat more detail than this is done for the other societies relevant to this chapter.

Warring-States China: the so-called *Mohist Canon* from ca. 300 BCE.[20] Referring to one particular passage in that text, Joseph Needham claimed that "this excerpt on the lever and balance [...] shows that the Mohist must have been essentially in possession of the whole theory of equilibria as stated by Archimedes."[21] Yet, on the background of the analysis of the cultural evolution of practical and theoretical knowledge given above, it seems quite possible that the Chinese example constitutes a genuinely alternative pathway. While this pathway did not lead to Archimedes' theory of equilibria, it does show that theoretical reflection on the functioning of mechanical devices emerged independently more than once in history.[22]

The idea of an independent emergence of a theoretical tradition does not, of course, preclude foregoing and contemporary processes of knowledge exchange. In fact, with regards to China and the part of Eurasia that lies west to it, there is ample evidence for a long history of intensive technology exchange since very early times.[23] Further, given its importance in trade, it is plausible to assume that weighing technology was particularly amenable to diffusion.

We still lack a comprehensive understanding of the early development of weighing technology in China. The earliest, albeit indirect, evidence for the presence of weighing technology in China is given by inscriptions on bronze wares from the time of the Western Zhou dynasty (eleventh to eighth century BCE) which display characters for units of weights, such as *jun* 鈞 and *lüe* 鋝.[24] The oldest extant bronze weights stem from the Spring and Autumn Period (770–476 BCE), and the oldest extant (wooden) beams and scale pans from the Warring States Period (475–221 BCE).[25] These finds all indicate the use of balances with equal arms.

What appears to be closely related to the emergence of the science of mechanics in ancient Greece are balances with unequal arms, such as bismars or steelyards. Different proposals have been made about the first appearance of this kind of balance in China. The earliest proposed date is from the Warring States Period (475–221 BCE). The main ostensible evidence invoked in favor of this early date is 1) the above-mentioned passage from the *Mohist Canon*, which Needham describes as being "on the lever and balance"; and 2) two bronze beams excavated in Anhui province and bearing the character "Wang 王," which are interpreted to represent a transitional form between equal-arm and unequal-arm balances (Liu 1979; Qiu 1992, 302–303, 2005, 32). We will discuss the passage in the *Mohist Canon* in more detail below. As will become clear, most probably it does not refer to an unequal-

[20] The *Mohist Canon* is contained in chapters 40–43 of the Mohist corpus, a compliation of texts that accummulated in the time ca. 500–300 BCE. Mohism owes its name to its legendary founder Mo Di. The Later Mohist texts, to which the *Mohist Canon* belong, stand out among sources from ancient China for their logical sophistication and topical breadth, see Graham (1989).

[21] Needham (1962, 23). According to A.C. Graham's numbering, the passage Needham refers to is Canon B 25b and its coordinated Explanation; see Graham (1978).

[22] See the discussion in Renn and Schemmel (2006).

[23] For a recent discussion, including many references to the literature, see Potts (2012).

[24] See Qiu et al. (2001, 33).

[25] See, e.g., Qiu (1992, 286–300).

arm balance and not even to any kind of weighing apparatus. In any case, it does not prove the existence of unequal-arm balances in Warring-State China.

As concerns the Wang bronze beams, their interpretation as weighing instruments also turns out to be untenable (and their dating must probably also be revised) (Renn and Schemmel 2000, 20–21). These beams are clearly rulers whose scales display ten equal intervals, each indicating precisely the length of a unit measure introduced in Qin times (221–207 BCE). They are very similar and almost equal in measure to bronze rulers from Eastern Han times (25–220 CE), which were also excavated in Anhui province.[26] The only main difference between the two groups of artifacts relates to the position of the eyelet, which possibly served to hang and store the instruments. There is thus no convincing evidence for the existence of unequal-arm balances in Warring-States China.[27]

It has been argued that the first clear evidence for the existence of unequal-arm balances in China stems from late Eastern Han times (late second, early third century CE) (Guo 1993, 29–32). The argument is mainly based on archeological finds, namely, the proliferation of weights that are not integers of any known standard unit of weight and are therefore believed to be moving weights of lost steelyards.[28] But there is also textual evidence that hints at an origin in late Han times (Guo 1993, 30). The earliest pictorial evidence for the use of steelyards is said to be from the time of the Southern and the Northern Dynasties (fifth to sixth centuries CE).[29] Textual evidence for a very elaborate tradition of steelyard production in Song times is given by two specimens of steelyard produced in 992 (Guo 1994). Extant steelyards stem from much later times, namely from the Ming Dynasty (Wanli Period, 1573–1619).[30]

Given the fact that there is no independent evidence for the existence of unequal-arm balances in Warring-States China, let us examine the above-mentioned passage in the *Mohist Canon*. It reads[31]:

> The beam: If you add a weight to one of its sides [that side] will necessarily drop down. This is due to the effectiveness and the weight matching each other. If they are made level with each other, then the base is short and the tip is long. Add equal weights to both sides, then the tip will necessarily go down. This is due to the tip having gained effectiveness.

---

[26] See, e.g., items 58 and 59 in Qiu (1992, 34–35).

[27] Guo (1993, 29), arrives at a similar conclusion.

[28] Steelyards contemporarily produced in China do have standardized moving weights.

[29] At least one of the invoked paintings does not hold up to closer scrutiny: The weighing instrument depicted in the painting on the northern wall of cave 254 in Dunhuang, a picture dated to the Southern and Northern Dynasties and often referred to as the earliest depiction of a steelyard in China (Guo (1993, 31); see also Renn and Schemmel (2000, 22)) is in fact an equal-arm balance, as is appropriate in the context of the Buddhist story represented in the drawing, in which a bird is balanced with a person's flesh.

[30] The steelyards are preserved at the National Museum of China, Beijing; Qiu (1992, 484–485; 2005, 151). For a brief discussion, see also Renn and Schemmel (2000, 8–9).

[31] This is the 'Explanation' (*shuo* 說) of Canon B 25b in Graham's numbering (Graham 1978). The translation presented here was prepared in cooperation with William Boltz. It is part of a more comprehensive project on the scientific sections in the *Mohist Canon*, which is currently being prepared for publication by Boltz and Schemmel.

The beam under discussion is of a kind where weights can be added on both sides. Its turning point (fulcrum) can also be moved so that the beam can be balanced, that is, the beam's own weight is taken into account. As for weighing apparatuses, the first property would speak for an equal-arm balance, while the second for a bismar-type balance. Such a weighing apparatus is unknown and does not appear to be particularly practical. As has been pointed out, there is no other evidence—either textual or material—that would indicate its historical existence. More probably, the *Mohist* passage refers to another type of implement. The shoulder pole is a plausible candidate, since its practical use entails exactly the kind of mechanical experience described. If something is loaded on one side, this side goes down. One naturally compensates this movement by shifting the heavy side closer to the shoulder, thus making one side (the "base") shorter and the other (the "tip") longer. The observation that seems remarkable to the Mohist author is that, by adding equal weights to both sides, the longer side goes down, despite the equality of the applied weights.

Although apparently inspired by an observation in a practical context, the text is obviously not a practical text but genuinely theoretical. The observed phenomenon is considered in need of an explanation, and the explanation is given by introducing technical terminology. The term 'weight' (*zhong* 重) is complemented and paired with another term, *quan* 權, translated here as 'effectiveness', which serves to explain the diverging effect of the two equal weights. In this respect, the Mohist text displays an astonishing similarity to the ancient Greek *Mechanical Problems* discussed above: An unexpected mechanical effect, which seems to violate the intuitive idea that equal causes have an equal effect, is explained by introducing a further quality in the discussion (such as the speed of motion, the position in the arrangement or the effectiveness or power of something in a given position). The Mohist passage in fact relates to an arrangement that would later be explained with reference to the lever principle. The term *quan* 權 is also used in one other passage that deals with a mechanical arrangement involving different behaviors of weights, this time an arrangement that from a modern perspective is unrelated to the lever principle.[32]

The *Mohist Canon* documents a reflective abstraction from practical knowledge, thus bringing about new conceptual structures that are externally represented by means of written language and make consistent use of technical terminology. It therefore presents the same kind of theoretical knowledge discerned in the Greek *Mechanical Problems*. A comparative study of the societal and cultural conditions that brought about this new form of theoretical knowledge in Greece and China has yet to be carried out. But there are conspicuous similarities between the contexts for the emergence of theoretical mechanics in the two societies. One such similarity is the existence of vivid cultures of disputation, which were not primarily related to questions of technology but rather to questions of ethics and worldview.

It is here that potentially relevant differences between the two intellectual contexts also come to the fore. In the Greek case, reflections on mechanical devices

[32] Section B 26, see Graham (1978, 390–392). The passage appears to be concerned with an arrangement involving a pulley, or a curtain being pulled over a rod.

took place against a background of comprehensive natural philosophies, while in the Chinese case, elements of natural philosophy, like the *Yin-Yang* 陰陽 and Five-Agents (*wuxing* 五行) concepts, had not yet developed into coherent and all-encompassing systems.[33] In consequence, the Mohists deal with a contradiction between mechanical effect and intuitive expectation: that equal weights have different effects appears paradoxical and must be explained. The Aristotelians, in contrast, are confronted with a contradiction between mechanical effect and a system of physics. That equal causes have equal effects is not just an intuitive expectation but has become a fundamental tenet in the context of Aristotelian natural philosophy. Their explanation of the mechanical effect must therefore be as fundamental and as general as the proposition that it will complement in the Aristotelian system of physics.

Another difference between the Greek and the Chinese ancient traditions of mechanics relates to sheer quantity. In the *Mohist Canon,* the mechanical passages are part of an ordered sequence of definitions and propositions on correct reasoning, ethical conduct, the relation of spatial, temporal, material and kinematic concepts, and phenomena brought about by mechanical and optical devices. Of the about 180 Canons and coordinated Explanations, only six are on mechanics. Comparing the sections on mechanics in the *Mohist Canon* to the peripatetic *Mechanical Problems*, one is thus confronted with a huge difference in the quantity of text. Moreover, while both texts are systematically structured, the balance-lever model is not a structuring principle in the case of the Mohist text. As explained above, there are only two sections for which such a claim could be made, both of which employ the term 'effectiveness', *quan* 權. Leaving aside the different status of physical system-building mentioned above, in the context of this chapter it is intriguing to speculate about the extent to which this disparity between Greek and Chinese traditions may be related to the probable absence of unequal-arm balances in Warring States China.

Without doubt, the most obvious difference between the two traditions lies in the fact that, in the later course of history, what in hindsight looks like the seeds of a Chinese tradition of theoretical mechanics did not germinate. In China, the new abstract concepts were not taken up and further developed. We have no evidence, for instance, that the Mohist concept of the effectiveness of a weight, *quan* 權, has ever been used in this technical sense in later texts. We also have no evidence for a development of a concept of center of gravity as a generalization of the fulcrum of a balance. No science of simple, force-saving machines or any other sort of theoretical mechanics developed in ancient China. While sciences like mathematics and astronomy and a tradition of philosophical writings dealing with natural processes flourished, a mathematical science of mechanics did not develop. This is a clear expression of the fact that the emergence of a certain intellectual tradition by no means guarantees its long-term survival. Specific socio-cultural conditions must be given in order to sustain such a tradition, and the history of mechanics in the West itself provides ample evidence that this is not a matter of course. Nevertheless, as

---

[33] Boltz and Schemmel (2016). On the early Chinese tradition of cosmos-building, see Graham (1989, 315–370).

concerns the Chinese case it is again interesting to ask to what extent the absence of a mathematical tradition of theoretical mechanics can be explained by the timing of the development of the discursive-theoretical tradition on one hand and the technological tradition on the other: When the cultural conditions were favorable for the emergence of a tradition of theoretical mechanics, as documented in the few sections on mechanics in the *Mohist Canon*, no technological device embodying quantitative relations connected to the lever principle was present.

When Western theoretical mechanics and the law of the lever were introduced to China nearly two millennia later in the context of the Jesuit missionary effort in the seventeenth century, the existence of a Chinese tradition of mechanical engineering, and of Chinese practical experience with unequal-arm balances in particular, were obviously instrumental for the success of this knowledge transfer.[34] Just as it is hardly conceivable that the science of weights in the Arabic and Latin Middle Ages could have flourished without the material basis of widespread weighing practices, the transmission of theoretical texts on mechanics to China was not independent of the existence of a material culture supporting important theoretical insights. This material culture had obviously evolved against the background of much earlier instances of the multi-directional transmission of technological and mathematical knowledge on the Eurasian continent.

The introduction in the early modern period of Western knowledge of theoretical mechanics to China shows at the same time that the technological and intellectual conditions enabling a successful transfer of knowledge do not automatically propel that knowledge so that it becomes dynamic or inevitably plays a fundamental role in the knowledge system of the target culture.[35] Theoretical mechanical knowledge remained marginal in China (Schemmel 2012, 287–288) at the very time when the same knowledge experienced an accelerated transformation in Europe, a transformation of its mathematical foundations (consider the invention of the calculus) and its role in the overall knowledge system (consider its takeover of natural philosophy). It was not before the second half of the nineteenth century that this transformed theoretical mechanics, which had now become an integral component in the knowledge systems of industrialized societies with global outreach, was newly introduced to China. Under drastically changed socio-political conditions, this reintroduction became part of a more fundamental transformation of the Chinese knowledge system, eventually leading to China's participation in the development of present-day global science and technology.

[34] See Schemmel (2013) for further discussion. For an annotated edition and analysis of the *Qiqi tushuo* 奇器圖說 from 1627, a Chinese work on mechanics through which this introduction took place, see Zhang et al. (2008) (in Chinese).

[35] On the relation between natural philosophy and technical knowledge in seventeenth-century China, see Cullen (1990).

4.6 Technical Innovation and the Evolution of Knowledge

To sum up the previous theoretical discussion, we recognize an iterative process in which:

- cognitive structures based on foregoing actions with physical objects are externally represented by artifacts, language or writing
- the exploration of actions with these external representations (such as the fabrication and usage of new devices or the composition of texts) opens up new possibilities for a reflective abstraction leading to new cognitive structures,
- the exploration of the options for actions is canalized at each step by historically specific contexts constraining the actors.

Being dependent on contingent boundary conditions, this process is highly path-dependent, that is, present structures can depend on antecedent contexts that are no longer necessarily given. The contributing actors form a network of interactions that is regulated by their internal cognitive and external social structures. The cognitive structure is shaped by and can also be changed by material culture. More than merely providing a selective, independent context for the activities of the actors, material culture incorporates the external representations of the very structures that regulate the actors' actions.[36]

In this scheme of knowledge development, cognitive structures and external knowledge representations thus develop in an iterative process. Can this scheme also be applied to technological innovation processes? Technological devices are external representations of the institutional and cognitive regulative structures of the societies that invent, produce and use them, and they shape these structures in turn by creating spaces of action that determine what people can and cannot do with them under the given historical circumstances. For technological development, we can furthermore distinguish features of the development of theoretical knowledge as described above. A given generation of technological devices acts as a precondition for the creation of the next generation where, as a rule, it is the exploration of the potential of the preceding generation that provides the means to enable the creation of novelty.

Subsequent layers of technology do not completely replace earlier ones, which instead continue to act as scaffolding on which to build, albeit in a modified form. This is particularly true for the practical and technological knowledge associated with devices, which in this respect is comparable to theoretical knowledge. In the realm of theoretical knowledge, we first have to learn for instance how to count before we can understand number theory. In the realm of technological knowledge, sophisticated balances with variable arm length for example rely on standard weights which in turn are a product of equal-arm balances.

An important difference between theoretical and technological knowledge, however, is the relation between the external representation of the knowledge and the

[36] See Laubichler and Renn (2015).

underlying cognitive and institutional structures. Theoretical knowledge can typically be appropriated by an individual through texts that are understandable with certain prior knowledge and under certain external circumstances; in particular the meaning of the main concepts used in a text have to be part of the shared knowledge of the group to which the individual belongs. The individual may also be required to have accumulated certain specific experiences prior to being able to understand a text. Practical and technological knowledge, in contrast, may not even pertain to an individual but may involve the distributive knowledge of a group cooperatively solving a technical problem—without any single individual intellectually mastering the entire process. A typical way for an individual to appropriate practical knowledge is by participating in joint working processes, which involves joint attention, observing others, imitating them, taking on their perspectives, gaining and articulating experience with the same tools they are using and taking up hints and learning from corrections.

Practical knowledge is often characterized as "implicit knowledge" because its verbal expression provides for only a very limited aspect of its transmission. In reality, such knowledge is characterized by typically requiring an even broader array of media and structured information for its communication than theoretical knowledge. Its external representation may comprise samples, a variety of tools and demonstration of their usage, verbal explanations (possibly involving technical terminology), drawings or models and a specific distribution of labor, as well as social and material contexts that may not be made explicit but taken for granted in a particular culture. As a rule, this renders the transmission of practical and technological knowledge much more context-dependent than the communication of theoretical knowledge through texts, as is witnessed by the difficulties of reverse engineering, that is, sucessfully reproducing a technology using only its material manifestation. This context-dependency—and hence locality—of practical and technological knowledge is often reinforced by the fact that technical solutions, at least until the pre-modern period, were themselves mostly attuned to specific contexts.

The dependence of technological knowledge on multiple forms of external representations, each connected with its own regulative structures, also accounts for the stability of this kind of knowledge, at least as long as the relevant contexts for its transmission do not substantially change. If they do change, however, technological knowledge is much more easily irrefutably lost than theoretical knowledge.[37] It is also much harder to reflect on and to successfully alter such a wide-ranging array of external representations than on the operations of a single device, or on symbolic representations of theoretical knowledge.

A general account of technical innovations along the lines sketched above suggests a number of distinctive features that should be identifiable in the early history of weighing technology: a superposition or co-existence of various stages, some

[37] In the history of the steelyard, this tendency is nicely illustrated by the apparent loss of the ability to produce fully functional steelyards with two or three fulcra in the Merovingian period. See Werner (1954).

serving as the scaffolding for others; a relative scarcity of innovations due to canalization; the crucial role of additional regulatory factors for the occurrence of larger innovations; and a transformation of the "inheritance system" underlying the transmission of technology. The superposition of layers is a consequence of the iterative evolutionary process described above. The scarcity of innovations follows from the fact that, at each step, the space of evolutionary possibilities is circumscribed by the available means and external representations. The crucial role of additional regulatory factors follows from the fact that the effect of a variation is not random, but may be small or large according to its role in the context of regulatory structures. Major innovations are due to changes upstream in the regulatory apparatus. The transformations of inheritance patterns are a consequence of the fact that the boundaries of technological systems are not fixed.

4.7 The Co-Existence of Different Lineages

The results of our investigations into the early history of weighing technology as they have been pursued so far are in agreement with the theoretical explanation of innovation sketched above.[38] In the history of weighing technology, various types of balances emerged, were widely spread and continue to coexist until today. This is evidently the case for the equal-arm balance and the Roman balance. It also applies, with some reservations, to the bismar. Aristophanes' passing allusion to the instrument suggests that his audience would have been familiar enough with it so as to understand the pun he makes in the play. In the *Mechanical Questions*, it is stated that bismars were used to weigh meat. It thus seems that the bismar was a rather common weighing instrument from at least the end of the fifth century BCE. This conclusion, however, does not seem to be supported by the archaeological record. Whereas the absence of any artifacts from the early period could be potentially explained by the fact that, at least initially, bismars were made out of wood and thus usually would not have been preserved, the lack of pictorial representations suggests moreover that the bismar persisted next to weighing with equal-arm balances only as a somewhat marginalized technology (Fig. 4.4).[39]

Yet, a second linage in the history of the bismar can be discerned which has mostly been ignored. In India, bismars are attested to in writings as well as pictorially possibly as early as the end of the fourth century BCE.[40] The archaeological records suggest that the bismar enjoyed a continued tradition and can still be found

[38] The theoretical explanation presented here is informed by a theory of extended evolution as laid out in Laubichler and Renn (2015).

[39] As the abundant representations of weighing with equal-arm balances that have been preserved show, the lack of pictorial representations of bismars cannot be explained by the fact that weighing as an everyday technology was not the subject of such representations.

[40] See Jenemann (1994). A bismar is mentioned in chapter XIX of the *Arthashastra,* an ancient Indian treatise on state governance. See Kautalya (1992).

Fig. 4.4 Greco Buddhist bas-relief from the ancient region of Gandhara (today a border region between Afghanistan and Pakistan) from the second to the third century showing a scene from one of Buddha's former lives as King Sibi (British Museum, Inv. Nr. 1912,1221.1). In the center, flesh cut from the king's leg is weighed with a balance, which is obviously a bismar. © Trustees of the British Museum

in use today.[41] It is certainly possible that we are concerned here with independent developments. Yet, acknowledging the fact that balances had been in use in both cultural areas, the Aegean and the Indus valley, for more than 2000 years, the emergence of the very same modification of weighing technology at apparently more or less the same time must be taken as a strong indication that this is the result of a transfer of technology.[42] Further research is required to answer this question and, should this be the case, decide in which direction the transfer of technology actually took place.

4.8 Canalization and Scaffolding: The Case of the Bismar

The mechanical lever balance constitutes a physical system with a limited design space for arranging load, fulcrum and standard weight.[43] Yet even this limited design space has not been fully exploited since historically the case of a moveable load has

[41] See Dikshit (1957, 1961).

[42] For the earliest evidence of weighing in the Indus valley culture, see Kenoyer (2010).

[43] For the concept of design space, in particular in relation to the formation of specific "body plans" or "dominant designs," see Murmann and Frenken (2006).

not played a role.[44] Given one of the basic "body plans" for balances, further innovations took the form of exploring possibilities of inherent optimisation. Larger-scale innovations associated with changes in the body plan were at best only indirectly linked to the existence or introduction of new regulatory structures related to weighing.

The basic invention of the first balances with unequal arms (i.e. balances of the bismar type) thus presupposed the firm establishment of sets of standard weights as the cognitive and social regulations needed to empirically gauge such a balance and make this gauging acceptable to its users, without requiring any further sophisticated knowledge. The bismar is indeed a rather simple instrument; palpably its most complicated feature is its non-linear scale where the scale intervals representing equal weight differences follow a harmonic division. It can be confidently stated, however, that in antiquity bismar scales were not theoretically established but empirically constructed by gauging.[45] If this is taken into account, it becomes manifest that the construction of a bismar indeed poses only minimal requirements for the underlying mechanical knowledge. The law of the lever governs the operation of a bismar, just as that of many other instruments. The reverse assumption that knowledge of the law would have been required to construct a bismar is as pointless as the assumption that such knowledge would be required to construct a set of pliers.

The allusion in Aristophanes' play to the transformation of an everyday item such as trumpet into a bismar underscores the comparatively scant requirements posed by the construction of a bismar. One could, of course, dismiss the passage in Aristophanes as literary fiction with little or no informative value concerning the actual feasibility of the transformation of an everyday item into a bismar were it not for a peculiar object found in Pompeii.[46] In this case, an ordinary kitchen casserole (hundreds of them were found in the Vesuvian town) has been transformed into a bismar. To this end, a slit was made in the handle of the casserole to which a suspension was linked and a load attachment adjoined to the handle. This bears witness to the relative ease with which a bismar could be fabricated.[47]

The advantages of the simple construction of the bismar stand in contrast to a certain shortcoming in its application for weighing purposes. Since the fulcrum of a

---

[44] See Jenemann (1989).

[45] A statistical analysis of the deviation of the scales' marks from their ideal positions in Roman steelyards has shown that the scales of these balances were produced by gauging at regular intervals. The intermediate marks were placed by dividing the distances into an appropriate number of equal parts. As concerns the bismar, a similar gauging routine for the establishment of the scales has been assumed, see Damerow et al. (2002) and Jenemann (1994,) but awaits confirmation by a detailed examination of the objects.

[46] Aristophanes' description, however, is not merely based on a superficial similarity of a trumpet and a bismar, as the instruction to fill the bell with lead corresponding to construction knowledge illustrates.

[47] A summary of the discussions of this particular object can be found in Damerow et al. (2002). Jenemann takes the Pompeian bismar as an indication that bismar technology was still in widespread use in the second half of the first century, Jenemann (1994). In view of the simplicity of the construction the argument is not very cogent.

bismar can move arbitrarily close to the load suspension, its weighing range is potentially infinite, limited only by the stability of the beam. However, in practice the resolution and accuracy of a bismar necessarily decrease with higher loads so that weighing becomes less expedient as small displacements of the fulcrum correspond to great differences in weight. A bismar is thus particularly suitable for determining small weights. Small weights and weight differences are especially relevant when dealing with valuable goods, where accuracy is key. Thus, it is in the weight range most applicable to the bismar that it is in direct competition with equal-arm balances, which are more accurate by design. This may provide an explanation of why, as suggested above, the bismar was a somewhat marginalized technology compared to equal-arm balances, of which there is ample evidence in the relevant period.

4.9 Canalization and Scaffolding: The Case of the Steelyard

We have emphasized that larger-scale innovations associated with changes in "body plans" require the existence or introduction of new regulatory structures in the role of scaffolding on which to build. In the case of the bismar, we have seen that the pre-existing level of knowledge and of societal regulations involving the abstract concept of weight and its representation by a series of standard weights took on the role of such scaffolding, enabling a new type of balance to be improvised by empirically gauging its scale. As we will see, the introduction, spread and development of the steelyard placed much higher demands on the underlying structures. The earliest evidence for the steelyard comes from a number of archeological finds that can be dated approximately to the middle of the first century BCE. Vitruvius mentions the Roman balance a little later in his *De architectura*. Here, he refers to it as a *statera* and explains its function by a qualitative law of the lever.[48] Only a handful of steelyards can be dated with certainty before the Common Era and, for this early period, the finds remain somewhat confined to the Roman core territory. From the middle of the first century CE onwards, a rapid growth in the number of preserved artifacts is observable. Clearly, the steelyard became widely used and produced all over the Roman Empire.[49]

Roman steelyards with two or even three fulcra were much more sophisticated than the bismar. Their introduction and spread required the articulation and transmission of sets of rules within an appropriate societal infrastructure. In contrast to the case of a bismar, the construction of such a steelyard cannot be achieved by

[48] Peculiarly, Vitruvius' description is the only unambiguous reference to a steelyard that can be found in textual sources from antiquity and late antiquity, see Rohmann (2017).

[49] The spread of the steelyard over a vast geographical area, and its persistence over time can be characterized as a complex innovation process in which different subtypes of steelyards emerged and replaced each other. This innovation process is being studied in detail by a Topoi junior research group (https://www.topoi.org/project/d-5-5/). The new findings concerning unequal-arm balances presented in this article result from this research agenda.

improvisation; the mechanical knowledge required to successfully design and construct a steelyard is much more intricate. In order to manufacture a steelyard suitable for weighing purposes, a number of non-trivial boundary conditions have to be satisfied. Many, but far from all, of the problems to be solved in fabricating a steelyard are related to its dead weight: the instrument has a weight, which unlike the case of the equal-arm balance, influences its equilibrium.[50] In a certain way, the steelyard can be said to weigh itself. Using two examples, the problems encountered in successfully designing a steelyard shall be briefly indicated below (Fig. 4.5).

In the simplest case, a steelyard is a beam divided into two unequal parts by a fulcrum. The longer arm, referred to as the scale-arm, carries the counterpoise and the scale; the shorter arm, referred to as the load-arm, carries some sort of load suspension. The scale of such a balance should ideally start at zero and run to a certain maximum weight, corresponding to the expected weight of the largest load. Optimally, the zero point of the scale should be close to the fulcrum so that the full length of the scale-arm is exploited, resulting in a better resolution and a greater ease of weighing.

In realizing such a simple construction, a number of constraints need to be observed. The zero point of the scale, that is, the position at which the counterpoise must be hung so that the unloaded balance is in equilibrium, can be on the scale-arm if and only if the center of gravity of the instrument without the counterpoise is located on the load-arm. Due to the unequal division of the beam, this will usually not be the case and the position of the center of gravity needs to be varied. Varying the weight of the load suspension usually does this. The largest weight determinable

Fig. 4.5 Roman steelyard from Pompeii (British Museum 1772, 0319.1). Photo: Jochen Büttner

[50] In equal-arm balances, the weight of the instrument itself influences its operation too, although not the equilibrium configuration. For the bismar, the influence of the dead weight of the instrument is handled with the gauging of the scale.

with the instrument obviously depends on the division of the beam by the position of the fulcrum, as well as on the weight of the counterpoise. The variation of each of these two design parameters in turn affects the zero position of the scale. The division of the beam affects the position of the center of gravity with respect to the fulcrum, and the heavier the counterpoise, the closer to the fulcrum the zero point of the scale will be for a given position of the center of gravity.

The complications are dramatically enhanced when, as is the case for the majority of the finds from the period in question, a second fulcrum and thus a second scale is introduced.[51] Here, the additional condition comes into play that the scales have to be harmonized, that is, that the minimal weight that can be determined using the second fulcrum should be slightly smaller than the maximum weight of the load determinable with the first fulcrum. Hence, in such balances, an optimal position for the second fulcrum exists. If this position is exceeded and the second fulcrum moves closer to the load suspension, a non-functional balance will result, with a gap in its weighing range. From a modern perspective, the optimal position for the second fulcrum depends on many, factors, including the division of the beam, the position of the center of gravity and finally, the ratio of the weight of the balance to the weight of the counterpoise.

An analysis of the steelyards from the Roman period has shown that their makers were able to solve the ensuing problems, such as the ones mentioned above, in a consistent and remarkably ideal fashion. Steelyards could be systematically produced with the position of their counterpoise in unloaded equilibrium at the very beginning of the scale-arm, and whose second fulcrum is positioned such that the two scales harmonize perfectly. We are only now beginning to understand how this and a great range of additional problems were solved and which mechanical knowledge this embraced. Suffice to say for the present purpose, the solutions were not, and indeed could not have been, obtained by trial and error with the individual objects. Rather, the preserved steelyards obey certain general principles, that is, they show repeating complicated patterns in the relations between the relevant design parameters. These regularities can be straightforwardly interpreted as the result of recurrent similar actions in their production, which themselves are the external consequence of the sequential execution of procedural rules. Even today, the manufacture of steelyards in some parts of the world is regulated in such a fashion.

Fieldwork in China has yielded some insights into different ways of determining the relevant design parameters, in particular, the position of the two fulcra.[52] In contrast to the Roman case of cast steelyards whose fulcra have to be determined before the casting of the beam, the beams of the Chinese steelyards whose production was investigated are made of wood so that the steelyard-makers can fix the fulcra at any determined position. One way of determining the fulcra was empirical: the steelyard-

[51] Whereas the earliest preserved steelyards all have two fulcra, instances of later types tend to have three. For a typology of steelyards, see Franken (1993). A new catalogue is in preparation and will be published soon. A prototype can be accessed via the webpage of the Topoi junior research group D-5-5 (http://www.topoi.org/project/d-5-5/), accessed August 23, 2017.

[52] See Renn and Schemmel (2000).

maker fixes the hook and scale pan and adds standard weights corresponding to the maximum weight of the scale. The counterpoise is then moved to the position at the end of the beam that will be the end of the scale. The beam is then balanced on a knife. The position of the knife when the beam is in equilibrium is marked as the position of the fulcrum.[53] Such an empirical method of determination is not possible in the Roman case with cast or forged beams. While the steelyard-makers described the empirical determination of the position for both fulcra as the most reliable, there are also rule-based procedures for their positioning. In particular, once the position of the first fulcrum is given, the position of the second fulcrum is found by the following rule: if the distance from the load suspension to the first fulcrum is divided into three equal parts, the second fulcrum lies on the first division counted from the load suspension.[54] Once the positions of the fulcra have been determined for one type of steelyard, they may be measured and handed down. Thus, the steelyard-makers divided the beam into 21.5 sections and placed the second fulcrum on the first section mark.[55]

Another way of determining the position of the fulcra involved calculation: The sum of the maximum weight to be measured and the weight of the scale pan, divided by the weight of the counterpoise equals the number of sections to be measured on the beam. The fulcrum then lies on the first section mark. The total weight is thus measured in units of the counterpoise, and from this the number of sections on the whole beam is determined. The calculation does not agree with the law of the lever. It deviates from that law in precisely the way in which the law of the lever would have to be corrected in order to account for the weight of the beam.[56]

Similar to this example of rules for determining the position of the fulcra in the Chinese case, generally there are rules for determining the components of the balance. Among the rules that were applied historically are those that embrace the variations that are contingent on prior choices, and that themselves depend on the specific purpose of the balance to be built. In consequence, the relation between different objects produced according to the same set of rules is not immediately apparent: they are in particular not geometrically similar. While it is conceivable that individual steelyards could have been produced by copying, the fabrication of a broad range of application-specific steelyards, as is evidenced in the Roman Empire at remote distances and time intervals, is only possible if these rules were explicated and diffused in the context of production. It can be argued that the steelyard as a technological innovation could only successfully take hold once the conditions for

[53] This method was performed by a steelyard-maker in Tongzhou (near Beijing) and described by a steelyard-maker in Changsha, see Renn and Schemmel (2000, 17).

[54] This was explained to us by the master steelyard-maker at the workshop at Tongzhou, see Renn and Schemmel (2000, 19).

[55] The value of 21.5 for determining the position of the second fulcrum was used both in Tongzhou and in Changsha, see Renn and Schemmel (2000, 18, 32). For the first fulcrum, the number of sections used in Changsha was 7.2.

[56] Renn and Schemmel (2000, 18–19). The method was explained to us by a steelyard-maker in Tongzhou. The numbers used at the workshop in Changsha are compatible with the application of the same method.

diffusing and transmitting such complex knowledge across larger geographical areas and over longer periods of time were given, that is, against the backdrop of the production infrastructure of the Roman Empire. This production infrastructure constitutes the additional regulatory module that made the success story of the Roman steelyard possible in the first place.

4.10 The Role of Infrastructures as Inheritance Systems for Technology

The Roman production infrastructure which, as argued above, enabled the introduction and spread of the steelyard, had itself emerged at a high point of technical capabilities and could benefit from the resources connected with them, which had developed since the fourth century BCE in the Mediterranean region. The Hellenistic period, in particular, can be seen as an era of technological boom: technological innovations such as the gearwheel, the water pump, the cylindrical screw with bolts or the dioptra can all be dated back to this period. Alexandria and its network, to which the Syracuse of Archimedes' time also belonged, was the basis and the institutional reference point for all these innovations, and also for the engineers and scholars who investigated them.[57] Technology was then further boosted by the need to realize and spread a powerful infrastructure for the military, political and economic maintenance of the Roman Empire.

Roads and water supply systems could only be built using reliable machines. Machine technology, however, received most attention because of mining activities, which became part of a self-reinforcing mechanism since the metal technology also underlying the production of weapons and machinery (including pumps and steelyards) would be unthinkable without it. The Roman Empire saw something akin to the creation of a military-industrial complex centered on metal production, which also became a presupposition for the widespread production and use of steelyards.[58] From an economic perspective, machines vastly increased the possibilities for more income because they enabled mining activities in regions and contexts that were otherwise not exploitable. In his in-depth analysis of several techniques of hydraulic mining and ore-processing applied by Romans in the Iberian peninsula and what is now French territory, the economic historian Andrew Wilson speaks about peaks of mining production that did not reach the same level again until the Industrial Revolution, and in particular, that are only rivaled in the modern era for their dependency on advanced technology.[59] We can thus see that the Roman steelyard, like

[57] For Hellenistic technology, see especially Chap. 4 of Russo (2004). See also Schürmann (1991).

[58] See Sommer (2013, Chap. 4).

[59] See Wilson (2002). Wilson's paper is an effective response to the widespread idea that technological innovation and economy were not linked in antiquity. This idea was diffused by M. I. Finley. See, in particular, Finley (1965). For a large study on the relation between technology and economy in antiquity, see Lewis (1997).

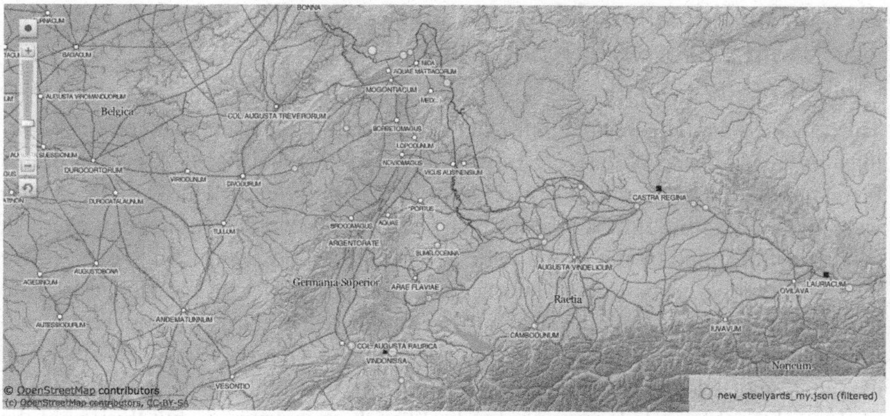

Fig. 4.6 Find-spots of iron steelyards of the first and second centuries CE mapped onto the *Digital Atlas of the Roman Empire* (http://dare.ht.lu.se). Forged iron steelyards have been found only north of the Alps. Their spatial as well as temporal distribution implies a close connection of their production and use to Roman military infrastructure

every technical product, is associated with a network of production conditions that constitute its variable inheritance structure. Changes in this network may change the product, and vice versa. The study of these mutual influences is ongoing research (Fig. 4.6).

In summary, bismar-type balances could be improvised, while the production of Roman steelyards required a rather elaborate societal and cognitive infrastructure. Compared to the bismar, the steelyard posed higher requirements regarding the knowledge needed to produce it, but this was compensated by the evident advantages in its use: it is more adaptable to different weighing purposes and generally more accurate and simpler to handle than a bismar. That the steelyard, despite these advantages, could apparently not be established against the bismar in India may thus be explained by the lack of an infrastructure capable of transmitting and diffusing the required complex production knowledge. One of the open and challenging questions in the evolution of weighing technologies is how the Roman steelyard could nevertheless survive the decline of the Roman Empire and become an important asset in the Islamicate empires, in China, in the European Middle Ages and during the Renaissance, and continue to be produced and used essentially until today.[60]

[60] For the production of steelyards in modern-day China, see Renn and Schemmel (2000).

4.11 The Unexplored Interaction between Technical and Theoretical Knowledge

The transition from the bismar to the steelyard also left traces in the theoretical writings on mechanics. When referring to unequal-arm balances, Aristotle in his *Mechanical Problems* and later authors such as Vitruvius or Hero of Alexandria were indeed treating different instruments. If their theoretical treatments were in some way informed by the technological knowledge associated with the respective instruments, one needs to take into account the fact that such knowledge is quite distinct for the bismar and the steelyard. Because of the relatively simple construction of the bismar, the passage in the *Mechanical Problems* (besides giving an explanation of the operational principle) could also thus function as a representation of the technical knowledge required to build such an instrument, but this is not the case for Vitruvius' treatment of the steelyard. Hero squarely addressed the difference between theoretical principles and the actual construction of a balance:

> Some people believe that when, in balances, the weights are in equilibrium to the weights, the weights have to be in said inverse proportion to the distances. Generally, this can not be maintained … .[61]

In the rather ingenious proof following this passage, Hero shows that a simple form of the law of the lever does not always adequately describe equilibrium when the dead weight of the balance beam is accounted for. The configuration discussed in his proof clearly corresponds to a steelyard, even if it is not explicitly referred to as such. Thus, here Hero addresses what has been identified above as one of the central problems to be overcome when designing and fabricating a steelyard. Whether scientific thought actually influenced the technological transition from the bismar to the steelyard cannot currently be answered and the question, due to the scant evidence, may in fact remain unanswerable.

Our studies of the innovation processes of early weighing technologies have drawn attention to yet another question regarding the relation between science and technology in antiquity. The rules applied in producing steelyards are entirely different from the modern physical formula that enable us to express the conditions a steelyard has to meet in order to actually function as a weighing instrument, yet the application of these rules in the production of steelyards gives rise to what, from a theoretical perspective, must be considered as an almost optimal result. Were these rules established on a purely empirical basis or did theoretical knowledge play a role in their formulation? We have reason to be believe we will be able to answer such fundamental questions in our further research.

---

[61] See Hero (1900, 86), authors' translation.

References

Boltz, William G., and Matthias Schemmel. 2016. Theoretical reflections on elementary actions and instrumental practices: The example of the Mohist Canon. In *Spatial thinking and external representation: Towards a historical epistemology of space*, ed. Matthias Schemmel. Berlin: Edition Open Access. http://edition-open-access.de/studies/8/5/index.html.

Brentjes, Sonja, and Jürgen Renn. 2016. The Arabic transmission of knowledge on the balance. In *Globalization of knowledge in the postantique Mediterranean, 700–1500*, ed. Sonja Brentjes and Jürgen Renn. New York: Routledge.

Büttner, Jochen. 2008a. Big wheel keep on turning. *Galilaeana* 5: 33–62.

———. 2008b. The pendulum as a challenging object in early-modern mechanics. In *Mechanics and natural philosophy before the Scientific Revolution*, ed. Walter Roy Laird and Sophie Roux, 223–237. Dordrecht: Springer.

———. 2013. The lever, the balance and a beautiful proof. In *Archimedes: The art and science of invention; Rome, Musei Capitolini, May 31, 2013–January 12, 2014*, ed. Giovanni Di Pasquale and Claudio Parisi Presicce, 85–91. Florence: Giunti.

Corti, Carla, and Nicoletta Giordani. 2001. *Pondera: pesi e misure nellantichità. Grandezze e misure nella storia*. Campogalliano: Museo della bilancia.

Cullen, Christopher. 1990. The science/technology interface in seventeenth-century China: Song Yingxing on qi and the wu xing. *Bulletin of the School of Oriental and African Studies* 53: 295–318.

Damerow, Peter. 1996. *Abstraction and representation: Essays on the cultural evolution of thinking*, Boston Studies in the Philosophy of Science. Vol. 175. Dordrecht: Kluwer.

Damerow, Peter, and Jürgen Renn. 2010. The transformation of ancient mechanics into a mechanistic world view. In *Transformationen antiker Wissenschaften*, ed. Georg Toepfer and Hartmut Böhme, 243–267. Berlin: de Gruyter.

Damerow, Peter, Jürgen Renn, Simone Rieger, and Paul Weinig. 2002. Mechnical knowledge and Pompeian balances. In *Homo Faber: Studies on nature, technology, and science at the time of Pompeii*, ed. Jürgen Renn and Giuseppe Castagnetti, 93–108. Rome: L'Erma di Bretschneider.

Di Pasquale, Giovanni, and Claudio Parisi Presicce. 2013. *Archimedes: The art and science of invention*. Florence: Giunti.

Dikshit, Moreshwar G. 1957. Narji: The one-pan scales in ancient India. *Bulletin of the Deccan College Postgraduate and Research Institut* 18: 5–7.

———. 1961. Two Pakrit verses. *Indian Historical Quarterly* 37: 189–191.

Finley, M.I. 1965. Technical innovation and economic progress in the ancient world. *The Economic History Review, New Series* 18 (1): 29–45.

Franken, Norbert. 1993. Zur Typologie antiker Schnellwaagen. *Bonner Jahrbücher* 193: 69–120.

Geller, Markham, ed. 2014. *Melammu: The ancient world in an age of globalization*. Berlin: Edition Open Access. http://edition-open-access.de/proceedings/7/.

Graham, Angus Charles. 1978. *Later Mohist logic, ethics and science*. Hong Kong: Chinese University Press.

———. 1989. *Disputers of the Tao: Philosophical argument in ancient China*. La Salle, IL: Open Court.

Guo, Zhengzhong. 1993. *San zhi shisi shiji zhongguo de quan heng du liang*. Beijing: Zhongguo shehui kexue chubanshe.

———. 1994. The Deng steelyards of the Song Dynasty (960–1279). *Cahiers de Métrologie* 11–12: 297–306.

Hero. 1900. *Herons von Alexandria Mechanik und Katoptrik*. Leipzig: B.G. Teubner. http://catalog.hathitrust.org/Record/008693868.

Jenemann, Hans R. 1989. Zur Geschichte der Waagen mit variablem Armlängenverhältnis im Altertum. *Trierer Zeitschrift für Geschichte und Kunst des Trierer Landes und seiner Nachbargebiete* 52: 319–352.

———. 1994. Die Besmer Waage im Altertum. *Jahrbuch des Römisch-Germanischen Museums Mainz* 41: 199–229.

Kautalya. 1992. *The Arthashastra*. New Delhi: Penguin Books India.

Kenoyer, Mark J. 2010. Measuring the Harappan world: Insights into the Indus order and cosmology. In *The archaeology of measurement: Comprehending heaven, earth and time in ancient societies*, ed. Iain Morley and Colin Renfrew, 106–122. Cambridge: Cambridge University Press.

Knorr, W.R. 1982. *Ancient sources of the medieval tradition of mechanics: Greek, Arabic, and Latin studies of the balance*. Florence: Istituto e museo di storia della scienza.

Laubichler, Manfred D., and Jürgen Renn. 2015. Extended evolution: A conceptual framework for integrating regulatory networks and niche construction. *Journal of Experimental Zoology (Molecular and Developmental Evolution)* 324B: 565–577. http://onlinelibrary.wiley.com/doi/10.1002/jez.b.22631/pdf.

Lewis, Michael Jonathan Taunton. 1997. *Millstone and hammer: The origins of water power*. Hull: University of Hull Press.

Liu, Dongrui. 1979. Tan Zhanguo shiqi de budengbi cheng 'Wang' tongheng. *Wenwu (Cultural relics)* 4: 73–76.

Malkin, Irad. 2011. *A small Greek world: Networks in the ancient Mediterranean*. Oxford: Oxford University Press.

Murmann, Johann Peter, and Koen Frenken. 2006. Toward a systematic framework for research on dominant designs: Technological innovations and industrial change. *Research Policy* 35 (7): 925–952. https://doi.org/10.1016/j.respol.2006.04.011.

Needham, J. 1962. *Science and civilisation in China. Vol. 4: Physics and physical technology, Part 1 (physics)*. Cambridge: Cambridge University Press.

Nenci, E. 2011. *Bernardino Baldi's in mechanica Aristotelis problemata exercitationes*. Berlin: Edition Open Access. http://www.edition-open-sources.org/sources/3/index.html.

Potts, D.T. 2012. Technological transfer and innovation in ancient Eurasia. In *The globalization of knowledge in history*, ed. Jürgen Renn, 105–123. Berlin: Edition Open Access. http://www.edition-open-access.de/studies/1/8.

Qiu, Guangming. 1992. *Zhongguo lidai du liang heng kao*. Beijing: Kexue chubanshe.

———. 2005. *A concise history of ancient Chinese measures and weights* (Zhongguo gudai jiliang shi tujian). Hefei: Hefei gongye daxue chubanshe.

Qiu, Guangming, Qiu Long, and Ping Yang. 2001. *Zhongguo kexue jishu shi: Duliangheng* (History of science and technology: Metrology). Beijing: Science Press.

Renn, Jürgen, ed. 2012. *The globalization of knowledge in history*. Berlin: Edition Open Access. http://www.edition-open-access.de/studies/1.

Renn, Jürgen, and Matthias Schemmel. 2000. *Waagen und Wissen in China: Bericht einer Forschungsreise*. Preprint 136. Berlin: Max Planck Institute for the History of Science. https://www.mpiwg-berlin.mpg.de/Preprints/P136.PDF.

———. 2006. Mechanics in the Mohist canon and its European counterparts. In *Studies on ancient Chinese scientific and technical texts: Proceedings of the 3rd ISACBRST; March 31–April 3, 2003, Tübingen, Germany*, ed. Hans Ulrich Vogel, Christine Moll-Murata, and Gao Xuan. Zhengzhou: Elephant Press.

Renn, Jürgen, and Peter Damerow. 2012. *The equilibrium controversy: Guidobaldo Del Monte's critical notes on the mechanics of Jordanus and Benedetti and their historical and conceptual backgrounds*. Berlin: Edition Open Access. http://www.edition-open-sources.org/sources/2.

Renn, Jürgen, Peter Damerow, and Simone Rieger. 2001. Hunting the white elephant: When and how did Galileo discover the law of fall? In *Galileo in context*, ed. Jürgen Renn, 29–149. Cambridge: Cambridge University Press.

Rohmann, Dirk. 2017. Ungleicharmige Waagen im literarischen, epigraphischen und papyrologischen Befund der Antike. *Historia* 1 (66): 83–110.

Russo, Lucio. 2004. *The forgotten revolution: How science was born in 300 BC and why it had to be reborn*. New York: Springer.

Schemmel, Matthias. 2012. The transmission of scientific knowledge from Europe to China in the early modern period. In *The globalization of knowledge in history*, ed. Jürgen Renn, 269–293. Berlin: Edition Open Access. http://www.edition-open-access.de/studies/1/15.

———. 2013. Stevin in Chinese: Aspects of the transformation of early modern European science in its transfer to China. In *Translating knowledge in the early modern low countries*, ed. Harold J. Cook and Sven Dupré, 369–385. Berlin: LIT.

Schürmann, Astrid. 1991. *Griechische Mechanik und antike Gesellschaft: Studien zur staatlichen Förderung einer technischen Wissenschaft*. Stuttgart: F. Steiner.

Sommer, Michael. 2013. *Wirtschaftsgeschichte der Antike*. Munich: C.H.Beck.

Valleriani, Matteo. 2009. The transformation of Aristotle's *Mechanical questions*: A bridge between the Italian renaissance architects and Galileo's first new science. *Annals of Science* 66 (2): 183–208.

———. 2010. *Galileo engineer*, Boston Studies in the Philosophy of Science. Vol. 269. Dordrecht: Springer.

———. 2012. Galileo's abandoned project on acoustic instruments at the Medici court. *History of Science* 50 (1): 1–31.

———. 2013. *Metallurgy, ballistics, and epistemic instruments: The Nova scientia of Nicolò Tartaglia – a new edition*. Berlin: Edition Open Access. http://www.edition-open-access.de/sources/6/.

———, ed. 2014. Appropriation and transformation of ancient science. *Nuncius*, 29 (1). Leiden: Brill.

Vitruvius Pollio, Marcus. 1999. *Vitruvius: Ten books on architecture*. Cambridge: Cambridge University Press.

Werner, Joachim. 1954. Waage und Geld in der Merowingerzeit. Vorgetragen am 5. Februar 1954. In *Bayerische Akademie der Wissenschaften München, Philosophisch-Historische Klasse. Sitzungsberichte 1*. Munich: Verl. der Bayerischen Akademie der Wissenschaften.

Wilson, Andrew. 2002. Machines, power and the ancient economy. *The Journal of Roman Studies* 92: 1–32.

Zhang, Baichun, Miao Tian, Matthias Schemmel, Jürgen Renn, and Peter Damerow. 2008. *Chuanbo yu huitong: 'Qiqi tushuo' yanjiu yu jiaozhu* (Transmission and integration: Qiqi Tushuo (Illustrations and descriptions of extraordinary devices) New research and annotated edition). Nanjing: Jiangsu kexue jishu chubanshe.

Chapter 5
The Balance, the Lever and the Aristotelian Origins of Mechanics

Jürgen Renn and Peter McLaughlin

Abstract The *Mechanical Problems* traditionally attributed to Aristotle is a short problem collection that also contains an ambitious project of reduction, which traces various mechanical devices back to the lever, the balance and the radii of a circle. This work is thus not just a collection of problems, but also the first theoretical mechanical treatise that has come down to us: Basic concepts of technical mechanics such as force, load, fulcrum are abstracted from an analysis of simple technology, and the workings of this technology are explained by arguments cast in syllogistic form. This chapter traces the origins of mechanical theory in this work and analyzes the form and structure of its argument. The key steps in the concept formation of basic mechanics carried out in this treatise are analyzed in detail. We focus on the special role of the balance with unequal arms in the early development of mechanics, on the interaction of various forms of explanatory practice and on their integration into systems of knowledge in the Peripatetic school.

Keywords *Mechanical problems* · Aristotle · The balance · The lever · Peripatetic school

5.1 Introduction[1]

One of the truisms of the traditional story of the Scientific Revolution of the sixteenth and seventeenth centuries is that mechanics was not part of Aristotelian physics, whereas it is constitutive of modern physics. The artificial motions of mechanics and

[1]The basic interpretation of the *Mechanical Problems* presented here was first developed in a research seminar at the University of Constance conducted by McLaughlin, Renn and the late Peter Damerow in 1997. Among the participants were Markus Asper, Elke Kasemi and Paul

J. Renn
Max Planck Institute for the History of Science, Berlin, Germany

P. McLaughlin (✉)
Philosophy Department, University of Heidelberg, Heidelberg, Germany
e-mail: peter.mclaughlin@uni-hd.de

R. Feldhay et al. (eds.), *Emergence and Expansion of Preclassical Mechanics*,
Boston Studies in the Philosophy and History of Science 270,
https://doi.org/10.1007/978-3-319-90345-3_5

mechanical technology could not be the subject matter of Aristotle's science of natural motions. In fact, it is the rejection of the distinction between violent motion and natural motion that marks the difference between Newtonian physics and Aristotelian natural philosophy. In opposition to this characterization some scholars have construed a greater continuity between medieval thought and modern science, and many recent scholars of sixteenth- and seventeenth-century thought even reject the existence of a scientific revolution.[2] But this one last bastion seems to resist razing. However, there is an alternative to the traditional story of early modern science as a journey along the high road from Copernicus by way of the law of fall to Newton and celestial mechanics. We can also tell a story that follows the low road from the Renaissance technicians to Guidobaldo, Galileo and terrestrial mechanics. In this case Aristotelianism provides not just the focus of attack of the new science but also one of its major resources, a work called the *Mechanical Problems*, in which not only is the law of the lever formulated, but also basic concepts of technical mechanics are abstracted from the study of simple machines. When Italian engineers began to reflect philosophically on their professional activities, they had a model of how to move from technical practice to "philosophical" theory—marketed under the name of Aristotle. If we follow the low road, we find that the contribution of the Middle Ages to modern science lay not just in the mathematical methods of fourteenth-century Franciscans or in the admission of mechanics to science through the back door in the *quadrivium* of Gundisalvus, but also in the technology employed in Dutch windmills and Italian harbor installations. In the opening lines of the *Discorsi* (1638/1974) Galileo praises the Venetian Arsenal for opening up "a large field to speculative minds for philosophizing" about mechanics.[3] The *Mechanical Problems*[4] spoke to Renaissance engineers not just as the source of the law of the lever and some comparatively primitive analyses of mechanical instruments, but also as a model of how to turn reflection on technical know-how into scientific knowledge.

Even the greatest heroes of twentieth-century historiography of science, pursuing the high road to celestial mechanics, often succumbed to what we may call the utilitarian fallacy on the relation of science and technology: If a theoretician is interested in technology, it must be because he wants to apply the theoretical knowledge sought, or already acquired, in some useful practice. But, so the argument goes, seventeenth century technology did not need and could not readily apply seventeenth century science; and if science is not pursued for the sake of technology, it is independent of technology. However, when Francis Bacon

---

Weinig, whose subsequent contribution to this project cannot be measured in footnotes. In the intervening years we have profited from collaborations with István Bodnár, Brian Fuchs, Malcolm Heiman, and Mark Schiefsky, as well as Albert de Puis and, for the current version, Joyce van Leeuwen. From the beginning we have benefited from Fritz Krafft's analysis (1970).

[2] See Duhem (1913–1959) and Crombie (1961) for the continuity thesis. See also Shapin (1996).

[3] Galilei (1638, 11).

[4] There is no accepted critical edition of the *Mechanical Problems*. We use the Greek text given in Hett 1936, cited, as is customary, according to page and column of the Bekker edition of 1831. All translations from the Greek are our own. They are intended to be as literal and interpretatively open as possible.

famously proclaimed, that "Nature to be conquered must be obeyed," the notion, that technology (conquering nature) obeys natural laws, is not limited to cases where a law is known and followed obediently in application.[5] It also holds for those cases in which a successful technology is studied in order to learn about and thus to formulate the natural laws that were already being unconsciously obeyed in practice. And given that modern science was not systematically applicable in technology until the nineteenth century, this version of Bacon better explains the almost universal interest of seventeenth-century scientists in technology.[6] Such a study of technology, looking for the general theoretical principles embodied in practice, is basically what the author of the *Mechanical Problems* is also doing—with much simpler machinery.

In the following we shall read parts of the *Mechanical Problems* as a project of abstracting general theoretical concepts from the analysis of concrete technical practice. The treatise formulates a number of basic mechanical concepts, connects them in an embracing conceptual scheme or mental model, and presents its theoretical explanations in the form of deductive arguments. In the context of the peripatetic project of systematizing various areas of knowledge, Aristotle—or one of the best members of his school—analyzes weighing practices, with their techniques of establishing *equality* of weights, and lifting practices, with their techniques for *outwitting* nature so as to raise greater loads with smaller forces. The balance is confronted with the lever and then with a wide range of technical devices. These disparate practices are integrated into a comprehensive system of knowledge by reducing the lifting devices to the measuring devices. It will be further argued that a key prerequisite for this project of reflection on mechanical practice was an innovation in weighing technology: the invention in (or introduction to) Greece of the balance with unequal arms. A by-product of the analysis will be that the law of the lever, taken as a strict proportionality, as it is formulated in this treatise, may be just an afterthought of the reduction program, or even a later interpolation.

In this paper, we will first briefly discuss the history and structure of the text of the *Mechanical Problems*. In Sect. 5.3 we will explicate the work's program of the reduction of technical devices to lever and balance. Section 5.4 illustrates this program through a reconstruction of two paradigm problems. Section 5.5 then presents a detailed analysis of the explanation of the balance with unequal arms in Problem 20. In the final section we analyze the historical conditions under which such an asymmetric balance could contribute to the origin of theoretical mechanics.

---

[5] Bacon (1858, §3). For the classic one-sided reading of Bacon, see Koyré (1943); Koyré (1961, 308) also argues that since technology in human history *precedes* science, it cannot be relevant to an explanation of science.

[6] Merton (1939, 5); Westfall (1993, 65).

5.2 The Text of the *Mechanical Problems*

The peripatetic *Mechanical Problems* presents us with the first documented example of a sustained theoretical reflection on mechanical knowledge. Compiled at Aristotle's Lyceum, in part perhaps during his lifetime[7] and passed down as authentically Aristotelian, it seems at first glance to belong to the genre of problem collections that arose out of the question and answer contests of the sophists in the fifth century and was still popular in the school of Aristotle.[8] The treatise was generally taken to be a genuinely Aristotelian work up to the early nineteenth century; but sometime between 1830 and 1870 the opinion of most classical philologists shifted from acceptance to denial of its authenticity—for reasons that probably have more to do with the constitution of classical philology as a discipline than with any characteristics of the text itself. In particular French and German classicists of the later nineteenth and early twentieth centuries tend to be adamant that the work is not Aristotelian—based primarily on a misguided judgment about the scientific quality of the work: Valentine Rose influentially spoke of the "triviality and confusion of the questions" and Paul Tannery considered it "a collection without order or method."[9]

The earliest still surviving manuscript of the *Mechanical Problems* was made in Constantinople around 1300 and all extant manuscripts have been traced back to one archetype from tenth-century Byzantium. Based on the same manuscript tradition, the Byzantine scholar, Georgios Pachymeres (1242–1310) wrote a Greek compendium on Aristotle called *Philosophia*, whose twelfth and last book contains an almost verbatim paraphrase of the *Mechanical Problems.* Some of his very minor emendations found their way into later Aristotle manuscripts and into the first printed edition. The only possibly independent tradition of the work lies in extracts from the Introduction and Problem 1 that were translated into Arabic at the end of the eleventh century. The usual lists of works by Aristotle attribute a work on mechanics to him. In the first century BCE Athenaeus Mechanicus reported that Aristotle's work in mechanics could only be of interest to "younger friends of knowledge" and dismissed any professional benefit for mature practitioners. Those who deal explicitly with some of the actual subject matter of the *Mechanical Problems* (Vitruvius and Hero) mention neither the work nor its author.[10] No

[7] There is no consensus on authorship or dating. Based on the way letters are used to locate points and figures, the work predates the Euclidean reform. On such formal questions see Heiberg (1904) and Netz (1999). Euclid's *Elements* are generally taken to have been compiled shortly after 300 BCE; Aristotle died in 322 BCE. If we don't want to resolve the question of Aristotle's possible authorship by stipulation, we have to date the work at some time between 330 and 270 BCE. Recent commentators tend to favor the later date.

[8] Flashar (1961, 297–316). On the culture of mathematics in this period see Asper (2008, 107–112).

[9] Rose (1854, 192); Tannery (1915, 33).

[10] For the manuscript tradition see van Leeuwen (2012, 2013, 2016); for the Arabic translation see Abattouy (2001). On the ancient lists see Flashar (2004, 189–191) and Hein (1985, 304). On Athenaeus, see Whitehead and Blyth (2004, 44) and Bodnár (2011); Vitruvius (1931–1934, Bk. 10,3); Hero of Alexandria (1900, Bk. 2.8; 2.33, pp. 114, 170).

medieval Latin translation of the work has been found. Although number of tantalizing hints at such a translation have surfaced,[11] none of them indicates an influential tradition of reception nor even provides hard evidence for the existence of a Latin text. A broad reception of the work begins only after it was first printed in the *Corpus aristotelicum* by Aldus Manutius in Venice in 1497. Between 1517 and 1629 four complete and two partial translations into Latin as well as some complete or partial vernacular translations were undertaken and more than a dozen extensive commentaries were produced.[12]

The text consists of an introduction and 35 sections, or "problems" generally merely one short paragraph in length. The problems (almost) always begin with the phrase *dia ti* ("Why is it that") and generally ask why a particular technical device performs as it does. This opening question is fairly consistently answered by a rhetorical counter-question *hê dioti* ("Is it because"), which usually identifies (some part of) the device and asks if it cannot be viewed as a lever. Then, in many instances, a general principle is invoked, under which the particular case can be subsumed. This principle can then function as the major premise in a syllogism-like argument with the counter-question as the minor premise. The conclusion of the argument repeats the statement posed as a question at the start. Sometimes the problem then closes with an evaluation. The ideal form of a problem, actually instantiated by a number of them, displays an analysis in the following form: (1) Why are Greeks mortal? (2) Is it because they are men? (3) And all men are mortal: Thus, since Greeks are men, they are mortal.

The *Mechanical Problems* departs from the usual form of the problem genre in as much as the introduction to the work announces a theoretical program of reduction.[13] All technical devices are to be reduced to the lever, the lever is to be reduced to a balance and the two arms of a balance are to be reduced to two radii of a circle, from whose dialectical properties all these devices get their remarkable abilities. The first three problems after the introduction systematically introduce the basic theoretical concepts and principles needed for the reduction that is to be carried out in the subsequent problems. However, many of the "problems" that actually follow have no recognizable connection to the reduction program. Although the text that has come down to us is to a certain extent a hodge-podge of disparate topics thrown together, nonetheless, parts of the work pursue an ambitious program of theoretical investigation of technical devices: the formation of theoretical concepts in the study of technology.

With two notable exceptions the problems are short: the mean length is 24 lines in the Bekker edition (1831). Problem 1 (115 lines) and Problem 24 (89 lines) are both significantly longer than any of the other sections—between three and two times longer than the next longest problem. Alone among the problems these two do not begin with the standard phrase *dia ti* but rather with a different (though

[11] Clagett (1959, 71, n.5 and 75–76, n. 6).

[12] See Laird (1986) and Rose and Drake (1971). The Latin translations were by Fausto (1517), Tomeo (1525), Bechio (1560) and Monantheuil (1599).

[13] The reduction program was already noted by Duhem (1905, 8).

equivalent) formula. Their content and terminology is also significantly more technically mathematical in character than that of other problems, except for the third longest (Problem 23): These three (a third of the text not counting the introduction) deal not with mechanics at all, but with kinematics—or perhaps better: the geometry of moving points. Importantly, in Problem 1 a simple mechanical discussion of why balances with longer arms are more sensitive than those with shorter arms, which carries out the first step of the reduction by comparing the arms of a balance to the radii of a circle, is interrupted by a long, unsuccessful and never relevantly used mathematical proof of the composition of motions. After this interruption the argument continues where it left off without noticeable regard for the proof.

Fifteen of the 32 later problems follow the program of reduction presented in Problems 1–3 and display the 'standard' pattern to some degree, although the reduction most often skips over the balance and goes straight from the lever to the circle—or even skips the lever, too.[14] However, there are a number of other problems, besides the "kinematic" ones (23, 24), that have no recognizable connection to the program of reduction: five problems (8, 10, 12, 19, 31) deal with why it is easier to move objects that are already in motion or tending to motion, and three of the last four problems deal with projectile motion (32, 33, 34). With the exception of the reduction program in Problems 1–3, there does not seem to be any systematic principle structuring the order of presentation. Thus the *Mechanical Problems,* as it has come down to us, seems to have been an open-ended compilation used in a school context and may have collected a number of different projects from different times at the peripatetic school. It displays aspects both of a tight programmatic treatise on mechanical theory and of a desultory collection of disparate questions about mathematics, mechanics and everyday experience. The analysis presented here applies only to those parts of the text that carry out the reduction program of the introduction—about a third.

[14] Problem 4: The oars of a ship are identified as levers.
Problem 6: The mast of a ship is identified as a lever.
Problem 9: The wheels of pulleys are identified as levers.
Problem 13: Handles of spindles and windlasses are identified as levers.
Problem 14: A piece of wood broken over the knee is identified as a lever.
Problem 15: Pebbles at the beach rotated and worn down by water are identified as levers.
Problem 16: Wooden planks raised are identified as levers.
Problem 17: Wedges are identified as consisting of levers.
Problem 18: A pulley is identified as a lever.
Problem 20: An asymmetric balance is identified as a lever.
Problem 21: The forceps of a dentist are identified as a pair of levers.
Problem 22: A nutcracker is identified as a lever.
Problem 26: Wooden planks carried on the shoulder are identified as radii.
Problem 27: Wooden planks raised up to the shoulder are identified as radii.
Problem 29: A plank carried by two men is identified as a lever.

5.3 The Program of Reduction

The Introduction to the *Mechanical Problems* opens with an affirmation of the distinction between things that occur in accord with nature (*kata phusin*) and things that are contrary to nature or perhaps only 'beyond' nature (*para phusin*) and with an exclamation about the "wonders" of technology, the trenchant negation of which by Simon Stevin (*Wonder en is gheen wonder*) in 1586 now graces the frontispiece and binding of every volume of the *Dictionary of Scientific Biography* (1970–1980). The treatise begins:

> It evokes wonder that there are things that occur according to nature, the cause of which is not known, and that there are things that occur contrary to nature, which are produced by art [*technê*] for the benefit of humans (847a12–13). (Hett 1936)

Phenomena of the second kind, produced by technology, in which the weaker seems to move or master the stronger and which are thus beyond or contrary to nature, are called "mechanical problems" by the author. It is these technical devices that are of special interest, for they seem to exhibit effects that are incompatible with everyday experience (and with central doctrines of Aristotelian physics).

The source of everything wondrous is rather quickly located in the circle, which, we are told, is the unity of opposites and thus has various dialectical properties. The principle of the wonders (*archê tôn thaumatôn*) will turn out to be the fact that points on a moving radius at different distances from the center trace greater or lesser circles depending on how far they are from the center. This principle, which we shall call the *circular motion principle,* refers back to what must have been a very well known passage in Plato's *Laws* (Bk. 10), where Plato engages in an extensive discussion of the wonders of the circle concluding:

> For [this circular motion] is the source of all the wonders bringing about what was supposed impossible, making slowness and speed at the same time harmonize [*homologeîn*] in greater and smaller circles.[15]

A similar principle, formulated numerous times in the course of the *Mechanical Problems,* states in its first formulation (see below) that each different point on a radius moves at a different speed: the farther from the center, the greater the circle traced in the same time, and thus the faster the motion.

After introducing the wonders of the circle the author then formulates the theoretical program of the treatise: The reduction of mechanical devices to the lever, the reduction of the lever to the balance and of the balance to the circle. Just as the end points of radii of different lengths move at different speeds, so too do weights on balances of different lengths move at different speeds. The balance can be interpreted and explained in its workings by the properties of the circle. And just as weights in the pans of balances with longer arms move more swiftly than in those

[15] Plato (1962, Bk. 10, 893b-e). It is significant that Plato has an exact mathematical grasp of the relation of the circumference (and motion along it) to the length of the radius but makes no reference in this context to any mechanical problems. See Berryman (2009, 61) for a different view.

with smaller arms so, too, does a weight/force applied to a longer lever move the load (at the other) end *more easily* than one applied to a shorter lever. Thus the lever can be interpreted and explained in its workings by the properties of the balance. Finally, every mechanical device can be interpreted and explained in its workings in terms of the lever. Thus, in each of the sections following this presentation of the reduction program a particular mechanical problem is to be solved by reducing the device by way of the lever to a *principle of circular motion*—which is then formulated explicitly:

> What pertains to the balance is traced back [*anagetai*] to the circle, and what pertains to the lever [is traced back] to the balance; and nearly all that pertains to mechanical movements [is traced back] to the lever. But no one of the marks on a line drawn from the center travels equally fast as the other. But always that which is farther opposite the resting [point] is faster. (848a11–17)

Thus we have a program of reduction of mechanical problems ultimately to the circular motion principle, which states that different points on a rotating radius cover different distances in the same time. That this is the explanation of the wonders mentioned in the opening sentence (we are then told) will "become evident in the subsequent problems" (848a19).

The details of the program of reduction are developed in the first three problems where, alongside the center and the radius, the basic concepts of the peripatetic mechanics are introduced. These three sections introduce the basic elements of a balance-lever system or model and begin to turn them into theoretical concepts: lever, balance, fulcrum, suspension point ("cord"), force/mover, load/weight, and the material beam.

Problem 1 begins with the question of why larger balances are more accurate than smaller ones—by which the author seems to mean, why balances with longer arms are more sensitive to weights placed on them than those with shorter arms. This state of affairs is explained by the circular motion principle, which receives its second formulation:

> The reason for this is that in the circle the line removed farther from the center moves faster than the smaller one, which is close, when moved by the same force. Faster has two senses: When something traverses the same space in less time, we say it is faster, and if more [space] in the same [time]. But the greater [line] in the same time describes a greater circle, for the outside is greater than the inside. (848b3–6)

The longer radius ("line from the center") covers more space than a shorter one in the same time even if both are moved by the same force or "strength" (*ischus*).[16] The inference that immediately suggests itself in this context, namely that the longer balance arm, like the longer radius, is moved more swiftly (has a greater displacement in the same time), is not made. Instead the text embarks on a lengthy mathematical

[16] There has been some disagreement in the literature on the meaning of *ischus* (force) here, Duhem (1905 and 1913), De Gandt (1982), and Krafft (1970). See the discussion in Schiefsky (2009, 59–61). For our purposes it suffices to note that motion presupposes a force and that wherever the force is applied to the rotating radius or beam: outside points move faster than inside points.

argument about the composition of two motions of a point.[17] After 85 lines on the geometry of motion, the text resumes the course of the mechanical argument by rehearsing the circular motion principle (this is the third version): "the more distant … the more quickly" (849b21). It then repeats the original question almost verbatim and draws some conclusions rather quickly, thus finally carrying out the first step of the reduction. The suspension point of the balance, the *cord*, is identified with the center of a circle, and the arms of the balance are identified with radii of the corresponding circle:

> For the cord becomes the center (for it stays the same), and each part of the beam is a radius. (849b24)[18]

The form of this reduction step will be repeated later like a formula: Each of the three steps in the program of reduction is carried out by just such an identification: *A becomes B*. Here the cord (*sparton*)[19] becomes (*ginetai*) the center (*kentron*) because it stays the same, (*menei gar touto*). The phrase 'it stays the same' or 'it stays put'—just like the center of a circle stays put—is used expressly by the author (or some later editor) to mark each of the three steps of the reduction. It is used as a formula in Problem 1 where the cord or suspension point is identified with (*becomes*) the center, in Problem 3 where the fulcrum is identified with (*becomes*) the suspension point and in Problem 4 where, in the first application of the reduction program to a concrete problem, an oarlock is identified with (*becomes*) the fulcrum. In these three first steps, as most often in the later problems, the identification of a device with a lever or a pair of radii or the identification of a part of such a device with the fulcrum or center is expressed by *ginetai,* not simply by *esti,* the verb "to be".[20]

The argument being made here is that just like a point farther from the center is moved more/farther, so too, the arm of a balance farther from the suspension point is moved more/ farther by the same force/weight and thus a motion that is scarcely

[17] Since Krafft's analysis in *Dynamische und statische Betrachtungsweise* (1970); this argument about the geometry of a moving point has been the focus of attention of scholars dealing with the *Mechanical Problems*. See also Mark Schiefsky, (2009) and De Groot (2009, 2014). No convincing argument has so far been advanced as to why the later problems should need this proof.

[18] Reading *phalanx* (beam) for *plastinx* (pan). Not only does the sense of this text argue for this reading, but also the next problem, which deals with a material beam balance (without actually calling it a beam), would be much easier to understand if the material beam had already been introduced. No modern translation takes the author to mean 'pan'. Forster (1995, 1302) and Hett (1936, 347), (both take the author to mean "balance" and not "scale pan." The Arabic translation (Abbatouy 2001, 114–115) renders whatever was in the original Greek text all three times as *beam* or *pole*. We thank Sonja Brentjes for advice on the Arabic.

[19] The term *spartion* (the diminutive of *sparton*: cord) becomes a technical term meaning *suspension point* after its identification with the center of a circle here. The two later occurrences of the non-diminutive *sparton* (in Prob. 3 and 20) may be copying mistakes, since in each case the text refers back to a cord previously mentioned in the diminutive form. In Problem 1, on the other hand, we are dealing with *larger* balances, where the larger cord might actually be meant.

[20] The phrase *menei gar touto* otherwise occurs only in Problem 27, where it plays exactly the same role as in 1, 3, and 4. The verb *ginetai* (it becomes) is used to express the identification of a part of a device with the fulcrum or center in Problems 1, 3, 4, 5, 6, 12, 15, 16, 19, 20, 26, 29. In Problems 9, 13, 14, 17, 21, 22, and 27 only the verb *to be* or no verb at all is used.

noticeable if it is near the center/suspension point (because it is so short) will be more easily noticed if it is far from the center/suspension point (because it is longer). That the argument—to the extent that it is plausible—has nothing to do with differences in the effects of a weight (nor with the lever) can be seen from the fact that it applies just as well to the *horizontal* motion of a turnstile.[21] Under the same force (applied to turning a horizontal bar) a point or part of the bar far from the center would move farther (and more noticeably) than one near the center. That is, the question is not really about how sensitive the balance is to a weight placed on it, but about how long and thus how noticeable the segment of the circle is that the weight describes in its motion. What the author needs, however, is an argument that takes him from 'more swiftly' (*thatton*) to 'more easily' (*rhaon*). Furthermore, no comparison is yet made between the motion of a long arm and the motion of a short arm: the comparison is between the motions of the arms of two different symmetric *balances*, one of which has long arms and the other short arms. The additional steps are taken only in Problem 3.

Problem 2 begins with the analysis of the behavior of a balance. We start out with a balance suspended from above its middle by a cord and inclined to one side by a weight in one of the pans. When the weight in the pan is removed, the depressed side of the balance rises. If, on the other hand, the balance is supported from below and the weight is removed it will remain as it is. Both observations are correct for the case that the balance in question is a homogenous material beam. If such a beam is supported from below in the middle and tipped to one side, it will tip still further down as far as it can (presumably until it hits the ground) because more than half the volume of the beam lies on the tipped side of the vertical plane through the fulcrum. Thus in the course of the discussion, it becomes clear that the balance is a material beam balance—although the treatise's term for the material beam (*phalanx*) is not used here. The question as formulated at the beginning of Problem 2 reads:

> Why is it that, if the cord [*spartion*] is from above, when the weight inclined below is removed, the balance rises back up. If however [the cord] is placed below, then it [the balance] does not rise, but stays. (851a3–6)

The author then points out that the vertical plane through the suspension point above or below (which divides a horizontal beam in two equal parts) divides a deflected beam balance into unequal parts: If suspended from above, the upper part of a deflected material beam is larger and heavier than the lower end and will therefore descend, whereas if supported from below, the lower part of the deflected beam is larger and heavier and thus stays where it is (on the ground). For an understanding of the mechanical knowledge embodied in this treatise, it is important to note that the third case, in which a (panless) beam is fixed neither from above nor from below, but in the (vertical) middle of the beam, is not discussed. We know, but the author apparently didn't, that if such a beam is supported in the middle (at its center of gravity), it will remain stationary in any given position. This follow-up question,

[21] The author discusses the horizontal motion of a suspended beam balance in Prob. 10.

which literally makes the fulcrum and the suspension point the center, is obvious only in hindsight.

The task of this problem is to extend the scope of the cord or suspension point of the balance (which has just been identified with the center of a circle) to cover cases in which the balance is supported not from above, as is the rule, but from below: The "cord" is now below the balance. The technical term for fulcrum (*hupomochlion*) is not yet used and the one mention of a "support" uses a term otherwise used for a substrate or substance (*hupokeimenon*).

Problem 3 then extends the identification of the suspension point of the balance with the center of the circle to the fulcrum of a lever: center = cord = fulcrum. The official question is: "Why is it that small powers can move great weights/loads by means of a lever?" The answer is that the lever is a kind of balance that has its suspension point below it and also does not have equal arms, but rather is longer on one side than on the other.

> Is it that the reason is the lever, which is a balance having the cord [*spartion*] below and being divided into unequals? For the fulcrum [*hupomochlion*] becomes [*ginetai*] the cord, for both of them stay the same [*menei gar amphô tauta*] just like the center. (850a34–35)

This is the second step of the reduction: The analysis of the material beam balance is extended to the lever, which—unlike the device considered in Problem 2 with its support below—is no longer equal-armed but rather is longer on one side than on the other. The fulcrum can be identified with the suspension point of an apparently unequal-armed balance. A balance of this kind is studied in detail in Problem 20 (see below). The further reduction step from the balance to the circle is continued immediately as prefigured in Problem 1: "For under the same weight, the greater radius moves faster" (850a36). In the formulations of the circular motion principle in the Introduction and Problem 1, two different-sized balances were compared. In Problem 3 different-sized arms are considered—but there is only one balance (divided into unequal parts). These different arms are reduced to two unequal radii that have different speeds (with the same angular velocity). Similarly the two ends of the same lever/balance move at different speeds. Recapitulating this argument, the text then specifies the various elements of a lever system: besides the bar itself, a lever system consists of a fulcrum, a force and a load. When the system is considered as a balance, however, both the force and the load are *weights*, and the fulcrum is equated with the suspension point (and of course, the center). After the analysis of the lever system into its components, the law of the lever is formulated without any preparation or argument; and only after that, is the inference made to a relation between distance and effect. On a balanced lever the weights or loads are inversely proportional to the distances from the fulcrum:

> There are three [things] concerning the lever: the fulcrum—the cord and center—and two weights [*barê*], the mover and the moved. Thus the weight [*baros*] moved is to the mover as the length to the length inversely. For always the more distant [the mover] is from the fulcrum, the more easily it will move. The reason for this is the one already given, that the line more removed from the center describes the greater circle. (850a37–b3)

Not only is this passage preceded by a brief formulation of the circular motion principle (the fourth version); it is also ends with another longer formulation of the same principle, which is said to be the reason for it all. The formulation of the law of the lever here seems to be an afterthought and may in fact be a later interpretative interpolation.[22] In any case, the real work for the goals of the treatise is being done not by the law of the lever but by the proposition that immediately follows it here—before the circular motion principle is again affirmed (for the fifth time). This proposition for the first time makes the transition from moving *more swiftly* (or moving something *more swiftly*) to moving it *more easily.* There is, so to speak, a corollary to the principle of circular motion: not only do points farther from the *center* of the circle move more swiftly (*thatton*), but forces farther from the *fulcrum* of the lever move a given load more easily (*rhaon*). The formulation of the law of the lever is thus something of a distraction on the way to this assertion, which is the real point of Problem 3. Only after the inference from distance from the fulcrum to ease of motion is made, is it even possible to specify the relation further as a proportion. Thus, the law of the lever would seem to have been inserted one sentence before it makes any sense to formulate it.

The inverse proportionality of weights and lengths formulated here is in fact never used or referred to in any of the later problems—which instead appeal directly to the circular motion principle ('more swiftly') or to the corollary derived from it ('more easily') to explain whatever it is they purport to explain. The circular motion principle, as Plato used it, had already formulated proportionalities: the circumferences of various circles are proportional to their radii. And motions along these circumferences in the same time should display similar proportions. Plato, in the passage cited above (893b–e) speaks explicitly of ratio (*logos*) and proportion (*analogon*)—as does the *Mechanical Problems*, too, occasionally when it is doing geometry instead of mechanics. However, the explanatory principle, as formulated in this treatise, never demands a proportionality between distances from the center and speeds or effects, just an increase in speed or effect with increasing distance from the center.

The formulations in the later problems explaining technical devices are always qualitative, comparative and generally adverbial: the greater the force, the faster something moves; the farther removed from the center, the more something is moved; the closer … the more slowly; the greater … the easier, etc. Often, easily quantified expressions are avoided: the greater force generally doesn't move more *weight,* it moves weight *more* or *more easily*. With the exception of the law of the lever (expressed using a technical term for inverse proportionality, *antipeponthen*) the relations of distances and weights or of speeds and forces are formulated merely qualitatively. Often the term *hosô* is used to suggest the coupling of two properties

[22] Stevin (1586, 65, 509) condemned Aristotle for this internally contradictory formulation. The strict proportionality of lengths and weights holds only in equilibrium, that is, in that case in which there is neither a moving weight nor a moved weight. When one weight moves the other, they are not in equilibrium. If we assume the author was aware of this fact, we have an explanation for his avoidance of the language of proportions.

or their increments. And in some of the problems a standard contrast form, *hosô* ... *tosoutô* (*as x so y*)[23] is used—a formulation that is certainly a bit more formal but still entails no quantification (similar to the Latin *tantum* ... *quantum*, which can but need not express a proportionality). Such formulations could later be interpreted as proportions, whether or not they were originally intended as such—or the *tantum-quantum* formula could have been added later to make the text clearer to someone who already knew the law of the lever. The functions are so to speak formulated as monotonic but also happen to be linear. In fact outside the mathematical arguments of Problems 1 and 24 on the composition of motions, we encounter neither the concept of proportion (*analogon*) nor even of ratio (*logos*).[24]

5.4 The Application Paradigm: Problems 4 and 6

Problem 4 in spite of some internal inconsistency presents the paradigm of the third step of reduction: the working of a technical device—in this case, an oar—is explained as an instance of a lever system. The three elements of the lever system just analyzed (in addition to the bar of the lever itself) are identified in a concrete device. The oar is the lever bar, the oarlock is the fulcrum, the sailor on one end of the oar is the moving force and the sea on the other end is the load:

> Why do those in the middle move the ship most? Is it because the oar is a lever? Thus the oarlock becomes [*ginetai*] the fulcrum (for it stays the same [*menei gar touto*]) and the sea is the load which the oar pushes. The mover of the lever is the sailor. The farther the mover of the load is away from the fulcrum the more it always moves the load. For thus the radius is greater, and the oarlock, which is the fulcrum, is the center. In the middle of the ship the most oar is inside. For the ship is broadest at that point, so that it is possible, that on both sides a greater part of the oar is inside the ship. (850b11–18)

Apollonius Rhodius tells us that when Jason and his Argonauts first boarded the ship, the seats on the benches running fore to aft were assigned by casting lots—except for the two seats on the bench in the middle, which were reserved for the heroes Heracles and Ancaeus.[25] So how could the two strongest men contribute most by sitting in the middle?

The *Mechanical Problems* gives an answer: The phenomenon to be explained is considered as an instance of the application of a lever to a load. The various elements of the lever are identified in concrete objects. A general principle is adduced and a compelling answer to the question is given. The text illustrates the typical three-step argument:

---

[23] For instance: Problems 16, 20, 29. As we shall see in Problem 20: With a given lever and a given force, the closer to the load the fulcrum is put, the greater the load that can be moved, but the increase in effect is not proportional to the change in distance from the fulcrum.

[24] There are two uses of *logos* in a different sense in Prob. 19 and 23.

[25] Apollonius Rhodius (1967, 31: Bk. 1, 394–401). See van Cappelle (1812, 199–200).

1) Question:	Why is it that the rowers in the middle move the ship most?
2) Counter-question:	Is it because the oar is a lever?—The elements of the lever are identified and the oarlock is said to *become* the center-cord-fulcrum, which stays put.
3) Principle:	A corollary to the circular motion principle is adduced. A force farther from the fulcrum effects more. This is a further step in relation to the corollary formulated in problem 3. There we learned that the same force moves a load more easily; here we learn that the same force moves a greater load.

The three steps can then be read backwards as a quasi-syllogistic argument. The principle is the major premise, the counter-question is the minor premise and the transformation of the original question into an assertion is the conclusion.

Major premise:	Forces farther from the fulcrum effect more.
Minor premise:	Rowers in the middle of the ship are forces farther from the fulcrum.
Conclusion:	Rowers in the middle effect more.

This presents an argument of the following form: [All] Bs are C. [All] As are B. Thus [All] As are C, which is a good approximation of a *Barbara* syllogism.

This kind of argument is made again and again in the *Mechanical Problems*. The important point is that, within this scheme, given that the identifications are plausible and the principle is accepted as certain, the answer generated is in a certain sense compelling. The identifications can of course be wrong and thus the answer may be false—even absurd, but the solutions given are not just empirical assertions, they are the logical consequences of the identifications, given the principles. The answer to the question is thus the conclusion of a compelling argument.[26]

After the appeal to the qualitative lever principle (the farther, the more) the text just quoted then takes a further step from the lever to the radii of a circle and from the fulcrum to the center. The next two sentences explain to us why the sailor in the middle sits farther from the fulcrum: the ship is wider there and since the benches are in a line down the middle of the ship, the sailor sitting at the place where the ship is widest has more oar inside the ship than at any other point. Presumably the oars all have the same length outside the ship.

[26] In this particular case the rower in the middle does not in fact effect more, he just has an easier time of it. Moreover, if the arm-length of the rowers is the same, the rower on the longer lever moves it the same length as the rowers on the shorter levers (oars) and thus actually effects less. Problem 3 asserted: *the farther, the more easily*; Problem 4 asserts: *the farther, the more effective*. Whereas Problem 3 held that the *same* effect is achieved with *less* force, Problem 4 wants to assert that a *greater* effect is achieved with the *same* force.

The discussion (see below) then continues, repeating the argument, but it is inconsistent with the previous course of argument. The identification of the elements of the lever is redone with different results; it is also not schematic, or even explicit. The lever appealed to now becomes a lever of the 'second kind'. In a lever of the *first kind* the fulcrum lies between the force and the load as in the ship construction just described, and the force applied pushes in the opposite direction of that in which the load is supposed to move. In a lever of the *second kind* the load is located between the fulcrum and the force—normally one end of the lever is on the ground and the load is between it and the raising force. The force at the end of the lever pushes in the *same* direction as the intended motion of the load. As the text continues, the point in the sea where the oar is placed is now seen as the fulcrum and the ship attached to the middle of the oar at the oarlock is the load:

> So the ship moves because, when the oar is fixed [*apereidomenês*] in the sea,[27] the end of the oar that is inside goes forward, and the ship which is bound to the oarlock also goes forward in the direction in which the end of the oar is [going].

The text then recapitulates the argument (ignoring the second lever model):

> For where the oar raises up the most sea, there the ship is necessarily pushed forward most. And it raises the most where the part of the oar from the oarlock is greatest. That is why those in the middle of the ship move it most. For in the middle of the ship the part of the oar from the oarlock inside is the greatest.

Thus, in order to solve a mechanical problem, we must be able to identify among the material bodies that we are dealing with something that can be viewed as a lever, something that can be viewed as the fulcrum—and of course the force and the load. Given the corollary to the circular motion principle, the principle of the lever, that the force farther from the fulcrum effects more, the problem can be solved.

This procedure, illustrated by Problem 4, can also be seen equally well in Problem 6, which also better illustrates how the scheme can go wrong. Problem 6 inverts the following quasi-syllogism:

Major premise:	Forces farther from the fulcrum effect more (Bs are Cs).
Minor premise:	Sails on higher yard-arms (blown by the wind) are forces farther from the fulcrum (As are Bs).
Conclusion:	Ships with higher yard-arms sail faster (As are Cs).

The text of Problem 6 follows the standard form.

Question: Why is it that As are C?

---

[27] The verb *apereidein* means to *fix* or *support* and is related to the term *peisma* for the ship's cable used to tie down the ship to land. This sentence may be a later insertion since the text then continues with a renewed identification of the sea as the load and a repetition of the explanation of why the lever (now of the first kind again) is longer in the middle. There is clearly some corruption in the text. Renaissance authors often pointed out that Aristotle should have used a lever of the second kind in his analysis. See Galileo's letter to Giacomo Contarini, March 22, 1593, in Galilei (1968, vol. 10, 55–57); Biancani (1615, 159); Baldi (1621, 41).

"Why is it that the higher the yard-arm the faster the ship [moves] with the same sail and the same wind?"

Identification: Is it because As are B? The four elements of the lever are specified.

"Is it because the mast becomes [*ginetai*] a *lever*, the base in which it is fixed [is] a *fulcrum*, the *load* to be moved [is] the ship, and the *force* [is] the wind in the sail?

Principle: Bs are C. The farther from the fulcrum the force is, the more easily and swiftly it achieves its effects.

"If, then, the farther away the fulcrum is, the more easily and more swiftly does the same power [*dunamis*] move the same load, then the yard-arm drawn higher makes the sail also farther away from the base, which is the fulcrum." [28]

In this case the argument, given the identifications, is compelling, but the result of the argumentation is absurd: The higher the yard-arm, the more likely the ship is to tip over, not to sail more swiftly.

5.5 The Asymmetric Balance and the *Mechanical Problems*

The crucial step in the program of reduction carried out in Problem 3 identified the lever as a beam balance supported from below and divided into unequal parts. The treatment of the question in this passage left it open whether this identification was a purely conceptual speculation or whether the author had some specific technical device in mind. Later in Problem 20 the author makes it clear that he did have a particular technical device in mind that instantiates this identification, a balance that is in fact divided into unequal parts, though it is still suspended from above. The asymmetric balance seems to have played a role analogous to that of Thales' Theorem, which was a major step in the development of geometry but in Euclid's *Elements* (Bk. 3, prop. 31) is just a trivial special case of an angle inscribed in a circle on an arbitrary chord. The unequal armed balance may thus represent a special case that made a cognitive advance possible.

The technical device that occasions wonder and needs an explanation in Problem 20, as we shall analyze in detail below, is what the author calls a "half-balance": If we wanted to weigh out a side of beef at the market, we could take a very large beam balance and put a number of large lead cylinders in the scale pan on one side and the beef on the other side. On the other hand, we could (and the Greeks apparently did) put the meat on a pan or a hook at the end of a beam that is shorter, and perhaps thinner and lighter, than the large balance and has a relatively small counterweight at the other end. We could then suspend the whole construction from a point quite

[28] The three sentences quoted make up the entire Problem 6, 851a38–b6. Note that the principle invoked is that the *same* force moves the *same* load more easily and thus more swiftly—and thus has a greater effect.

close to the hook hanging up the beef and "wonder" how this device can do the same job as a much larger symmetric balance with very large measuring weights. The author asks why we can weigh out *large* quantities of meat with a *small* counterweight by means of a device that seems to be only half a balance, seeing that it has only one (real) arm and one pan (or hook):

> Why do beams weigh out meat (great loads) by means of a small attachment [*artêma*], the whole being a (small) half-balance?[29] For the scale pan is suspended only there where the load [*baros*] is put in, but on the other [side] is only the beam. (853b25–28)

This half-balance (with only one pan) can nonetheless perform the same task as a full balance of even greater size. How can that be? What sort of device is this?

The scale balance is ancient; it was in widespread use in Egypt and Mesopotamia by the beginning of the third millennium BCE. Systems of weights are undocumented in Mesopotamian sources before 3200 BCE; but after 3000 they prevail over other measuring systems. All balances known from the two and a half millennia after its invention were of the same type: balances with equal arms. These do not seem to have given rise to any generalized mechanical concepts other than weight.[30]

The first theory of mechanics, attempted here, seems to have arisen in conjunction with the invention of the balance with unequal arms. This new type of balance differs from the earlier ones in that two *different* weights at *different* distances from the suspension point are compared with one another. In this practice weight and distance had to be coordinated such that unequals were in equilibrium. The theoretical quantity that emerged from thinking about this practice came in the course of time to be called variously *moment* or *positional gravity*. The Greek term adapted to denote it was *rhopê*. It is not known exactly when or where the asymmetric balance was invented, but the balance with unequal arms, called a *steelyard* in English,[31] was quite widespread in the early Roman Empire. The Roman steelyard (in modern Italian called *statera*)—normally made of iron or bronze—was so common at the end of the first century CE that literally hundreds of them have been preserved in the ruins of Pompeii as well as in many other places. Greek balances, on the other hand, were typically made of wood, and evidence is much more sparse. Greek vase paintings picture only balances with equal arms, but there is literary evidence of asymmetric balances that shows them to have been quite well known in Athens of the late fifth-century BCE.

There are two fundamentally different kinds of asymmetric balance. Best known is the Roman steelyard, which has a fixed pan or hook, one or more fixed suspension points and a moveable counterweight. The vast majority of steelyards preserved

---

[29] *hêmizugiou*. This is the only occurrence of the term in the classical Greek corpus: it could either be an adjective (*hêmizugios*) meaning "forming a half-balance," Liddell et al. (1996) or (as Markus Asper has suggested to us) a diminutive noun (*hêmizugion*) meaning "small half-balance."

[30] On the history of the balance in general, see Robens et al. (2014).

[31] From the end of the Anglo-Hanseatic War (1474) until 1598, Hanseatic merchants (who apparently used such devices) had a compound on the north bank of the Thames near London Bridge called the "Stalhof" (steel yard).

from Roman antiquity are of this type. On the other hand, both the pan and the counterweight can be fixed, in which case a balance can weigh out only one particular measure; or, if the suspension point is moveable, it can weigh out a number of different weights. A few examples of this moveable type of balance from the early Roman period have been preserved. This kind of balance, sometimes called the *bismar* (*bessmer*, *besemer*) or *Danish steelyard*, was however common enough in Athens in the later fifth-century BCE for Aristophanes to be able to make a joke by referring to it. In the comedy *Peace,* first performed in 421 BCE, Aristophanes alludes to an unequal-armed balance of the simple bismar type.

It is significant that the first documented reference to the asymmetric balance apparently describes a merely one-portion measuring instrument, which evidently presupposes no knowledge of the law of the lever and from which it is also highly unlikely that the law could be just read off. Let us take Aristophanes' example: An arms merchant, who has been complaining about not being able to sell off his wares because of the new peace, asks what he should do with a particularly expensive war trumpet. He is given some suggestions about what to do with it, including the following: "Well, here's another idea. Pour in lead [into the bell of the trumpet] as I said, add here a pan hung on strings, and you will have something to weigh out the figs for your slaves in the field."[32] A sample representing the amount of figs that each slave is to receive can be placed in the pan hung from the mouthpiece of the trumpet, and the suspension point at which the lead-filled bell end balances the figs can be found empirically. Any predetermined weight portion can be built into this weighing instrument without any theoretical knowledge just by attaching a cord where the trumpet hangs horizontally. The invention and retainment of such a one-pan balance presupposes at most a weighing practice common and specialized enough to want one-portion weighing devices. And the introduction of standards for weights and coins accelerated by the reforms of Solon at the turn of the sixth century BCE would seem to accommodate such practices. As a matter of fact, to build such a device, the standard weight in the weighing pan of a normal symmetric balance could simply be built into the beam of the balance—even before it was learned that a different suspension point would balance a different standard weight.

A balance found in Pompeii, which in the past was wrongly interpreted as a Roman steelyard turns out to be just such an improvised *bismar* balance. It was produced from a vessel that looks like a saucepan, using the pan itself as the fixed counterweight, while its handle serves as the beam. A slit cut down the length of the handle makes it possible to hang the pan from a chain at different points along the handle. The object to be weighed was then hung from the end of the handle, where a rivet is fixed. Along the slit in the handle a scale is marked, evidently produced by empirically gauging each mark, that is, independent of any theoretical knowledge of the law of the lever.[33] Given the evidence, it would seem that the asymmetric balance with a moveable suspension point was the earlier device and was then replaced by the much more efficient Roman steelyard with a moveable counterweight.

[32] Aristophanes (1998, lines 1245–49, pp. 58, 304).

[33] See Damerow et al. (2002).

With this background, let us return to the text: why can a simple beam (with a counterweight) do so much? The answer begins with the standard form of identification in a counter-question, in which the beam is identified as a balance and a lever at the same time. In his exposition the author then explains (1) why the "half-balance" is in fact a (full) balance, (2) why it is not only one balance, but many different balances, (3) why it is also a lever and then (4) what is the specific difference of this balance to the lever. Let's look at the first step. In answer to the question, how a small counterweight can balance a great load, the author asks:

> Is it because it happens that the beam is a balance [*zugon*] and a lever [*mochlos*] at the same time? A balance inasmuch as each of the cords becomes the center of the beam. Only that on the one [side] it has a scale pan, but that on the other [side] instead of the scale pan [it has] the counterweight [*sphairôma*][34] which lies on the balance, as if one were to put the other scale pan and the measuring-weight [*stathmos*][35] on the end of the beam.[36] For it is clear that it draws just as much load laid in the other scale pan. (853b28–35)

A load or heavy thing has a weight, which can be measured by placing the load in one scale pan of a balance. In the other pan we place a standardized measuring weight, or several of them. But the measuring weights and the pan holding them could also simply be fastened to the beam of the balance instead of hanging loosely from it—or they could even be replaced by an equivalent counterweight built into the beam, as long as the counterweight draws the same load as the pan and measuring weights did. This would also explain why this device is indeed a sort of half-balance, since it has only one scale pan. The cord by which the beam (*phalanx*) is suspended and which divides it in half at the center turns it into a balance (*zugon*). In fact as long as the beam with the counterweight still balances out the load placed in the one pan, the cord suspending the device could actually be moved off-center to a different place on the beam—in which case each of the new suspension points would constitute a new "center" of the device. But except for the as yet uncalled-for plural ("each of the cords") the argument thus far merely describes an equal-armed balance with only one pan. The second step justifies viewing this new balance as many different balances:

---

[34] *sphairôma*: literally "round thing"

[35] The term *weight* can be used in two quite different senses: Like *length* it names an abstract physical quantity, but it also can name a concrete material thing such as a measuring weight. If a piece of silver weighs three talents, the silver *has* a weight, the talent *is* a weight. Greek of Aristotle's time sometimes distinguished the two senses by the gender of the noun: *ho stathmos* (masculine) could refer to the abstract quantity measured and *to stathmon* (neuter) to the standardized measuring weight placed in the pan [see Liddell et al. (1996)]. The author of the *Constitution of Athens* (presumably Aristotle) uses the terms in this manner (Aristotle 1995, pp. 2346, 2373; ch. 10.1–2 and 51.3). However, Aristotle at one prominent place (*Metaphysics N*, 1087b37) uses the masculine form for the measuring unit. The standard published versions of Problem 20 use both forms of the word: masculine twice and neuter once. Of the relevant manuscripts, all but the one on which the first print edition was based have only the masculine form. We thank Joyce Van Leeuwen for checking the manuscripts for us.

[36] Reading (with Cappelle) *phalanx* (beam) for *plastinx* (pan).

> And in the same way as the one balance is many balances, many such cords lie in such a balance, in each of which that [part] on the side of the counterweight is half of the beam, and the weight [*stathmos*] [behaves] in the same way as the cords are moved with respect to one another so that it is measured how much load that which is placed in the scale pan draws; so that, whenever the beam is straight [= horizontal], one knows from which cord [is used] how much load the scale pan holds, as has been said. In short, this is a balance having one scale pan in which the load is weighed out, but [having] the other in which the measuring-weight [*stathmos/on*] is in the beam. And therefore the counterweight is the beam on the other side. Being such, there are many balances, as many as there are cords. (853b35–854a7)

The cord by which the beam is suspended and which divides it in half at the center could also be moved to a different place on the beam—in which case the balance would weigh out a different load. Each time we move the cord, the counterweight weighs out a different load. We can have as many different (one-portion) balances as we have different places to attach the cord. These new balances are asymmetric in two ways: one side has a scale pan with a load in it while the other has only a counterweight and furthermore the two *halves* of the beam, divided at the *center* where the cord is fastened to the beam now have different lengths. This makes it clear that the balance under consideration is a *bismar* that can effectively be treated like a symmetric balance with a weight fixed to one side of the beam. To a certain extent this is just an empirical description not an explanation: as a matter of empirical fact the balance is horizontal. If two different objects placed successively in the pan, using the same suspension point, each make the beam horizontal, then they are of the same weight.

It is clear that the half-balance is very much like a real balance. But it is still necessary to show that it is also a lever since this will explain why this kind of balance can do what it does. This is the third step in the text. The asymmetric balance can be viewed as a kind of lever: If the length of the beam is given, then the closer to the load the cord is set, the farther the counterweight is from the cord and the greater is the load that it can balance. The reason for this, the author tells us, is that the whole construction is a sort of lever that is upside down in the sense that the fulcrum, instead of supporting the beam from below, is actually the cord suspending it from *above* and that the load to be moved is not on top of the beam, but is rather located in the pan hanging *below* it. We know (from Problem 3) that the farther from the fulcrum the force is placed, the more easily it moves a load.

> Always, the nearer the cord is to the scale pan and to the weighed-out load, the greater the load it draws because of the fact that the whole beam becomes a lever, which is upside down, for each cord is a fulcrum [*hupomochlion*] being from above, and the load is that which is in the scale pan; but the greater the length of the lever from the fulcrum, the more easily it [the lever] moves there, but here it makes counterpoise [*sêkôma*] and weighs out the load of the beam toward [in relation to] the counterweight. (854a7–15)

As we noted in the discussion of Problem 3, the law of the lever plays no explanatory role in this text. It would indeed have been difficult to extract this law either from mere technical practice with the lever or from handling a bismar type of balance as treated here. In contrast to an idealized steelyard with a weightless beam, the scale of the bismar as it is here described is nonlinear and also depends on the qualities of the material beam that is used. Accordingly the text makes use of functional descriptions that can hardly be directly translated into mathematical relations. It is true, as the text formulates, that the closer to the load the fulcrum is pushed, the greater is the load that can be moved by a given force because the other side of the lever, where the force acts, gets longer. And the closer to the pan the cord is placed, the farther from the cord is the counterweight located and thus the greater is the weight that it can balance out. However, it is not true that the load moved or the weight balanced increases merely proportionally with the distance of the force/counterweight from the fulcrum/cord—since the length added to one side is also subtracted from the other side.[37]

The conclusion (fourth step) drawn in the last clause above returns to the original question of why the asymmetric balance can do what it does. On a lever, the farther from the fulcrum the force is placed, the more easily it moves a load. However the counterweight does not actually *move* a load, but rather *balances* a load ("makes counterpoise" as the author puts it), that is, induces equilibrium. Thus, because the half-balance is like a lever, the farther the counterweight (which plays the role of the force) is from the cord, the greater the load it can weigh out. Problem 20 thus explains the asymmetric balance in terms of the lever, which has already been reduced to the symmetric balance.

The fact that this text deals with a *bismar* not a Roman steelyard is significant, and it was scarcely recognized in the early modern reception. Almost all Renaissance commentators (one notable exception is Biancani 1615) interpret the balance as a Roman steelyard and many identify the round counterweight at the end of the beam (the *sphairôma*) with the moveable counterweight of the steelyard called a *romanum* or *marcum*.[38] The prevalence of the Roman steelyard led these commentators to misunderstand the text. In fact, given the comparative rarity of the *bismar* once the Roman steelyard came into use, it is possible that already in the Hellenistic period the argument had become effectively unintelligible even to the Greeks, which might account in part for the corruption of the text.

Let us return briefly to the reduction program and see how it works in this particular case. In the few cases analyzed so far we have seen the reduction of a technical device to the lever, then to the balance, and then to the circle. In this case, however, we have a balance to begin with, which purportedly only works the way it does because it is simultaneously a lever. The argument is not circular since the balance to be reduced is of a different kind. Moreover, the example even suggests that the origin of the entire reduction program depended on the identification of balance and lever enabled by this device. The asymmetric balance embodies this

[37] In the text just quoted, the appeal to the lever principle of Problem 3 "the greater the length … the more easily (*rhaon*) it moves" is marked by the *tantum-quantum* formula *hosô … tosoutô*. On the other hand, the state of affairs that this principle is supposed to explain "the nearer the cord … the greater the load it draws" is not so marked.

[38] Biancani (1615, 183); cf. Piccolomini (1547, 42v); de Monantheuil (1599, 147): "… aequipondium, Graecis dictum *sphairôma*, nostris Marcum vel Romanum…." Even Baldi (1621, 134), who mentions that "we could use the steelyard in a different manner" and then describes a *bismar*, does not suggest that this might have been intended by Aristotle himself.

identification; it really does work simultaneously as a lever and a balance. It thus seems plausible that the invention or use of such balances constituted the historical starting point for the reduction program itself.

5.6 Historical Conditions for the Origin of Theoretical Mechanics

The aim of the *Mechanical Problems*—at least of that part of it that exemplifies the mechanical reduction program—is to explain technical devices and to show that their 'wonderful' success is compatible with other aspects of everyday experience as well as with certain principles of Aristotle's physics, in particular, with the principle of the correspondence of force and effect. This is taken up early in the Introduction:

> For as Antiphon, the poet, said, this is so: "By means of art we master that by whose nature we are conquered." Of this kind are those in which the lesser master the greater and things possessing little moment [*rhopê*] move heavy weights, and all similar devices, which we term mechanical problems. (847a20–24)

The cornerstone of this project is the equation of the balance and the lever. What is shown again and again in the work is that technical devices can be made intelligible by assimilating them to a *balance-lever model*. Technical devices can now be analyzed in terms of levers, loads, fulcrums and forces, which are no longer just directly given concrete things but also instantiations of abstract concepts, which can stand in logical and even quantitative relations to one another. What the author does is systematically to derive these abstract concepts in the study of technical devices or, perhaps better, to articulate and to reflect on the identifications that actually occur in mechanical practice. Viewing an iron bar and a small stone as a *lever* and a *fulcrum* involves a process of abstraction—even if the concrete objects and the abstraction products retain the same names. Viewing a wheel as a circle involves a similar abstraction. And when parts of technical devices are identified with ('become') parts of the balance-lever system, a further abstraction is made. What we can see in the text is scientific theory arising from reflection on a particular kind of technical practice.

As we have seen, the asymmetric balance, which is both a balance and a lever, plays a special role in the construction of the *Mechanical Problems,* providing the point of departure and the model for the cognitive development realized in the treatise. Why, of all the possible technical devices of the period, could the asymmetric balance play this special role? And why of all places did this happen in Greece?[39] As we have already mentioned, there had been around 2500 years of weighing practice, before the *bismar* was invented without any indication of mechanical theorizing,

[39] On the possibility of comparable developments in China in the same period, see Renn and Schemmel (2006).

and ancient civilizations also had knowledge of many devices that display the lever effect in a simple and perspicuous form. However, relatively soon after the invention of the bismar, we see the beginning of theoretical mechanics.

In the course of three millennia in the construction of the temples and monuments in Mesopotamia and Egypt all the devices known to us as "simple machines" came to be employed. Studying the passageways in the Pyramids archaeologists can reconstruct from the putlog holes in the walls where the pulleys were placed to lower the sarcophagus of the Pharaoh. Thus it is clear that the machines were not only used, but also discussed and reflected on beforehand by the construction planners and overseers. But there is no evidence that this kind of planning practice led to any abstract mechanical theory in connection with these devices.

How does the situation in Greece compare to these highly developed civilizations? Due to the military and architectural activities of the Greeks, practical mechanical knowledge of the simple machines and the planning knowledge needed for construction was widespread in Greece as well. More specifically, such devices as the *shadoof* (or sweep-beam, *kêlôn*), which is discussed in Problem 28 of the *Mechanical Problems,* and the counterbalanced long spear (*sarissa*) of the Macedonian army indicate that counterweights, which practically embody the complementarity of weight and distance, were a common aspect of ancient Greek culture. Any one of these could in the right context have provided the occasion for theoretical questions. Just as Galileo much later used the Venetian Arsenal as an occasion for philosophical reflection, so, too, could Aristotle have used Alexander's exercise yards or the nearest temple construction site. But there are good reasons why the asymmetric balance is a much better candidate than these others for initiating the development that led to the foundation of mechanics. The main reason is simply that a balance is a measuring instrument that by its very nature establishes equivalence between cause and effect.

One foundational experience of practitioners' knowledge since antiquity has been the equivalence of the weight of a body and the force needed to raise it. This equivalence is prototypically embodied in a real instrument, namely the balance with equal arms, which lies at the center of a social practice of weighing and grounds an associated shared mental model that we can call the *balance model*. The heavy body placed in one scale pan that keeps the balance horizontal is equal in weight to the body in the other scale pan. The force applied to one scale pan that keeps the balance horizontal (in equilibrium) is equal to the weight in the other scale pan.

However, the practical knowledge of the technicians of antiquity also involved other basic experiences, in particular, the experience that the constraints set by the equivalence of weight and force could be overcome. In fact, the art of the mechanical technician consisted precisely in overcoming the natural course of things with the help of instruments such as the lever and other machines. In this practice a mechanical instrument achieves (with a given force) a greater effect than could have been achieved without the instrument. That a weight can be moved by a smaller force seemed to involve a "wonderful" trick against nature, called *mêchanê* in Greek and from which the term *mechanics* is derived. This unnatural trick of mechanics is prototypically embodied in a real instrument, the lever, whose use unifies a different practice and grounds a different mental model we can call the *mechanics model*.

Both forms of practice and the mental models they embody could (and did) peacefully coexist without a problem as long as the areas of application remained separate. And even where some practices overlap, no contradiction need occur. Any water-carrier balancing two buckets at the ends of a pole across his nape and shoulders who was unfortunate enough to get different amounts of water in his buckets had to make adjustments in the position of the pole. Neither the water carrier nor even the makers of asymmetric balances have to possess knowledge of the law of the lever or of any similarly specific quantitative rule. They need only know what must be done to keep the beam horizontal. Even though the asymmetric balance seems to instantiate two incompatible mental models or forms of practice, there is no conflict so long as they are not exercised at the same time. It is of course a completely different state of affairs when the two potentially conflicting practices are made the subject of philosophical reflection with a view to integrating them into an embracing system of knowledge.

The appearance of the asymmetric balance in Greece some time before Aristotle made it possible to combine the *mechanics* model with the *balance* model, producing the *balance-lever* model of the *Mechanical Problems* and integrating theorizing on the two areas of practice. The balance-lever model can be understood as a generalizing transformation of the balance model, associated with the ordinary balance with equal arms. In the case of an equal-arms balance, weight differences are balanced by adding weights; in the case of an unequal-arms balance, they are balanced by changing the distance of the counterweight by changing the position of the suspension point. This necessarily extends the simple balance model: Not only can heavy things be moved by a small force, but weights can be compensated not only by weights but also by distances. This is not just practical knowledge gained by using the devices but also a kind of theoretical knowledge gained by systematizing.

As far as explanatory power is concerned, there is a difference between the balance and simple machines like the lever. The balance ascertains equality of weight. The machines apply force. Both deal with weight since weight is what demands the application of force to raise it; but the machines make it possible to raise a weight with less force than is needed without them. Machines make the mover stronger; balances ascertain equality. Equating the lever with the balance, as is done in *Mechanical Problems* on the basis of the asymmetric balance, is a singularly strong explanatory move. Insofar as they are levers, asymmetric balances partake of the general property of mechanical devices by achieving large effects with small forces (great loads, small counterweight). But insofar as they are balances, they must establish equivalence between cause and effect, which is just what is required for an explanation within the Aristotelian framework. This equivalence is achieved by appealing to the compensation of weight by length as it takes place in the asymmetric balance. Thus, it was the practical knowledge related to balances with unequal arms that ultimately provided the empirical basis for the formulation of the law of the lever.

While this is certainly one very important element in an explanation of why mechanics became possible in Aristotle's Greece, there is more to the story: The development of a theoretical science would not have been possible without a pre-existing theoretical tradition concerned with the explanation of natural phenomena. The emergence of mechanics, the realization of the theoretical possibilities of the lever-turned-balance, took place in a particular social setting in which theoretical reflection was already pursued in a number of other areas of experience and in which there existed a strong culture of disputation and persuasion by (public) justification in the form of arguments. The political and juristic customs of the Greeks (at least in Athens) made the success of someone's projects often depend not just on the usual mechanisms of common interest, bribery and nepotism, but also on the effective persuasion of others by rhetoric and argument. Proof in court and in mathematics had a peculiar relevance in Aristotle's Greece. Furthermore, throughout the Mediterranean world with the spread of trade and coined money, weight standards and the associated practices also spread. In Athens one of the feats for which the semi-legendary Solon would be praised by Aristotle was the establishment and reform of standards for weights and coins.[40] Finally, the formation of abstract concepts such as length, and weight already had a firm place in Aristotelian natural philosophy, which also possessed the means to treat simple functional relations such as proportionality and sought to bring ever more areas of human practice into its system of knowledge.

The articulation of the practical knowledge of the balance-lever model in this theoretical context thus constituted an important step in the integration of mechanical practice into systematic knowledge, which led (somewhat later we think) to the formulation of the law of the lever. The *Mechanical Problems* represents one of those rare instances in the history of science where an intermediate or transitional stage has been preserved, showing practical knowledge being turned into the theoretical foundation of science.

References

Abattouy, Mohammed. 2001. Nutaf min Al-Hiyal: A partial Arabic version of pseudo-Aristotle's 'Problemata mechanica. *Early Science and Medicine* 6: 96–122.

Aristophanes. 1998. *Peace*, ed. S. Douglas Olson. Oxford: Clarendon Press.

Aristotle. 1995. The complete works of Aristotle, ed. Jonathan Barnes. Princeton: Princeton University Press.

Asper, Markus. 2008. The two cultures of mathematics in ancient Greece. In *The Oxford handbook of the history of mathematics*, ed. E. Robson and J. Stedall, 107–112. Oxford: Oxford University Press.

Bacon, Francis. 1858. Novum organum. In *The works of Francis Bacon*, ed. J. Spedding, R. Ellis, and D.D. Heath, vol. 4. London: Longmans.

Baldi, Bernardino. 1621. *In mechanica Aristotelis problemata exercitationes: adiecta succinta narratione de autoris vita et scripti*. Mainz: Albinus.

Bechio, Philippo. 1560. *Georgii Pachymerii Hieromnemonis in universam fere Aristotelis philosophiam epitome*. Basel: Froben.

[40] See the already cited passage from the *Constitution of Athens,* ch. 10, in Aristotle 1995, p. 2346.

Bekker, Immanuel, ed. 1831. *Aristotelis opera, 2 vols*. Berlin: Reimer.
Berryman, Sylvia. 2009. *The mechanical hypothesis in ancient Greek natural philosophy*. Cambridge: Cambridge University Press.
Biancani, Giuseppe (Blancanus). 1615. *Aristotelis loca mathematica*. Bologna: Coch.
Bodnár, István. 2011. The pseudo-Aristotelian mechanics: The attribution to Strato. In *Strato of Lampsacus: Text, translation and discussion*, ed. M.-L. Desclos and W.W. Fortenbaugh, 443–455. New Brunswick: Transaction Publishers.
Clagett, Marshall. 1959. *The science of mechanics in the Middle Ages*. Madison: University of Wisconsin Press.
Crombie, Alistair C. 1961. *Augustine to Galileo*. Vol. 2. London: Mercury Books.
Damerow, Peter, Jürgen Renn, and Simone Rieger. 2002. Mechanical knowledge and Pompeian balances. In *Homo Faber: Studies on nature, technology, and science at the time of Pompeii*, ed. G. Castagnetti and J. Renn, 93–108. Rome: L'Erma di Bretschneider.
De Gandt, François. 1982. Force et science des machines. In *Science and speculation: Studies in Hellenistic theory and practice*, ed. J. Barnes et al., 96–127. Cambridge: Cambridge University Press.
De Groot, Jean. 2009. Modes of explanation in the Aristotelian mechanical problems. *Early Science and Medicine* 14: 22–42.
———. 2014. *Aristotle's emprицism: Experience and mechanics in the fourth century BC*. Las Vegas: Parmenides.
Dictionary of Scientific Biography. 1970–1980. Ed. Charles Coulston Gillispie. New York: Scribner.
Duhem, Pierre. 1905. *Les origines de la statique*. Vol. 1. Paris: Hermann.
———. 1913–1959. *Le system du monde: Histoire des doctrines cosmologiques de Platon à Copernic*. Vol. 10. Paris: Hermann.
Fausto, Vittore. 1517. *Aristotelis mechanica ... ac latinitate donata*. Paris: Badii.
Flashar, Hellmut, ed. 1961. *Problemata physica*. Berlin: Akademie-Verlag.
———, ed. 2004. *Grundriss der Geschichte der Philosophie. Die Philosophie der Antike. Bd. 3. - Ältere Akademie, Aristoteles, Peripatos*. 2nd ed. Basel: Schwabe.
Forster, E.S. 1995. Mechanics. In *The complete works of Aristotle*, ed. Jonathan Barnes. Princeton: Princeton University Press.
Galilei, Galileo. 1638. *Discorsi e dimostrazioni matematiche: intorno à due nuoue scienze attenenti alla mecanica i movimenti locali*. Leiden: Elsevir.
———. 1968. *Le opera: Nuova ristampa della edizione nazionale 1890–1909*. Vol. 21. Florence: Barbera.
———. 1974. *Two new sciences, including centers of gravity and force of percussion* (trans: Stillman Drake). Madison: University of Wisconsin Press.
Heiberg, Johann Ludwig. 1904. Mathematisches zu Aristoteles. *Abhandlungen zur Geschichte der mathematischen Wissenschaften mit Einschluss ihrer Anwendungen* 18: 3–49.
Hein, Christel, ed. 1985. *Definition und Einteilung in der Philosophie: Von der spätantiken Einleitungsliteratur zur arabischen Enzyklopädie*. Frankfurt/Main: Peter Lang.
Hero of Alexandria. 1900. *Heronis Alexandrini opera quae supersunt omnia, vol. 2, Mechanica et catoprica*. Leipzig: Teubner. (Reprint 1976. Stuttgart: Teubner).
Hett, W.S., ed. 1936. *Aristotle: Minor works*. Cambridge, MA: Harvard University Press. Loeb Classical Library.
Koyré, Alexandre. 1943. Galileo and Plato. *Journal of the History of Ideas* 4: 400–428.
———. 1961. Les philosophes et la machine. [*Critique* 1948]. In *Études de la pensée philosophique*, 279–309. Paris: Armand Colin.
Krafft, Fritz. 1970. *Dynamische und statische Betrachtungsweise in der antiken Mechanik*. Wiesbaden: Steiner.
Laird, Walter Roy. 1986. The scope of renaissance mechanics. In *Osiris*, 2nd Series, vol. 2, 43–68. Philadelphia: University of Pennsylvania.
———. 2013. The text of the Aristotelian *Mechanics*. *The Classical Quarterly* 63: 183–198.

———. 2016. *The Aristotelian mechanics: Text and diagrams. Boston Studies in the Philosophy and History of Science*. Vol. 316. Cham: Springer International Publishing.

Liddell, Henry G., Robert Scott, and Henry S. Jones. 1996. *A Greek-English lexicon: With a revised supplement. Compiled by Henry George Liddell and Robert Scott. Revised and augmented throughout by Henry Stuart Jones with the assistance of Roderick McKenzie and with the cooperation of many scholars*. 9th ed. Oxford: Clarendon Press.

Merton, Robert K. 1939. Science and the economy of seventeenth-century England. *Science and Society* 3: 3–27.

Monantheuil, Henri de. 1599. *Aristotelis mechanica graeca, emendata, latina facta, et commentariis illustrate*. Paris: Perier.

Netz, Reviel. 1999. *The shaping of deduction in Greek mathematics: A study in cognitive history*. Cambridge: Cambridge University Press.

Piccolomini, Alessandro. 1547. *In mechanicas quaestiones Aristotelis, paraphrasis, paulo quidem plenior*. Rome.

Plato. 1962. Laws. In *Platonis opera*, ed. John Burnet, vol. 5. Oxford: Clarendon Press.

Renn, Jürgen, and Matthias Schemmel. 2006. Mechanics in the Mohist canon and its European counterparts. In *Studies on ancient Chinese scientific and technical texts: Proceedings of the 3rd ISACBRS. International symposium on ancient Chinese books and records of science and technology*. March 31–April 3, Tübingen, Germany, ed. Hans Ulrich Vogel, Christine Moll-Murata, and Xuan Gao, 24–31. Zhengzhou: Elephant Press.

Rhodius, Apollonius. 1967. In *Apollonius Rhodius, the Argonautica*, ed. Seaton Robert Cooper. Cambridge, MA: Harvard University Press. Loeb Classical Library.

Robens, Erich, Shanath Amarasiri A. Jayaweera, and Susanne Kiefer. 2014. *Balances: Instruments, manufacturers, history*. Heidelberg: Springer.

Rose, Valentine. 1854. *De Aristotelis librorum ordine et auctoritate commentatio*. Berlin: Reimer.

Rose, Paul L., and Stillman Drake. 1971. The pseudo-Aristotelian *Questions of mechanics* in renaissance culture. *Studies in the Renaissance* 18: 65–104.

Schiefsky, Mark. 2009. Structures of argument and concepts of force in the Aristotelian mechanical problems. *Early Science and Medicine* 14: 43–67.

Shapin, Steve. 1996. *The Scientific Revolution*. Chicago: University of Chicago Press.

Stevin, Simon. 1586. *De beghinselen der weeghconst*. Leiden: Plantijn. (reprint 1955. *The principal works of Simon Stevin*, vol.1. Amsterdam: Swets & Zeitlinger).

Tannery, Paul. 1915. Sur les problèmes mécaniques attribués à Aristote. In *Memoires scientifiques*, vol. 3. Paris: privately published.

Tomeo, Niccolò Leonico. 1525. *Aristotelis quaestiones mechanicae: Opuscula nuper in lucem aedita*. Venice.

van Cappelle, J.P., ed. 1812. *Aristotelis quaestiones mechanicae*. Amsterdam: den Hengst.

van Leeuwen, Joyce. 2012. *The tradition of the Aristotelian mechanics: Text and diagrams*. Berlin: Diss. Humboldt University.

Vitruvius. 1931–1934. *On architecture*. Vol. 2. Cambridge, MA: Harvard University Press. Loeb Classical Library.

Westfall, Richard S. 1993. Science and technology during the Scientific Revolution: An empirical approach. In *Renaissance & revolution*, ed. J.V. Field and Frank James, 63–72. Cambridge: Cambridge University Press.

Whitehead, David, and Philip H. Blyth. 2004. *Athenaeus Mechanicus. On machines (Peri mechanêmatôn)*. Wiesbaden: Steiner.

Chapter 6
A Journey to the Center of the Earth: Cosmology and the Centrobaric Theory from Antiquity to the Renaissance

Elio Nenci

Abstract This chapter aims to throw light on the ways in which the concept of center of gravity interacted with some of the cosmological ideas conceived in antiquity and in particular with the idea of the figure of earth as presented in Aristotle's *De coelo*. Developing earlier research, this study provides a better understanding of the scientific discussion that took place during the crucial first stage in the development of modern science. The origins and earliest stages in the development of the concept of center of gravity in Ancient Greece was briefly studied by Duhem, whose cursory analysis of a few texts by Pappus and Archimedes was undertaken with the specific purpose of showing the supposed faults inherent in Greek statics. The chapter will begin with a discussion of these Greek sources and attempt to follow the intellectual recovery of this key concept in the Renaissance and the first decades of the seventeenth century, relying in particular on a thorough study of textbooks used for teaching astronomy in Jesuit schools, for example, *Commentaries on De sphera* of Sacrobosco.

Keywords Equilibrium · Balance · Centrobarics · Lever · Aristotle · *De coelo* · Johannes de Sacrobosco · *Sphaera* · Figure of Earth · Natural motion

6.1 Introduction

The theory of the center of gravity played an important role in mathematical and mechanical studies in antiquity. It was an essential part of Archimedes' geometrical works: its most thorough exposition is to be found in the treatise Ἐπιπέδων ἰσορροπιῶν (*On the Equilibrium of Planes*). The most important result of this work

---

The original version of this chapter was revised. A correction to this chapter is available at https://doi.org/10.1007/978-3-319-90345-3_11

E. Nenci (✉)
University of Milan, Milan, Italy
e-mail: elio.nenci@unimi.it

R. Feldhay et al. (eds.), *Emergence and Expansion of Preclassical Mechanics*, Boston Studies in the Philosophy and History of Science 270,
https://doi.org/10.1007/978-3-319-90345-3_6

was the demonstration of the law of the equilibrium of the balance: this result was not an isolated case but was used to solve complex mathematical problems, such as, first of all, those concerning the area or volume of figures bounded by curved lines or surfaces, as can be seen in the book *On the Quadrature of the Parabola*, and further as a method for finding the ratios between the volumes of parts of regular solids inscribed and circumscribed within or without the solid for which the volume is sought. This second method using the centrobaric theory was utterly unknown before the discovery in 1906 of a letter written by Archimedes to Eratosthenes. This treatise, which is known as *On the Method* shows the great potential of such kinds of mechanical considerations for the solution of geometrical problems. But this potential does not seem to have been appreciated by later mathematicians. The concept of center of gravity, though peculiar to Archimedes' great mathematical talent, did not disappear from mathematical thought in later periods, but it was mostly and successfully employed in mechanics. During the Renaissance, the study of the so-called "simple machines" (balance, lever, pulley capstan or wheel-and-axle, wedge) became one of the main fields in which the centrobaric theory re-emerged.[1]

The thorough study made in the sixteenth century of the Greek and Roman writings on mechanics and technology played an essential role in the so-called Scientific Revolution and many recent scientific contributions by historians of science have correctly stressed the importance of the Archimedean tradition within this group of works. Compared to this great process of recovery of ancient texts, the slow and subdued penetration of the new learning into the works of the sixteenth century that were developed for teaching purposes has remained in the background. It has perhaps been often assumed that works of this kind were not open to new ideas, though in fact they also underwent important transformations. This for example is the case with the treatise *De sphera* by Johannes de Sacrobosco (John of Holywood). After having been the subject of several commentaries over four centuries, it experienced a substantial updating in the sixteenth century, which subsequently changed the way in which extensive sections of the text were understood. Of particular interest for this paper is the introduction in this treatise of the concept of 'center of gravity,' which, though it was already present in earlier times, was then completely revised and corrected.

In the following, I shall try to shed light on the ways in which the centrobaric theory interacted with some of the cosmological ideas conceived in antiquity and in particular with the idea of the figure of the earth presented in Aristotle's *De coelo*. This research, though not entirely new, is nevertheless necessary for a better understanding of the scientific discussion that took place during the crucial first stage of the development of modern science.

[1] To the study of the centrobaric theory in the field of mechanics was later joined a more strictly mathematical study concerning the determination of rigorous methods for finding the center of gravity of solids. The lack of ancient specific texts made this study more complex and difficult, entailing a discontinuity between the two stages of development.

6.2 Prelude: The Different Treatments of the Centrobaric Theory by some Historians of Mechanics

Though there are many and various writings by seventeenth-century authors that deal with the center of gravity in Archimedes' works, up until the last decade of the nineteenth century the complex development of the centrobaric theory has not been studied as an important topic by historians of mathematics or physics. Considering Pierre Duhem's numerous important studies on this subject, such a statement appears at first sight rather absurd: yet this was the situation that clearly emerged from reading some of the most important works in the field published before Duhem.

The state of affairs seems to have been a direct consequence of the new system of mechanical theory elaborate by Lagrange, who in the historical sections of the *Mecanique analytique* (1788) pointed out the main stages in the development of the mechanical principles and, in the case of statics, the three general principles at the foundation of the laws of equilibrium: the principle of the lever, the principle of the composition of forces and the principle of virtual velocities. From an historical point of view, the centrobaric theory had been closely connected with the study of the law of the lever, but this was not enough to assure the survival of the long tradition of the theory within the new development of mechanics outlined by Lagrange. The fragile connection linking the study of the centers of gravity with the research concerning the laws of equilibrium had in fact been broken forever. "The equilibrium results from the destruction of several forces that fight against each other and annihilate reciprocally the action they exert on each other."[2] In this brief definition, we find the main reason for the lack of interest in the centrobaric theory.

Studies on the center of gravity, which had been developed from antiquity to the earlier decades of the seventeenth century, were thus removed from the main branches of the history of mathematics. They did not disappear completely but left a trace of their history in the general principle of the lever. This faint trace, which the following historians of mechanics were inclined to ignore, came back into full view towards the end of the nineteenth century. This made it possible to gain a more precise picture of the development of this branch of mathematics and of its past achievements.

For the moment, the dominating view was that dictated by the *Mecanique analytique*, as can be seen from major nineteenth-century works on the history of mechanics such as Eugen Karl Dühring's *Kritische Geschichte der allgemeinen Principien der Mechanik* (1873), and Ernst Mach's *Die Mechanik in ihrer Entwicklung: Historisch-kritisch dargestellt* (Mach 1988) in which he presented a rather detailed analysis of what the Siracusan mathematician had written on equilibrium.

[2] "La Statique est la science de l'équilibre des forces [...] L'équilibre résulte de la destruction de plusieurs forces qui se combattent et qui anéantissent réciproquement l'action qu'elles exercent les unes sur les autres; et le but de la Statique est de donner les lois suivant lesquelles cette destruction s'opère. Ces lois sont fondées sur des principes généraux qu'on peut réduire à trois ; celui du *levier*, celui de la *composition des forces*, et celui des *vitesses virtuelles*." Lagrange (1811, 1–2).

From the study of Archimedes' text, Mach went on to make important epistemological reflections, but without making any reference to the concept of center of gravity. By now, reflections on equilibrium seemed to be disengaged from any of the fundamental concepts that determined the discussions on this subject during most of the previous centuries. This, however, was not exactly the case. Mach had also found fault with the Archimedean demonstration from another point of view: according to him, on the basis of just those assumptions given at the beginning of the text, it would have been impossible to deduce the law of the lever (Mach 1988, 38–39). The answer to this criticism prompted the recovery of the Archimedean text relating to the centers of gravity.

What seemed to have completely disappeared from the history of mechanics reappeared thanks to the Italian scholar Giovanni Vailati, who in a paper presented at the International Congress of Historical Sciences (April 1–9, 1903) tried to show that Archimedes could have been able to deduce the law of the lever from

> certain reflexions concerning the center of gravity, to which he briefly refers several times in his demonstrations without insisting at length on them, as if it were a subject already discussed in some previous treatise which was lost (Vailati 1904, 245).

Mach briefly discussed Vailati's argument in the seventh edition of his *Die Mechanik* (1907), and this discussion, reproduced in all the subsequent editions of Mach's work, made it imperative to go back and study the ancient texts.

That same edition of 1907 contained a long *excursus* summarizing the result of Pierre Duhem's research made in the first years of the twentieth century. Mach's interest was mainly focused on the studies collected in the first volume of *Les origines de la statique* (Duhem 1905), where the French physician had brought back to light the medieval *scientia de ponderibus* ('science of weights') and presented its development from its beginning to its late followers (Mach 1988, 101–8). But for the history of the centrobaric theory, Duhem's studies collected in the second volume of its work (1906) are much more interesting: there he adopted for the first time a method of investigation which was going to be systematically practiced in the ten volumes of his *Le système du monde.*[3]

We have thus come to the great revival of the study of the theory of the centers of gravity promoted by the French historian's unremitting researches, but before discussing in detail his works we must also remember the important contributions to that study given by the Italian historian, Raffaello Caverni.

Though Caverni is less famous than Duhem, his work in the history of science is no less valuable, and he was perhaps more rigorous in his analysis of manuscript and printed sources. In his *Storia del metodo sperimentale in Italia*, volume IV (Caverni 1895), Caverni discussed at length the development of the studies on the center of gravity and on the equilibrium of the balance from antiquity to the seventeenth century. Caverni's research greatly contributed to the resumption of the Archimedean tradition in the sixteenth century, and it presented the development of

[3]Volumes I–V, 1913–1917; volumes VI–X were published posthumously by his daughter, 1953–1959.

ancient mechanics in an original though not entirely new way. Caverni assumed a clear continuity between Archimedes' work and the pseudo-Aristotelian *Mechanical Problems*. The principle of the balance as established in this work was thus connected with the origin of the concept of the center of gravity. On what was this belief based? The author of the *Mechanical Problems* had pointed out that each point on the radius of a rotating circle moves with different speed according to its distance from the center, which is at rest. He then distinguished two components of that motion: a rectilinear natural movement downwards and a lateral violent motion towards the stationary center of rotation, which increased the nearer the points were to the center, was the cause of the lesser speed of the points nearer to the center. This explanation showed that there was a relation between the weight that was moved, the moving power and the different speed of the points at which the power was applied. This result was based on the analysis of different motions. How could this way of proceeding be connected with the demonstration of the law of the lever later established by Archimedes? In Archimedes' surviving writings there is no reference whatsoever to possible movement or speed. Caverni nevertheless thought that what was established in the first book of Archimedes' *On the Equilibrium of Planes* had somehow originated from the *Mechanical Problems*.[4]

Caverni's interpretation, based on an hypothetical reconstruction of Archimedes' lost work *περί ζυγῶν*, *On the Balance* was a bit strained and assumed that the treatment of questions relating to motion and the equilibrium of heavy bodies was

[4]The explanation of the functioning of the lever in the *Mechanical Problems* is vague. This fact was pointed out more than once by Renaissance scientists following the Archimedean tradition. The vagueness was explained as a first imperfect result of the earliest investigations on the lever, which were later to be fully developed only in Archimedes' work and in the centrobaric theory. This point of view is clearly presented in Guidobaldo del Monte's preface to his edition of Archimedes' work *In duos Archimedis aequiponderantium libros paraphrasis*, Monte (1588, 4). The same point of view is present in Archimedes' biography by Bernardino Baldi: "Since Archimedes (as it is probable and as Guidobaldo himself guessed in the preface to Book One of *On the Equilibrium of Planes*) regarded this Aristotelian work as being based on solid principles, but not being very clear in explaining them, he wanted to make it more explicit and more easily understandable by adding mathematical demonstrations to physical principles. Aristotle solved the problem of why the longer the lever, the easier it moves the weight, by saying that this happens because of the greater length on the side of the moving power; this was true according his principle, in which he supposed that the things that are at greater distance from the center move more easily and with greater force; the cause of which he saw in the greater speed with which the bigger circle moves compared to the smaller circle. This cause is indeed true, but lacks precision; for given a weight, a lever and a power, I do not know how I should divide the lever in the point where it turns, so that the given power balances the given weight. Archimedes accepted Aristotle's principle, but went further; he was not satisfied with saying that the force would be greater on the longer side of the lever, but he determined how much longer it should be, that is, what proportion it should have with the shorter side, so that the given power would balance the given weight. [...] He established this with a brilliant demonstration in Book One of *On the Equilibrium of Planes*, which, as Guidobaldo pointed out, was the book of *Elements* of the whole field of mechanics. In the preface of his paraphrase of Archimedes' work, Guidobaldo showed that Archimedes had followed Aristotle entirely, as far as the principles were concerned, but he had added his own exquisite demonstrations." Baldi (1887, 54–55).

uniform in the Ancient World. But this assumption is unfounded. However, it is on this assumption that Caverni described the origin of the centrobaric theory:

> We see that any body always falls by natural necessity when it lacks a support; and whether it falls freely or it is supported, it is always a very thin thread that marks the clear way or that obstructs the tendency to move. This observation, which is obvious and meaningless for common people, was the start of a scientific investigation for the philosophers who, considering how any weight could be prevented from falling by holding it by means of a very single thread, drew the conclusion that in the fall the weight's strength (*conato*) gathered in the vertical direction. This first important result was then further developed on the base of the experiences: seeing that the weight was keeping equally at rest from whatever point of its surface it was hanging, it was not difficult by means of geometry to draw the conclusion that all the strength of the falling body gathered not in a thin line, as it had seemed before, but at an invisible point which was determined by the intersection of two vertical lines that could be drawn across the hanging weight when it was taking now one now another position. (Caverni 1895, 101–102)

In this explanation Caverni was comparing different Aristotelian considerations drawn from *De coelo*: the definition of the center of gravity given by Pappus in Book Eight of *Mathematical Collections*, and Proposition One from the same Book. After the passage quoted above, Caverni made references to the balance and to Archimedes' treatment of this instrument. In his explanation Caverni seemed to regard all these passages as stages of a uniform development. But certainly things were a bit more complicated, as will be seen.

Pierre Duhem was without doubt aware of this greater complexity, but in his investigations he also assumed the existence of a close connection between the origin of the centrobaric theory and the philosophical considerations concerning the natural motion of heavy bodies. Therefore, I think it necessary to assess whether this assumption is correct in order to evaluate Duhem's general interpretation.

This chapter attempts to analyze this particular aspect of the history of the centrobaric theory by considering the use of the concept of center of gravity in the field of cosmography. The analysis is based on some works overlooked by previous historians, assuming that Pierre Duhem's studies on the philosophers belonging to the so-called Parisian school of the fourteenth century are already well known and do not need to be discussed.

I shall start by criticizing the interpretation given by the French physicist of some Aristotelian passages, and his attempt to see a first confused appearance in *De coelo* II, 14 of some ideas that were later combined to form the concept of center of gravity. First, I shall try to determine more exactly Aristotle's aim in that chapter of his work by specifying the function of the distinction between the center of the world and the center of the earth. This distinction was made in order to explain the process of formation of the earth, its shape, its position at the center of the world and its being there at rest. This distinction should not be related to the centrobaric theory but rather considered in a different way. At the center of my new interpretation is an explanation I will give of the *aporia* or difficulty contained in the chapter of *De coelo* mentioned earlier. My explanation will lead to a discussion of the role of the word ῥοπή in Aristotle. According to Aristotle, this word seems to abandon any reference to the movements of the balance and to the problem of the equilibrium to which it had been closely connected since Homer. Through a linguistic and conceptual analysis, I will try to explain the reason for Duhem's interpretation of *De coelo*'s passage, which will lead to an analysis of the origin of the word ἰσόρροπος before Archimedes.

In the second part of my essay I will show how the rediscovery of *De coelo* influenced the interpretation of Sacrobosco's *De sphera*, a textbook on cosmography which has been studied for centuries. I will also show how more and more space was dedicated to discussions on the natural motion of heavy bodies and how the model of the balance was increasingly used for the explanation of the earth being at rest at the center of the world. The change that took place after Simplicius's commentary on *De coelo* became known, which contained a direct reference to the centrobaric theory, will be discussed here only incidentally since it is a fact well known to historians of science such as Pierre Duhem, Giuseppe Boffito (1902) and Edward Grant (1984).

The last part of my essay will deal with the great changes that took place in the sixteenth century in those fields of science considered here. These changes were first introduced mainly as a consequence of geographic discoveries, which undermined previous ideas concerning the distinction between center of gravity and center of magnitude. To all this we should add the rediscovery of Pappus' *Mathematical Collections*, which in Book Eight presents a different treatment of the center of gravity. The author who best exemplifies the novelties is Francesco Maurolico, who in his *Cosmographia* introduced a radical change in the discussion of these questions in the Renaissance. Through Clavius' *Commentary on Sacrobosco's De sphera*, the new results obtained by Maurolico became part of the teaching of mathematics. Thus, towards the end of the sixteenth century, a new theoretical approach was introduced in the discussion of the role of the centrobaric theory within the science of 'cosmography.' This new role was connected to the development of the theory of simple machines, which was based more and more often on Archimedean principles. In all this movement of ideas, the Jesuits played an important role, as can be judged from the example of Giuseppe Biancani, who went far beyond his masters in his discussion of the question of the direction of movement of heavy and light bodies. He imagined that the different parts of a heavy body fall along parallel lines, a consequence of the fact that the center of gravity of the body moves along a line to the center of the earth. In this way one could discuss the question of the equilibrium of the balance in close connection with that of the movement of heavy bodies.

6.3 At the Beginning there Was Aristotle: The Early Stages in the Development of Centrobaric Theory According to Pierre Duhem

The origin and the earliest stages in the development of the concept of center of gravity in ancient Greece were briefly studied by Duhem at the beginning of the second volume of his *Les origines de la statique*. The cursory way in which he discussed the contributions of the ancients is probably the consequence of the fact that his main interest was in the scientific results achieved by a series of Parisian philosophers of the fourteenth century, who were relatively unknown before Duhem's research, figures such as Jean Buridan, Nicole Oresme, and particularly Albert of Saxony, all of whom formed the so-called "Parisian school of science". Duhem's special focus on the importance of this school of thought had a negative influence on his reconstruction of the history of science. His lack of interest in a precise evaluation of the different stages in the development of the concept of center of gravity by ancient Greek mathematicians and philosophers resulted in a simplified view of that development and in an inaccurate discussion of that concept, which would only be rigorously defined by Torricelli. Duhem's view derived once again from Lagrange, though he pointed to some incorrect judgements made by the Italian mathematician, which could have been avoided if the latter had played greater attention and 'respect' to the medieval *scientia de ponderibus* (Duhem 1905, 1–6).

As a consequence of these preconceived ideas, Duhem was satisfied with a brief analysis of a few texts by Pappus and Archimedes, chosen for the precise purpose of showing the supposed faults presented in Greek statics. He then proceeded, with little respect for chronology, to a brief and insufficient discussion of Chapter 14 of Book Two of Aristotle's *De coelo*. In this text concerning the determination of the spherical shape of the earth, Duhem saw a first and still confused idea of a doctrine that would have a long and successful history. He described this doctrine as follows:

> In each heavy body there is a point where its weight or gravity is concentrated: this is the *center of gravity*. In each heavy body the gravity is its desire to unite this center of gravity with the center of the world. If its center of gravity coincides with the center of the world, the heavy body is at rest. If the center of gravity is outside the center of the world, the first point aims at joining the second and, if it is not stopped from doing that, moves towards it in a straight line. The earth is a heavy body like all other heavy bodies; therefore it joins its center of gravity to the center of the world; and it is for this reason that the earth is at rest at the center of the world (Duhem 1905, 9–10).

But is this a correct way of putting it? Is it right to explain Aristotle's thought in the light of a concept that was introduced only later and, according to Duhem himself, was unknown to the Greek philosopher? Duhem's explanation should have been based on a deeper and more elaborate analysis of the Aristotelian text. It is therefore necessary to re-examine Chapter 14 of Book Two of *De coelo*.

6.4 How the Spherical Shape of the Earth Was Formed

In the chapter ending Book Two, Aristotle concludes the discussion he began in Chapter 13 of various questions concerning the earth: its position and shape, and whether it is at rest or in motion. He had started by reviewing previous theories, in particular, the doctrine developed by the Pythagoreans who held that the earth revolves like a planet round a center, which is occupied by fire: by proposing this picture of the world they were denying the idea that the earth was at rest at the center of world. But Aristotle also discussed at length many other theories that assumed this idea and explained why the earth was at rest at the center of the world.

This problem was closely linked with that of the shape of the earth, and was connected with Aristotle's doctrine of the natural motion of the elements: at this point he was breaking with the previous tradition of thought. What up to then had been explained by a violent action (such as, for instance, the motion of the heavens around the earth suggested by Empedocles) for Aristotle was a consequence of the nature of the elements: these moved in a straight line towards their natural places, the heavy bodies towards the center of the world, the light bodies towards the concave surface of the sphere of the moon; and both kinds of bodies stopped moving when they reached their natural places. Based of these principles it would be meaningless to ask why the earth was at rest in the middle of the world; likewise it would be absurd to ask the same question concerning fire being in its natural place (apart from the circular motion imparted to fire by the motion of the sphere of the moon.)

The earth was not at rest because, as Anaximander had argued, being at the center and in exactly the same relation to the extreme parts of the world around it, it had no reason to move in one direction rather than in another. This explanation would make the earth's immobility depend only on its position. If this explanation were true, then any other element, for instance fire, if placed at the center, would remain there like the earth. But the earth not only rests at the center but also moves towards it. This indicates that the doctrine of natural places, and not the argument from the position at equal distance from the extreme parts of the world, gives the true explanation. If we judge Anaximander's argument from the point of view of the doctrine presented in Aristotle's *Posterior Analytics*, it would be considered as a reasoning based on extremely general principles, rather than on the principles of the science of nature, where bodies are always considered as being either heavy or light.

Put in these terms, the question concerning the position of the earth at rest at the center of the world was closely related to the study of the motion of heavy bodies, which became an essential part of the argument together with the observation of the rising and setting of the constellations. One could positively state for certain that heavy bodies move naturally towards the center of the earth and that they do not fall in parallel lines, but so as to make similar angles with its surface. This statement was considered essential by Aristotle who repeated it twice in chapter 14. This indicated that the earth, towards which their fall was directed, was spherical and placed at the center of the world. The motion of heavy bodies was regarded as perpendicular not to the center of the earth, which was only accidentally the natural place towards which the heavy bodies tended, but rather to the center of the world, which was the true limit of all the motions of these bodies.

It was on the base of these premises, ignored by Duhem in his brief exposition, that Aristotle in chapter 14 of *De coelo* discussed in detail the question of the shape of the earth. The distinction between the two centers previously mentioned was here introduced in a dubitative manner in order to strengthen his criticism of his predecessors made in chapter 13, and at the same time to develop more clearly his arguments concerning the natural motions of elementary bodies. Beginning his historical reconstruction of the Aristotelian doctrine on the basis of this passage from the text of *De coelo*, the French scholar was moving from a completely different presupposition, which induced him to interpret the whole chapter in light of the medieval discussion on the different centers of the sphere of the earth and of that of water, and as a consequence, to distinguish between the center of magnitude and the center of gravity. Based on this interpretation, there was the idea that the concepts developed by Aristotle and other ancient philosophers could be applied to those medieval problems, but this assumption needed a rigorous analysis of the texts within an exact historical perspective. But what emerges from reading the texts tends rather to deny Duhem's assumption.

If Aristotle had kept to the principle of the eternity of the world, his analysis of the motion of natural bodies previously mentioned would have been sufficient to establish the spherical shape of the earth. But his argument had been developed in opposition to the cosmogonic theories of the Presocratic philosophers, who were incapable, according to him, of explaining in a non-contradictory manner the generation of the cosmos and its present shape. The validity of his doctrine of natural motions had therefore to be proved by a demonstration based on its application to a process of formation of the sublunar world taking place over time. How can the formation of a spherical agglomeration by the natural motion of earth towards the center of the world be explained?

In the case of a motion towards the center of similar parts coming from all directions and from equal distances, the problem was easy to solve because it seemed evident that by adding similar parts coming from all directions the result would have been a spherical body. As in the previous case, all the parts pushed by their weight would have continued to move towards the center of the world until they reached it. But what would have happened if a larger part pushed a smaller one? This question cannot be answered on the basis of the principles of modern physics: we do not have here two rigid bodies that collide in space but a completely different situation. It is not very clear what Aristotle means when he says that "when a smaller part is pressed on by a larger, it cannot surge round it [like a wave] (οὐχ οἷόντε κυμαίνειν), but is rather squeezed together with the other and combines with it (συγχωρεῖν) until they reach the center" (Aristotle and Guthrie 1939, 246–247/297a9–12).

It would seem that in this case the smaller part could still offer a certain resistance, and this would entail a reciprocal "action" between the two parts. The final result of this process of bidirectional buckling would be the formation of one single body, that is, the inclusion of the smaller part inside the larger. But it is not clear whether at the end of this process it would still be possible to single out in some way the two parts that have combined to form a new body.

It would seem that this explanation could be referring to the way in which the artisans made things from clay, but a "physical" explanation could not refer to an external cause such as the artisan, but must be based on principles relating to natural bodies, that is, the "weight" and the tendency to move downwards (i.e. towards the center of the world).

6.5 The Aporia in *De coelo* II,14

How can the process previously described be understood? What happens when the smaller part, which is already placed at the center of the world, comes into contact with the larger one? Being in its natural place, the smaller part, according to Aristotle, would have been able to offer a certain resistance to the larger one that was pushing it, and both buckled together would have formed a single spherical body, remaining then at rest at the center of the world. Should we perhaps think of the smaller part as a smaller immobile sphere stuck at the center? Or could we believe that the smaller part has been moved slightly away from the center of the world by the larger part that pushes it? These questions show that here we have a problem that must have already been considered in Aristotle's time, seeing that he discussed the following *aporia,* or perplexing difficulty, which is similar to our previous considerations.

> If, the Earth being at the centre and spherical in shape, a weight many time its own were added to one hemisphere, the centre of the Universe would no longer coincide with that of the Earth. Either, therefore, it would not remain at the centre, or, if it did, it might even as it is be at rest although not occupying the centre, i.e. though in a situation where it is natural for it to be in motion (Aristotle and Guthrie 1939, 248–249/297a31–297b1).

How must the expression "a weight added to one hemisphere" be understood? If we consider Aristotle's solution of the difficulty, this thing does not seem very important:

> It is not hard to understand [the difficulty], if we makes a little further effort and define the manner in which we suppose any magnitude, possessed of weight, to travel towards the centre. Not, clearly, to the extent of only touching the centre with its edge: the larger portion must prevail until it possesses the centre with its own centre, for its impulse extends to that point. It makes no difference whether we posit this of any chance portion or clod, or of the Earth as a whole, for the fact as explained does not depend on smallness or greatness, but applies to everything which has an impulse towards the centre. Therefore whether the Earth moved as a whole or in parts, it must have continued in motion until it occupied the centre evenly all round, the smaller portion being equalized by the greater under the forward pressure of their common impulse (Aristotle and Guthrie 1939, 248–251/297b2–14).

This is the passage where Duhem, following Simplicius, saw the first still confused appearance of what will be subsequently called the center of gravity. This interpretation will be discussed later; up until now we have considered the first part of Aristotle's statement concerning the motions of bodies towards the center. Do we face reasoning that implies the displacement of a rigid body caused by another rigid body placed on it? No, we do not. What could have caused this sudden change? Was it perhaps a different way of dealing with this kind of problem by the presumed author of the *aporia*? But it is probably Aristotle himself who is the source of this *aporia*.

To understand this part of Aristotle's reasoning we must return to what was said earlier. There we have a weight much larger than the earth. The earth takes the place of the smaller sphere immobile at the center of the world. Though offering some resistance, after a reciprocal deformation it will eventually merge with the larger body. When could the arrangement of the two parts—to form a single body—be considered to accomplish? And what form will this body have? Aristotle does not say it explicitly, but on the basis of the text of the *aporia* we may suppose that the final result is a spherical body since there is no mention of a possible different figure. But Aristotle adds that this *aporia* still holds valid since it is supposed that the formation of the new body ends when the edge of the larger part touches the center of the world. Why did he conceive such an idea? It may be that this is a naïve concept of the center as a support, almost a *floor*, which stops the weight from going further down. But in this case the new shape of the earth would be similar to a *pear*. But it seems correct to assume that Aristotle was thinking of something more subtle.

Let us therefore imagine that the *aporia* describes the meeting of two spheres of different magnitudes. We may then think of a deformation of the smaller one, that is, the earth, similar to squeezing its two hemispheres along the diameter parallel to the line passing through the tangent point. This would be a symmetric process, which would take place at both sides divided equally by the diameter, and which would come to an end when the external surface of the new body reaches the center of the world. If it were possible to single out the two different deformed bodies at the end of this process, they would appear as two bodies that are tangent to one another at the point previously occupied by the center of the smaller sphere, that is, the center of the world.

According to this reasoning, the end of the process of formation of the new body would be the final moment. But this would agree only partially with what Aristotle has said since in this case a fundamental aspect of the problem would be ignored, that the natural tendency of heavy bodies to move towards the center of the world, a tendency that does not cease to act until the bodies come to equal distances from the center. If, on the contrary, we consider the newly formed body only in relation to the center, without referring to its natural tendency, we get a situation exactly equal to that described in the *aporia*. We can then refer to the earth in its final placement, which is enclosed in the new body and therefore no longer in a central position, but actually displaced. Or we can consider the earth in relation to its original center, from which it has never moved away, despite being squeezed. But now this point is no longer its center since it is placed on the surface, and is to an even lesser extent the point towards which every part having weight tends to move.

Once we have established that the idea on which the *aporia* is based is wrong, how can we solve the problem? Going back to the process mentioned above, we can state that the body does not stop when its edge touches the center of the world; the larger body continues to act on the smaller and this would continue until the center of the new body coincides with the center of the world.

For Aristotle, it was of paramount importance to conceive a cosmogonic process, that is, a process for the generation of the world that was based on a rigorous doctrine of the natural motion of bodies. The main points of this doctrine have already been mentioned: heavy and light bodies have a natural tendency to move towards their natural places along a straight line, the heavy bodies towards the center of the world, the light ones in the opposite direction from the center towards the concave surface of the sphere of the moon. When they reach those places, the bodies no longer have that tendency and remain at rest. The reasoning by which Aristotle had tried to solve the problems of the placement and of the shape of the earth has already shown how important it was that all its parts were at equal distance from the center since only in this way was the earth at rest. But how must we understand this equal distance? Can we simply refer it to a 'mechanical' context, as Duhem did? Or is it necessary, also in this case, to be cautious? In other words, must we consider the parts of the earth placed at equal distances, from the point of view of the equilibrium of the balance? It is now necessary to analyze briefly the concepts of βάρος (weight) and ῥοπή (tendency of the heavy bodies to move downwards, that is, towards the center of the world) as they are employed in *De coelo*, Book Two, Chapter XIV.

6.6 The Different Ways in which the Terms βάρος and ῥοπή Are Used in *De coelo* II,14

We must first note that both terms are peculiar to Chapter XIV since they occur there more often than in any other Aristotelian text, and they are particularly frequent in the passages quoted above: they are employed (1) to determine the true natural place of the motion of heavy bodies and to show that the center of the earth only accidentally coincides with it; (2) to deny that heavy bodies fall along parallel lines and to prove that they move along lines that converge to the center of the spherical earth placed at the center of the world; (3) to show that parts of different dimensions meet to form a single body; (4) to solve the *aporia* of the weight placed on one hemisphere and to state the truth of what was said in point (3) for bodies of any dimension. For our problem, the last two points are particularly interesting.

The question raised in point (3) is mentioned twice by Aristotle: the first time at the beginning of the discussion concerning the shape of the earth; the second time after discussing the formation of a spherical body by the motion towards the center of equal parts coming from all directions and from equal distances, and just before enunciating and discussing the *aporia*. This repetition is not casual, given the general character of the explanation suggested, and it does not seem that there is a substantial difference between the two passages, though in the first case Aristotle emphasizes the way in which the two parts come together to form a single body, whereas in the second case he stresses the necessity for this process to take place. Aristotle employed the two terms, βάρος and ῥοπή, in a rather indiscriminate way. In the initial passage, the contact and the following reciprocal 'action' between the parts is possible because "every one of its parts has weight until it reaches the center (ἕκαστον γὰρ τῶν μορίων βάρος ἔχει μέχρι πρὸς μέσον)" (Aristotle and Guthrie 1939, 246–247/297a8–9); whereas in the later passage "a greater mass must always drive on a smaller mass in front of it, if the inclination of both is to go as far as the center (τὸ γὰρ πλεῖον ἀεὶ τὸ πρὸ αὐτοῦ ἔλαττον προωθεῖν ἀναγκαῖον μέχρι τοῦ μέσου τὴν ῥοπὴν ἐχόντ ων ἀφοῖν)" (ibid. 248–249/297a27–29). But perhaps in this second case the term ῥοπή was used to avoid linguistic confusion, since the parts were differentiated here by their weight rather than their dimension, as had been done before.

In the following passage, where the presuppositions of the *aporia* are criticized, the term employed by Aristotle is always ῥοπή, and this time without any possibility of misunderstanding. The supposition that the formation of a new body was completed when the edge of the larger part touched the center of the world was wrong; as a matter of fact, it was necessary for the center of the newly formed body to coincide with the center of the world since the larger part continues to act on the smaller part and both had an inclination to go as far as the center (μέχρι τούτου [μέσου] γὰρ ἔχει τὴν ῥοπὴν) (ibid. 297b6–7.)

For Aristotle, the determining factor in this process was not so much the size of the two parts, nor their weight, but rather what was common to all parts, that is, the impulse to go towards the center (ἀλλὰ κατὰ παντὸς τοῦ ῥοπὴν ἔχοντος ἐπὶ τὸ μέσον) (ibid. 297b–9). And whether the earth moved as a whole or in part "it must have continued in motion until it occupied the center evenly all round (ἀνανκαῖος μέχρι τούτου φέρεσθαι ἕως ἂν πανταχόθεν ὁμοίως λάβῃ τὸ μέσ ον)" (ibid. 297b11–12). It was therefore "under the forward pressure of their common impulse" that the smaller portions were *equalized* by the greater (ἀνισαζομένων τῶν ἐλαττόνων ὑπὸ τῶν μειζόνων τῇ προώσει τῆς ῥοπῆς) (ibid. 297b12–14).

The solution of the *aporia* and the whole process of formation of the earth were based on the concept of ῥοπή. But how can this Aristotelian concept be related to the idea of equilibrium? Can we with the help of Duhem come to the conclusion that Aristotle's reasoning contains a first vague idea of what was going to be called the center of gravity? Some further considerations induce us to utterly exclude such a possibility.

6.7 The Concept of ῥοπή before Aristotle

The term ῥοπή acquires a particular meaning in Aristotle's writings, which amounts to a deep change of this concept as compared to its traditional meaning. In texts written before the fourth century B.C., the concept of ῥοπή was most often related to the idea of a scale, particularly to the idea of the inclination of the scale and precisely to the idea of the heavier scale pan going down towards the ground. The scale was considered from the point of view of common experience, without any reference to its structure or to the motion of heavy bodies placed on it. What was considered important was the moving away of the beam from the position parallel to the ground, which disrupts the equilibrium. This image, when being metaphorically applied to human destiny, evoked decisive moments in a man's life. The use of the term ῥοπή in this context goes back to the earliest times of the Greek world, and it occurs already in Homer where it characterizes one of the meanings of the verb ῥέπειν.

In Book Eight of the "Iliad," Zeus placed the fates of the Achaeans and of the Trojans on a scale and, lifting it, he weights them. The scale pan holding the fate of the Achaeans then "inclined downwards" (ῥῆπε) and rested on the ground, whereas the Trojans' scale pan "rose to the sky," and so the battle, which had lasted. the whole morning with alternate fortune, was settled in the Trojans' favor:

The Sire of Gods his golden scales suspends, With equal hand: in these explore'd the fate
Of Greece and Troy, and pois'ed. the mighty weight,
Press'd with its load, the Grecian balance lies
Low sunk on earth, the Trojan striks the skies. (Homerus and Pope 1760, book 8/verses 88–92).

Similarly to Homer, at a date closer to Aristotle, Euripides employed the term ῥοπή in his "Helen" (1090) with the same meaning but without making a direct reference to the scale. At the time, the phrase must have been so familiar to the spectators of the tragedy that it could be used in the right context without further illustration. It is the moment in the drama when Helen and Menelaus, after persuading Theonoes not to reveal to his brother Theoclimenus that the news of Menelaus' death was not true, start to plan their escape from Egypt. They have to get away from Theoclimenus, who wants to marry Helen, but at the same time they need his help. Therefore they try to cheat him, and it is Helen who has to face this risky endeavor. Being aware of her situation, she then exclaims: "I see two outcomes (ῥοπάς). Either I must die, if my tricks are discovered, or I return to my fatherland and save your life" (Euripides and Kovacs 2002, 132–133). Once again, the scale could sink on either one side or the other, but this time the outcome would have decided not the destiny of two groups of warriors fighting against each other, but that of a single person. The term ῥοπή meant life on the one side, death on the other.

We could give further examples, and even more if we consider the various meanings of the verb ῥέπειν, but this would only confirm the variety of contexts in which this term has been used and which determine its meaning of 'inclination,' also in many modern languages.

Let us go back to our problem. A comparison between the concept of ῥοπή as employed by Aristotle and the concept of it as used by Homer and Euripides shows clearly that in the first case the "tendency to go towards the center of the world" results in reaching a state of rest, whereas in the second case the action described by the term is that of leaving the state of rest. For Aristotle, the action described by the term ῥοπή is the necessary condition for reaching a state of equilibrium, whereas for the other authors it is the cause of disruption of equilibrium.

It is true that the two concepts appear to be complementary; it could be said that they agree on an essential condition of the state of equilibrium, that is, that what is at rest must be equidistant from something that is without motion, that is, the center of the world for Aristotle or the ground for those refer to the scale. But is this condition sufficient to establish a satisfying doctrine of the equilibrium? Certainly not. Both conceptions in fact underwent further developments. In the end, it was the development of the studies concerning the scale that produced the greatest results, whereas the direction taken by Aristotle turned out to be rather sterile. Aristotle in fact gave up any reference to the scale and continued to analyze the concepts of rest and equilibrium simply in terms of equidistance. On the other side, however, the studies of these concepts were developed mainly in terms of equal ratios of weight and distances. What we state here is confirmed by the transformation that the concept of ῥοπή underwent in other passages in Aristotle's works.[5]

Already in *De coelo*, Aristotle did not hesitate to make use of that term also with reference to upward motions. From his point of view, the term ῥοπή could indeed be conceived as a general tendency of bodies to move towards their natural places and therefore it would have been more correct to define the meaning of this term without specifying any preferred direction; but, as we have seen, the common usage of that word was still prevalent in his writings. One of the clearest examples of the new broader meaning of the term can be found in Book Three, Chapter II of *De coelo*, where after showing that "every body has a natural motion performed neither under compulsion nor contrary to nature," Aristotle states briefly that "some bodies must owe their impulse to weight or lightness (ὅτι δ'ἔνια [τῶν σωμάτων] ἔχειν ἀναγκαῖον ῥοπὲν βάρους καὶ κουφότητος)" (Aristotle and Guthrie 1939, 276–277/301a22–23). To show this, he argued that "if that which moves has no natural impulse, it cannot move either towards or away from the center (εἰ δὲ μὴ ἔξει φύσει ῥοπὲν τὸ κινούμενον, ἀδύνατον κινεῖσθαι ἢ πρὸς τὸ μέσο ν ἢ ἀπὸ τοῦ μέσου)" (ibid. 276–277/301a24–26).

Here the meaning of the term ῥοπή has lost any connection with the inclination of the scale. Its meaning here is so general that it could be used not only to prove the spherical shape of the earth but also to explain the formation of the sphere of fire. But the process by which light bodies are placed in the space between the earth and the concave surface of the sphere of the moon is analyzed by Aristotle in a completely different way. Surely fire also moves towards its natural place along a straight line and meets the edge of that place at equal angles. But that is all. Moreover, this kind of reasoning would not even have been necessary for Aristotle: none of his predecessors had even conjectured that the world, or at least the sublunar region,

---

[5]A first stage in the transformation of the term ῥοπή can be seen in a passage from *De justo* ascribed to Plato, where the verb ῥέπειν usually refers to the motion of a heavy body downwards, seems to be used also with reference to the motion of a light body upwards. But the thing seems rather dubious. This brief text attempts to define what is *just* through a Socratic discussion, where the solution of the problem is found through a series of short questions and answers concerning less problematic fields of knowledge than the one discussed here. Among these is the doctrine dealing with the concepts of *heavy* and *light*. The question then is: How can we judge whether a body is heavy or light? By its weight. How can the weight of a body be assessed? By the art of weighing. Now the art of weighing makes use of the scale, and therefore "what on the scale inclines downwards is heavy, what *inclines* upwards is light" (τὸ μὲν κάτω ῥέπον ἐν τοῖς ζυγοῖς, βαρύ· τὸ μὲν ἄνω, κοῦφον). In the second part of this sentence, the verb is missing, but it seems correct to assume that the term ῥέπον should be repeated. But this not a question of a natural tendency of the light body to move upwards but rather of the simple observation of what happens on the scale: the equilibrium has been disrupted, one body goes downwards, the other goes upwards. We should not think of a scale for weighing the lightness, but of a determination of *heavy* and *light* as relative terms.

could have been formed through a process of *emanation* from a single point, as understood for instance by Robert Grosseteste (c. 1168–1253), who in his treatise *On light* perceived light, as the first form to be created in the prime matter, as being propagated from an original point into a sphere, thus giving rise to spatial dimensions and to everything else, according to optical laws.

Aristotle followed a different theory, which introduced a sharp distinction between absolutely heavy and light elements, such as earth and fire, and elements that were heavy and light only in relation to other elements, such as water and air. On the basis of this theory, he solved the problem of placing the elements in the sublunar region. This solution was of great consequence for later developments of the scientific thought.

6.8 Critical Discussion of Duhem's Interpretation. The Use of the Term ἰσόρροπος before Archimedes

Let us go back to Duhem and his interpretation of the *De coelo*. Without making any reference to its context, he simply translated the passage concerning the question of locating the true natural place to which heavy bodies move. In this passage, Aristotle had tried to explain that the center of the earth was only accidentally the point towards which they moved. After quoting this passage, Duhem proceeded immediately to the other passage that presents the *aporia* concerning a weight placed on one of the hemispheres of the earth, and focused his attention on the passage stating the need to place the center of the heavy body at the center of the world. Following Simplicius, and without considering the whole argument developed by Aristotle, Duhem concluded:

> Aristotle's doctrine is still vague on this point; the Philosopher does not define this center, τὸ μέσον, which in any heavy body tends to reach the center of the universe; he does not make it identical with the center of gravity, which he did not know (Duhem 1905, 11).

The French scholar, by interpreting Aristotle's text from the point of view of the historical development of the concept of center of gravity, regards it as an unsuccessful attempt to describe the conditions of equilibrium of a heavy body placed at the center of the world. All his analysis is based on the assumption that the Greek philosopher regarded the earth as a rigid body, and that those passages from his work were the earliest documents of a tradition of thought which lasted until the first half of the seventeenth century. The first assumption is a bit overstretched, but the second is the result of adopting without criticism the interpretation of the Aristotelian text by ancient and medieval authors. Our previous considerations lead us to quite different conclusions.

On the basis of what he states in the *De coelo*, it is probable that Aristotle regarded the earth as a very malleable body. To show the validity of the principles of his physical theory, he confronted earlier thinkers and developed a cosmogony based on a rigorous doctrine of natural motions, according to which one could explain the central position, the state at rest and the spherical shape of the earth. In the concept of ῥοπή, which in his work underwent a substantial change of meaning, Aristotle found a physical principle that explained the present shape of the earth, its place at the center of the world and it being at rest. As soon as a condition of rest/equilibrium is established, this active principle no longer holds: the condition is defined by the equidistance of each earthly part from the center of the world. From this point of view, each part having ῥοπή will converge towards the center and acquire a spherical shape. Any other solid geometrical figure, since the bodies are not rigid, will not remain unchanged at the center of the world but will change shape by effect of its ῥοπή and will always result in a spherical body having a center coinciding with the center of the world. In other words, the earth would behave like *a fluid body*. The interpretation of Aristotle's *aporia* based on a concept of equilibrium modeled on the idea of a scale is misleading. Such a concept emerged much later in developing the idea of equilibrium conceived in terms of equal ratios between weights and distances. It is imperative to put aside this idea, though it might not be so easy since it has influenced all subsequent interpretations and commentaries of *De coelo*.

A clear example of such a misleading interpretation is offered by Duhem's translation of the final passage of the *aporia*. Aristotle says that "whether the Earth moved as a whole or in parts, it must have continued in motion until it occupied the center evenly all round, the smaller portions being *equalized* by the greater under the forward pressure of their common impulse (Ὥστε εἴτε ὅλη ποθὲν ἐφέρετο εἴτε κατὰ μέρος, ἀναγκαῖον μέχρι τούτου φέρεσθαι ἕ ως ἂν πανταχόθεν ὁμοίως λάβῃ τὸ μέσον, ἀνισαζομένων τῶν ἐλαττόνων ὑπὸ τῶν μειζόνων τῇ προώςει τῆς ῥοπῆς)" (Aristotle and Guthrie 1939, 250–2551/297b10–14).

The analysis of the concept of ῥοπή makes this process of "equalizing" very clear, but for the one who could not free himself from the Archimedean idea of equilibrium, this was very difficult to see. For Duhem, the arrangement of the parts around the center would not result from the pressure of the bigger parts on the smaller, but from the fact that "the tendency of the different parts to move cause them to counterbalance each other." For the French historian, the meaning of the verb ἀνισάζειν became unconsciously identified with that of the verb ἰσορροπεῖν, and Aristotle's vague idea was replaced with the more precise Archimedes' concept. The Archimedean concepts of equilibrium resulting from an equal ῥοπή of heavy bodies placed at certain distances from the center of a scale was inserted into the Aristotelian text, where it assumed a new aspect. But Aristotle could not have conceived in this way the arrangement of the parts of earth at equal distances from the center, because with such an arrangement their tendency to go towards the center of the world would have completely ceased.

Was there any reason to justify such a change of meaning? Was there any passage in Aristotle's works that would support such an interpretation? In order to answer these questions, we must analyze the term ἰσόρροπος and discuss the ways in which it was used by ancient authors. This discussion will complete and confirm what we have said concerning the term ῥοπή.

The concept of ἰσόρροπος is closely related to the idea of a scale: it refers to the state of equilibrium and the absence of inclination. The prefix ἰσο– signifies the idea of equality between inclinations and their reciprocal elimination. The impossibility of one weight prevailing over the other results in the immobility of the scale and the equidistance of all its parts from the ground. In Homer's and Euripides' passages quoted above, the term ἰσόρροπος does not appear, but it would have stood for the first moment of the operation of weighing: if on a scale in the state of equilibrium two weights are placed and they are not equal, the scale would incline on one side. To this situation it could be added, as in the case of Homer, that the greater weight would come down to the ground.

If the idea of equilibrium could be tacitly assumed in the situation just described, a very different situation would occur if the initial state were one of non-equilibrium. It is in such a situation that the term ἰσόρροπος occurs for the first time in an ancient author, that is, in Aeschylus' *Persians* (346). In the battle between the Greeks and Persians, the destiny of the former seemed to be already decided: the number of Xerxes' ships was by far greater than that of the Greek's, but then some divinity destroyed the Persian fleet "weighting the scales so that even fortune did not fall out (τάλαντα βρίσας οὐκ ἰσορρόπῳ τύχη)" (Aeschylus and Sommerstein 2008, 137–138). Here the inclination is not going to take place in the immediate future, but is already present. The scale is not in equilibrium and the inclination must be reversed.

Aeschylus' image is very evocative, but I have recalled here it because of the earliest occurrence of the term and of its connection with the scale. Plato's use of the term ἰσόρροπος is much more significant for us because in this case it refers to the kind of questions which were later discussed by Aristotle, and the position taken by his master is very important in this context. The most interesting passage in which the term occurs is in the *Phaedo*, where Socrates presents his idea of the shape, place and dimension of the earth. He begins to describe the position of the earth in the cosmos in a way similar to that later followed by Aristotle.

> If the Earth is round in the middle of the heavens, it needs neither the air nor any other similar force to keep it from falling, but its own equipoise.
> (τῆς γῆς αὐτῆς τὴν ἰσορροπίαν) and the homogeneous nature of the heavens on all sides suffice to hold it in place; for a body which is in equipoise (ἰσόρροπον), and is placed in the centre of something which is homogeneous cannot change its inclination (κλιθῆναι) in any direction, but will remain always in the same position without inclination (ὁμοίως δ'ἔχον ἀκλινὲς μενεῖ) = (Plato and Flowler (1914, 374–375/109A).

Plato's last sentence is very similar to one in *De coelo* describing Anaximander's position as reported by Aristotle: "That which is situated at a position at the center and is equably related to the extremes (ὁμοίως πρὸς τὰ ἔσχατα ἔχον) has no impulse to move in one direction or either upwards or downwards" (Aristotle and Guthrie 1939, 234–235/295b12–14).

The fact that Plato insists here on using such terms as ἰσορροπίαν and ἰσόρροπον shows that for him the state of equilibrium of the earth was an essential supposition assumed as a basis of his argument. Unfortunately, he does not say anything about the way in which such equilibrium came about. He states it as a fact, which shows that the equidistance of the earth from all sides of the heavens is not sufficient to assure its remaining at the center of the world. The earth, as a scale, must be in such condition that all of its parts have equal inclinations, which eliminate each other. The beam of the scale remained equidistant from the ground as long as it remained in equilibrium, that is, as long as one weight could not prevail over the other.

If this was the true meaning of the term, it is clear why Aristotle avoided using it. He had based the process of formation of the earth on the concept of ῥοπή, the tendency of heavy bodies to move towards the center of the world, a tendency destined to disappear as soon as they reached their natural place, where they were arranged to form a spherical figure and remained at rest.

Our suggestion that Duhem's translation of Aristotle's passage is misleading can be further confirmed by the fact that the term ἰσόρροπος occurs rather infrequently in Aristotle's work and is used with the common meaning. The way in which this term is employed in *Nicomechean Ethics* (Book Nine, Chapter I) is surely less relevant to our enquiry: there concerning "those who have imparted instruction in philosophy" Aristotle says that "the value of their service is not measurable in money, and no honour paid then could be an equivalent (τιμή τ'ἰσόρροπος οὐκ ἂν γένοιτο)" (Aristotle and Rackham 1926, 520–521/1164b2–4). The term here has the metaphoric meaning of "well-balanced" or "well-matched" or "having the same value": no honor could have such a value as to adequately counterbalance the value of their service. In other words, the imaginary balance with which we could weight the value of wisdom cannot be put in a state of equilibrium by placing on one of the dishes either money or honor.

From the point of view of the history of mechanics, it is more interesting to consider the use of the term ἰσόρροπος in a passage from Aristotle's *Parts of Animals* where the structure of birds is analyzed in terms of equal distribution of their weight similar to the placing of equal weights on the scales of a balance:

> Quadrupeds have forelegs to support their forward weight; birds, however [...] nor have forelegs, because they have wings instead. By way of compensation, Nature has made the ischium long, reaching to the middle of the body,
> (μακρὸν ἡ φύσις τὸ ἰσχίον ποιήσασα εἰς μέσον προσήρεισεν), and has fixed it fast,
> while beneath it she has placed the legs, so that the weight may be equally distributed on either side and the bird enabled to walk and to stand still
> (ὅπως ἰσορρόπου ὄντος τοῦ βάρους ἔνθεν καὶ ἔνθεν πορεύεσθαι δύνηται καὶ μένῃ)
> (Aristotle, A. L. Peck and E. S. Forster. 1937, 414–415/695a6–13).

The concept of equilibrium derived from the idea of balance was essential for understanding how birds could walk and stand still. This sort of explanation would later lead to important developments in the study of animal anatomy. The body of a bird, supported by legs placed in a central position, did not keep the same 'inclination,' and could therefore not be immediately related to the idea of the beam of a balance equidistant from the ground. The body of a bird, usually in a slanting position, would continuously bend down, first on one side and then on the other, but in spite of that the animal would remain in equilibrium. Such a situation would have offered the opportunity for a deeper analysis of this phenomenon, but the complexity involved prevented the development of further research in antiquity. The advancement in the study of the anatomical structure of animals became possible only in the sixteenth and seventeenth centuries thanks to the revival and new development of the doctrine of the center of gravity.

Aristotle's passage quoted above shows a widening and transformation of the meaning of ἰσόρροπος, which had already been used in relation to human beings. It had been employed in the field of medicine in some of the works forming the *corpus hippocraticum*, but there the term was still connected to the idea of equidistance or setting in a line. For instance, in the case of a fracture of the thigh bone, it was advised to pull the limb strongly so as to avoid ending up with one leg shorter than the other. This kind of operation, if done unskillfully, would have made the patient wish that both his legs had been broken, for then at least he would be in equilibrium (ἰσόρροπος γοῦν ἂν εἴη αὐτος ἑωυτῷ), that is, he would not be lame. In this case, real bodies, and not simple distances, are equal, and, being used to support something placed above, they could cause an unwanted inclination. Such a lack of equilibrium would surely be of no help for normal walking.

6.9 The Multiplication of the Centers. The Study of Aristotle's and Sacrobosco's Texts from the Middle Ages to the Beginning of the Sixteenth Century

Johannes de Sacrobosco's *De sphera* enjoyed great renown during the Middle Ages and from the middle of the thirteenth century onwards it was taught at all the schools of Europe: during more than three centuries it served as the main textbook on cosmology for the students in medieval and Renaissance universities. The work presented in a simplified form some fundamental astronomical concepts and required only knowledge of elementary mathematics. Written probably around 1220, it was the subject of numerous commentaries that continued to be produced up until the first decades of the seventeenth century. *De sphera* was repeatedly enlarged with additions and developments based on Aristotle's *De coelo* and other mathematical and astronomical works by Greek and Arab authors, and on the new geographical discoveries that produced a better knowledge of the sky of the southern hemisphere. The observation of the new stars from 1572 onwards helped to raise doubts about the incorruptibility of the celestial spheres and, as a consequence, to overcome the traditional contraposition between the heavenly and sublunar world.

Within this complex process of assimilating new ideas and revising old ones, reviving the concept of center of gravity played a significant role: in particular, it was employed mainly to explain how parts of the terrestrial sphere emerged from water. This problem had been discussed in the previous centuries within the field of the exegesis of the Holy Scriptures, particularly "Genesis" and the "Psalms." But with the recovered knowledge of Aristotle's natural philosophy, this problem was seen in a new light. The whole question was the result of a sort of misunderstanding of Aristotle's texts, which led to suppose that in the sublunar world each element was entirely surrounded by the element next to it, and that on the basis of examples quoted by Aristotle of transformation of a thick substance into a thin one, the thinner element was taking up a space ten times larger than that taken up by the thicker element. This supposition, in connection with the doctrine of natural place, suggested the idea that the sphere of water was ten times larger than the earth.

Moving away from this supposition, two fundamental questions were raised that led to a revival of the doctrine of the center of gravity: what had caused the water surrounding the earth to shift away? What allowed the parts of the earth that had emerged to remain so without being again submerged by water?

The various answers given to these questions were discussed by several scholars, among them Pierre Duhem, Giuseppe Boffito and Edward Grant. It seems useless to discuss these questions again after all the work invested by these scholars. But both Boffito and Grant have scarcely considered the centrobaric doctrine, while Duhem, though focusing his attention on it, treated it within an idea of the historical development of mechanics which I intend to correct here. Thus it will be appropriate to discuss once again the authors who had tried to solve these questions with the help of the concept of center of gravity.

6.10 The Form, the Position and the Immobility of the Earth in Sacrobosco's *De sphera*

The question of the arrangement of the elements within the sublunar world is discussed by Sacrobosco at the beginning of Chapter One of *De sphera*, after giving two definitions of a sphere according to Euclid and Theodosius and analyzing the real sphere on the base of the concept of 'substance' and 'accident.' The division of the sphere in two separate regions, the heavenly or *ethereal* sphere and the sphere of the elements, is followed by a description of the elements which are concentrically arranged around the earth placed at the center of the world so that each of them completely surrounds the element below, with the exception of "that dry part of the earth which withstands the flow of water to keep safe the lives of living things."[6] Sacrobosco's brief sentence needs some explanation but he says nothing else on this point in his work.

This sentence is surely not related to a general doctrine of the motions of heavy bodies, which is not even mentioned in *De sphera* where the process of formation of the earth is not discussed. Furthermore, Sacrobosco shows no interest in the cosmogonic problem and is therefore satisfied with a very simple argument based on motion to prove the spherical shape of the earth. This argument is similar to that

[6] The original reads: "trium quorum [elementa] quodlibet terram orbiculariter undique circumdat, nisi quantum siccitas terre humori aque obsistit ad vitam animantium tuendam." Thorndike (1949, 78–79).

developed in *De coelo*: if equal parts of earth move from all directions and from equal distances towards the center of the world, they will form a spherical body around it. Except that here the tendency of the parts of earth to move towards their natural place is not considered only in relation to the center of the world, as was the case for the Aristotelian ῥοπή in Chapter XIV of Book Two of *De coelo*, but also with regard to their moving away from the rotating motion of the celestial spheres.[7] It is not clear what the author of *De sphera* had in mind, but this is all that can be found in this work concerning the process of formation of the earth.[8]

In Sacrobosco's work the earth is presented as a body already formed, and its spherical shape is proved entirely on the basis of the observation of the motion of the celestial bodies. In truth, there is still a brief mention of the natural motion of the earth, at the end of the section discussing the position of the earth at the center of the world, and it is made in order to prove its immobility. But this does not suffice to modify our judgment. Let us see, therefore, how the questions concerning the shape of the earth and of water are discussed in the first chapter of *De sphera*: these questions, it must be remembered, are not here related to what had been said concerning the size of the elements and are therefore treated separately.

That the earth is spherical from east to west derives from the fact that the stars do not rise and set at the same time, but at different times depending on the location. The stars rise first for those who live in the east and later for those who live in the west. This is evidently shown also by observations of lunar eclipses: the same eclipse observed by us around the first hour of the night (after sunset) is visible by those who live in the east around the third hour of the night. This shows the sun rises earlier in the east than in the west.

The mere observation of the rising and setting of stars is enough to prove the spherical shape of the earth from north to south: an observer moving from north to south at a certain point sees those stars rising that were previously always visible; the opposite happens to an observer moving in the opposite direction. All of these phenomena can be explained only if the earth has a spherical shape.

Naturally, the same arguments that were used to prove the spherical shape of the earth could have been used to find out if water is also spherical. But the difficulty in measuring distances and the position of ships during navigation prevented the application of this same method. Although it is almost impossible to sail for long distances in the same pre-established direction, this kind of experience could have been

---

[7] "Omnia etiam preter terram mobilia existunt, que ut centrum mundi ponderositate sui magnum extremorum motum undique equaliter fugiens rotunda spere medium possidet." Thorndike (1949, 79).

[8] Sacrobosco might have deliberately avoided a discussion of this type of problem in his treatise. It would seem that he abstained from referring to Aristotle's *De coelo*. At the time that Sacrobosco wrote his *De sphera*, Aristotle's work was already available both through Gerardo of Cremona's translation, and through the Pseudo-Avicenna's *Liber celi et mundi*, which was much more common. This work was a paraphrase of some sections of *De coelo* and at the time of Sacrobosco, generally regarded as an original work by Aristotle, Gutman (1997, 121–8). The tenth chapter of the *Liber celi et mundi* which is titled *Quod figura terre sperica est*, contains an interesting discussion of the problem I have been considering here. I will discuss the content of this chapter on another occasion. Gutman (2003, 181–183).

used to solve the question. It would have been enough to take into account an experience familiar to sailors: a sailor standing at the top of the mast can see the coast before it is visible to those standing on deck.

This experience was actually referred to by those who tried to prove the spherical shape of water. After placing a signal on the coast, the ship sailed away from the land until an observer on deck could no longer see the signal. After stopping the ship, the observer climbed to the top of the mast so that he could see the signal again. Since the distance between the observer and the signal was shorter when he was on deck, the only explanation that could be given for his not seeing it on deck was the interposition of the water, which could be eliminated by climbing to a higher position.

This observation could only prove that the surface of the water was curved, not that it was actually spherical. The author of *De sphera* must have been aware of this fact, as is apparent from the words he uses to introduce the argument: "Quod autem aqua habeat tumorem et accedat ad rotunditatem sic patet," that is, "that water has a swollen shape and is almost round is thus evident" (Thorndike 1949, 83). To this *proof* he added an argument that referred to the natural tendency of water to assume a spherical shape. This tendency, which can be observed in drops of water and in dewdrops, could also be ascribed to seawater because of its homogeneity and of the assumed identity of nature between the parts and the whole.[9]

Not based on a general physical principle, these arguments, being disconnected and contradictory, needed further development. It is possible that Sacrobosco intended to leave out of his work the doctrine of natural motions because it was a subject belonging to natural philosophy. With this decision, he would have confirmed the strict distinction between disciplines that were observed by medieval universities. But if this is true, his attitude was outdated. With the revival of the study of Aristotle's works on natural philosophy, and particularly of *De coelo*, the discussion of subjects thus far treated separately, increasingly involved a unitary approach. The subjects treated in *De sphera* and in *De coelo* were so close that it became impossible to discuss them separately.

[9] It is interesting to observe that even in this case Sacrobosco never refers to *De coelo* II, 4 which would have given him a good argument to prove that water takes on a spherical shape. In the passage concerned, Aristotle presents a geometrical demonstration of the type of *reductio ad absurdum*, in which the conclusion is not in contradiction with the assumed geometrical principle, but rather it conflicts with the evidence of sense experience: water runs by nature from higher places to lower places. It would have been impossible for Sacrobosco to refer to the proposition at the beginning of Archimedes' *On Floating Bodies*, which was later translated by William of Moerbeke. Only during the Renaissance were references to Aristotle and Archimedes made by those authors who wrote commentaries on the *Sphere*, thus showing the need for more rigorous demonstrations on this important point.

6.11 The Rediscovery of *De coelo* and Cosmography

The first document testifying to this new approach is the *Commentary on De sphera of Sacrobosco* (*Super auctorem Sperae cum questionibus expositio*)" ascribed to Michael Scot (†c. 1235), where this author's continuous referral to Aristotle's works turns his *expositio* into a proper and detailed *quaestio*.[10] Thus, for instance, the discussion concerning the elements, which had been only briefly mentioned by Sacrobosco, was developed by Scot into a thorough examination of the various positions taken by Aristotle on that subject in *Metaphysics*, *Physics*, *De coelo*, *De generatione et corruptione* and in *Meteorologica*. But when Michael Scot discussed the passage in *De sphera* concerning the arrangement of earth and water, he abandoned any reference to the Aristotelian text and relied instead on direct observation and the "Holy Scriptures." He strongly rejected what was stated at the beginning of the *De sphera* concerning the arrangement of the element in the sublunary world, and maintained that it is rather the earth that is placed above the water, in the same way as the islands are placed in the middle of the sea. This truth was confirmed also by Psalm 103 "Qui fundasti terram super aquas," that is, "You who firmly based the earth on the water" and by Psalm 23: "Quia ipse super maria fundavit eam," that is, "Because he himself firmly based it on the sea."

How can this change of approach be explained? Were there no passages in the Aristotelian works that could help to clarify this point? Perhaps some assertion that could confirm the literal meaning of the commented text? Before this, Michael Scot had in fact just written: "From one fistful of earth ten fistfuls of water are generated by rarefaction and by thinning. And inversely from ten fistfuls of fire one fistful of air is generated by condensation and by thickening."[11] How can this statement, based on a partial interpretation of some Aristotelian passages (mainly from *De generatione et corruptione*, Book Two, Chapter VI), be made compatible with the solution of the problem taken from the Holy Scriptures and based on observation? Unfortunately, it is not possible to give a clear answer and find out whether the commentator had intentionally avoided facing an apparent contradiction or whether he was able or not to understand the Aristotelian text correctly.[12] Pierre Duhem accu-

[10] This must be obviously related to Michael Scot's activity as a translator of Aristotelian works (with Averroës' *Commentaries*): he had translated into Latin an Arabic translation of *De coelo* before 1230. Another, perhaps not complete, translation was made by Robert Grosseteste soon after 1230. After these works became available, it was unlikely that those who wrote on cosmography could ignore Aristotle's work.

[11] The original reads: "ut ex pugillo terre per rarefactionem et subtiliationem fiunt pugille aque decem etc. Et econverso ex decem pugillis ignis per condensationem et inspissationem fit unus pugillus aeris et deinceps." Thorndike (1949, 263–264).

[12] Further on in the text just quoted (Thorndike 1949, 296), Michael Scot discussed the same question from the point of view of concept of place by investigating the reason for the placement of the earth within the sphere of the immediately higher element. The internal surface of the sphere of water, as it was not in contact with all parts of the sphere of the earth, could not be considered as the place of the earth, whereas in some passages in Aristotle's works it appeared that the internal part of the sphere of any element was the place of the element immediately lower. The solution

rately explained this difficult point of Aristotle's philosophy in his *Système du Monde*, where the reader could find further explanation (Duhem 1958, 91–97).

Concerning the question of the spherical shape of the earth, Michael Scot's commentary is more dependent on Aristotelian texts, although the proof of the spherical shape derived from the doctrine of the motion of heavy bodies is only briefly mentioned in his *Expositio*. This proof is presented within a discussion concerning the local motion of the elements seen as a process that leads to the complete actualization of the form.

The passages from *De coelo*, which are more frequently quoted in the discussion of the shape and position of the earth, are often given different interpretations that are not always true to the original. This is the case for the brief mention of the Aristotelian text contained in Bartholomaeus Anglicus' *De proprietatibus rerum* (Chapter I): the medieval author seems to use the term *equilibratus* with the same meaning as that of ἰσόρροπος, so that the passage concerning the equal arrangement of parts of earth around the center of the world caused by ῥοπή is interpreted. as stating a situation of equilibrium resulting from equal and opposite inclinations.

Such interpretation is clearly based on the idea of balance: the earth seems to be in a state of equilibrium because its heavy parts have a tendency to go towards the center of the world as a consequence of their weight; and the earth, because of the tendency and inclination of its parts, is *hanging* at the center in a state of equilibrium and remains at rest in the same place.[13]

6.12 The Acquaintance with Simplicius's Commentary on *De Coelo*: Through its Medieval Latin Translation, the Centrobaric Theory Becomes a Part of Cosmography

Around the mid-thirteenth century, the Aristotelian texts had already been given different interpretations, and their use to explain Sacrobosco's *De sphera* became a generalized practice. Only during the second half of the century were all those questions discussed more deeply and on the base of new texts. This development was made possible by the Latin translation of Simplicius' *Commentary on De coelo,*

given by Michael Scot was similar to that briefly mentioned by Sacrobosco: according to the form proper to each element, the earth should have been contained by water, but the world was not perfect and the great majority of the animals and of the plants could not live in water, therefore part of the earth had been cleared of water. No reasons are given to explain how this situation came about.

[13] Bartholomaeus Anglicus (1505, sign. [s 3v]) "Terra, ut dicet Philosophus, est propriis equilibrata ponderibus. Quelibet enim suarum partium suo pondere nititur ad mundi medium, quo nisu et inclinatione singularium partium, tota circa centrum equilibrata suspenditur, et equaliter immobilis retinetur." The first part of this passage seems to be in relation with Publius Ovidius Naso (*Metamorphoseon*, I, 1, vv. 12–13) "nec circumfuso pendebat in aere tellus ponderibus librata suis…"

made by William of Moerbeke in 1267. Through this work, the Western philosophers became acquainted with the doctrine of the center of gravity: with this new conceptual instrument they could give new answers to the traditional cosmological questions.

In truth, knowledge of the part of Simplicius' work that concerned these questions did not spread very rapidly. Thomas Aquinas, who was one of the first to know about Moerbeke's translation, did not use it in his *Lectio* XXVII on Book Two of *De coelo*, where he gave an explanation of the reciprocal action between parts of earth that is different from Aristotle's text and replaced the term *inclinatio* with *pondus*, in a way similar to that which we have already seen:

> (versio vetus) Plus enim semper quod ipso minus propellere est necessarium, usque ad medium inclinationem habentibus ambobus, et graviori propellente usque ad hoc minus grave.
>
> (comm. Thomas) Nam si versus unam partem terrae sit maior quantitas, ad hoc quod ipsa magis appropinquet medio, depellit minorem partem per violentiam a medio *quousque aequale pondus ex omni parte inveniatur*.[14]

At the beginning of the fourteenth century, the philosophers of the Parisian school approached questions of the position and shape of the earth from a new point of view. By making an exhaustive analysis of the Aristotelian concept of 'place' based on parallel passages from Book 4 of *Physica* and from *De coelo*, these authors put the study of the formation of the earth in second place and gave prominence to the problem of the arrangement of the "elements" and of the receding of water from land populated by men and animals. They were able to satisfactorily tackle these problems by introducing a distinction between center of magnitude and center of gravity of a body. This distinction, which had already been made by Simplicius in his commentary on *De coelo*, was employed by the Parisian philosophers in their discussion of the *aporia,* mentioned earlier.

In explaining the shifting of the center of gravity of the spherical body placed at the center of the world as a result of adding a great weight to one hemisphere, they were using an argument belonging to the Archimedean tradition, but said nothing about the place of the two centers, one of gravity and the other of magnitude, within one single non-homogenous spherical body. This problem was related. to another question in Simplicius' commentary, namely the question of the perfectly spherical shape of the earth, which he discussed by referring to Alexander of Aphrodisia's statement on the existence of non-homogeneous parts in heavy bodies. This observation necessarily involved a distinction in a body between center of gravity and center of magnitude: one center was important for determining the motion of a heavy body downwards, whereas the other was irrelevant from this point of view and had a purely geometrical meaning. In consequence, it was possible to place a heavy body differently with respect to the center of the world according to whether it was homogenous or not.

[14] *De caelo et Mundo*, Book. 2, lectio, 27, de Aquino (1866, 142–143). Author's italics.

The author, who in the Middle Ages developed this argument in the most rigorous way, was Albert of Saxony. In his *Questiones subtilissime in libros de celo et mundo* (1520), Albert of Saxony discussed the problem of the place of the earth in the world by making use of that distinction. There were thus two different ways of considering the position of the earth at the center of the world: either in relation to its center of magnitude, or in relation to its center of gravity. The fact that parts of land were not covered by water excluded the first case and involved the necessity of placing the two centers differently. The earth was therefore to be placed in the world with respect to its center of gravity because (to quote Albert's paraphrase of the Aristotelian passage) "the heavier part pushes" the less heavy one "until the center of gravity of the whole earth is at the center of the world," for "then the two parts will have the same gravity, though one is bigger and the other smaller with regard to their magnitude": the inverse proportion between the two parts is the same as that of two different weights placed on a balance in equilibrium.[15]

The transformation of the Aristotelian concept of ῥοπή could not be expressed more clearly. Here, it is no longer the question of several parts of earth of different size, which by interacting with one another end up making a spherical body, but rather of weights of different magnitude which, as in Proposition 3 of Book One of *On the Equilibrium of Planes*, are in equilibrium when their common center of gravity is placed nearer to the heavier weight.[16]

But in addition to this, we also have to take into consideration other factors that alter the weight of the different parts of the earth. The cause of the difference in weight between the two hemispheres is the sun, the heat of which makes the land not covered by water less dense, and therefore lighter. It is a cause which endures over time: the changing action, together with other changes taking place on the surface of the earth, determine a continuous shifting of its center, which therefore can move in a straight line.[17]

[15] "Quod terra est in medio mundi quo ad centrum sue gravitatis. Probatur: nam omnes partes terre tendunt ad medium per suam gravitatem, sicut dicit Aristoteles in littera; et verum est. Modo pars que esset gravior depelleret aliam tam diu quod medium gravitatis totalis terre esset in medio mundi; e tunc starent due partes eque graves; licet una maior et alia minor quantum ad magnitudinem contra se invicem; sicut duo pondera in equilibra." Albert of Saxony (1520, 40r, lib. 2, quaestio 25).

[16] We could even think that Albert refers to the law of lever as presented by Archimedes in Proposition 6 of Book one, but from the Latin passage quoted in the note above it would seem that Albert only points out that in this case the heavier weight is nearer the center of the world, and the lighter weight is further away from it.

[17] By referring to what Aristotle says in chapter XIV of Book Two of *Meteorologica*, Albert probably thought that the continuous changing location of lands, rivers and seas involved a continuous changing arrangement of the weighing parts of the earth.

6.13 The New Worlds and the Centrobaric Theory: Franciscus Maurolycus, Christoph Clavius, Giuseppe Biancani

Around 1540 the geographical discoveries made during oceanic voyages by Spanish and Portuguese sailors had changed the picture of the earth and raised doubts about many ideas on which medieval cosmology was based. The impossibility of reconciling old ideas with the knowledge of new facts made it necessary to introduce substantial corrections in the astronomy textbooks used in the schools. The passages in these textbooks, which needed corrections, were those concerning the inhabited lands rising from water, as a consequence of the discovery of new countries in the West Indies. Thus there was no longer any reason for referring to the concept of center of gravity in the commentaries on *De sphera,* though it was still considered necessary to go back to the Aristotelian text.

In the discussion concerning the position, the immobility and the shape of the earth, the situation remained unchanged and the doctrine of the motion of heavy bodies still played an important role in this context. But during the sixteenth century things began to change. In addition to the "old world" represented by Aristotle's works, a "new world" made its first appearance in the form of texts that were almost unknown up to this point: the *Mechanical Problems,* at that time ascribed to Aristotle; Archimedes' *On Floating Bodies* and *On the Equilibrium of Planes;* the works by Hero of Alexandria; and those by Pappus. These new texts offered a more complex idea of the motion of bodies, based on the observation of motion in certain machines. For the first time, scholars could familiarize themselves with works containing rigorous discussion of problems concerning the moving and lifting of heavy loads, and could learn how simple machines functioned. The concept of center of gravity was not used systematically in all these works but it did play a central role in Archimedes' texts. Through Book Eight of Pappus' *Mathematical Collections* scholars became aware of the only definition of center of gravity to be handed down from antiquity.[18]

[18] Here, I deliberately disregard the tradition of the medieval Latin *Scientia de ponderibus*, which greatly influenced discussions of the problem of the equilibrium of balances during the Renaissance. This is not because I assume there is no relation between the question I discuss and those treated in the works ascribed to Jordanus Nemorarius. On the contrary, the premises of this work treat explicitly the relation between *gravitas* and *rectitudo* of the path of descent of heavy bodies (with reference to the line that ends in the center), Moody and Clagett (1960, 128–129). I chose not to deal with this tradition because the problems of the centrobaric theory are not mentioned. For an extensive treatment of the dispute between followers and opponents of the *Scientia de ponderibus* in the sixteenth century, see Renn and Damerow (2012). In the Pseudo-Aristotelian *Mechanical Problems* there is also no reference to the concept of center of gravity, but in this case there is a tendency to adopt Archimedes' method, on the assumption of a continuity between Archimedes' work and the mechanical tradition of Pseudo-Aristotle.

6.14 The Rediscovery of Pappus' "Mathematical Collections": The Center of Gravity of Bodies: From Cosmography to Mechanics

The fact that all these new sources of information became available produced important changes in the use of the concept of center of gravity within the discussion on cosmology. This new development started with Franciscus Maurolycus, one of the major mathematicians of the sixteenth century, who in his *Cosmographia* (Maurolico 1543) introduced radical changes both in the content and form of exposition. Written in the form of three dialogues, this work included in a systematic way new ideas derived from the two "worlds" mentioned above and focused attention on many questions that were later discussed in similar works. Maurolycus' new discussion on the shape and place of the earth and of water was included in its entirety, without any mention of the source, in Clavius' *Commentary on De sphera*, a work that was used as an astronomy textbook by generations of scholars of the Jesuit Order.

On the model of Sacrobosco's *De sphera*, Maurolycus discussed the question of the spherical shape of the earth and of water just after presenting the proofs for the spherical shape of heavens.[19] The whole argument is developed by following, from the beginning, the line of reasoning traced by the Sacrobosco but it is made more precise and articulate in a similar way to that adopted by many commentators of Sacrobosco's work. In the case of the first proof for the spherical shape of the earth, Maurolycus specified that the lunar eclipse must be observed from two towns on the same latitude. This condition was essential to guarantee the truth of the reasoning, which ended with an observation made in different places on the same parallel: only then would it be possible to establish that the differences in time were proportional to the differences in places, and to prove rigorously the spherical shape of the earth (Maurolico 1543, 7r/v).

Concerning the shape of water, Maurolycus sensibly departed from the medieval author. At the beginning, he seemed to move in the same direction as Sacrobosco and to rely again on the common experience of sailors who, when approaching the coast, first see the tops of mountains, bell towers and high buildings. But immediately afterwards, Maurolycus considered the whole terrestrial globe, formed by earth and water, and discussed the proof of its spherical shape by observing the shape of the earth's shadow on the surface of the moon during the lunar eclipse. This proof was based on the fact that water has a natural tendency to move towards lower places and, owing to its instability, stops moving only when its surface has the same height everywhere: this situation occurs when the surface of water is equidistant

[19] Maurolycus' peculiar way of phrasing the question must be noted. For him, the problem was the need to explain "ut hanc terrae marisque congeriem conglobatam esse." Maurolico (1543, 7r/v). This seemed to go back to the tradition of discussions on the causes of why the earth was not completely covered by water. Sacrobosco instead discussed separately, one after the other, the question of the spherical shape of the earth ("Quod terra etiam sit rotunda sic patet," Thorndike (1949, 81)) and of water ("Quod autem aqua habeat tumorem et accedat ad rotunditatem sic patet," Thorndike (1949, 83)).

from the center of the world, which is the lowest point that a heavy body can reach (Maurolico 1543, 7v–8r).

Any other argument thus fell short of proving anything, including the presumed tendency of water to take on a spherical shape: the phenomenon observed in dewdrops and in drops of water could not take place in the whole element of water; this element took on a spherical shape in order to keep the equilibrium of its parts, which struggled equally to move towards the center (servet aequilibrium aequaliter ad centrum connitens), whereas dewdrops and drops of water had a spherical shape as a result of the action of their opposite, that is, the dry element (Maurolico 1543, 8r). The cause of the spherical shape of water could therefore be found in its nature as a heavy body.

This explanation was no different from that given by Aristotle for the spherical shape of earth, which, though it was not made to flow and have its parts uniformly arranged, could not have naturally taken any other shape but the spherical one (Maurolico 1543, 8r). It would never have a perfectly spherical shape like that of water but its lack of perfection was not so important. Mountains and valleys could be regarded as having no tangible dimensions compared to the great size of the earth, even though for us they appeared to be very large (Maurolico 1543, 8v).

The question of the shape of the earth and of water was thus inevitably related to the doctrine of the motion of heavy bodies and to the concept of natural place. Everything encouraged the scholars to believe that the center of the terrestrial globe, formed by earth and water, coincided with the center of the world. But what had been said so far was not sufficient to prove the truth of this conception. Other proofs were needed.

The proofs adduced by Maurolycus were various and also in this case went further than what had been said by Sacrobosco. Their starting point was the same, that is, the absurdities that would follow if the earth were placed outside the center of the world: the stars would appear to come close to and move away from the earth; it would have been impossible to see half of the celestial sphere and therefore to observe the rising and setting of six signs of the zodiac. All of these absurd consequences were contradicted by observation. To these arguments, which were the only ones presented by Sacrobosco, Maurolycus added other arguments based on astronomy and, more interestingly, on some thoughts concerning the weight and place of the elements in the sublunar region.

Both the arrangement of the elements on the basis of their being heavy or light and their motions *from* or *towards* the center of the world confirmed what had been said by Aristotle: heavy bodies move towards their natural place along lines meeting at one single point and tend to keep their tendency to move until they reach a place where all their parts are at equal distances from the center, thus forming a spherical figure (Maurolico 1543, 8r). The heavy bodies move along lines perpendicular to the surface of the earth and of water, as can be seen from observations made with a plumb-line; and this ensures that both elements, that is, water and earth, are placed around the same center (Maurolico 1543, 8r). That the lines along which heavy bodies fall are convergent is not immediately visible: to an observer they seem to fall along parallel lines, but this impression is due to the short distance they are observed to fall. As in the case of the walls of two buildings or of a well, the apparent parallel direction is the result of their small dimensions compared to the size of the earth: if we imagine the size of the two buildings to be increased enormously, they would appear to be more and more divergent, whereas the walls of the well would in the end converge at the center of the earth (Maurolico 1543, 15v–16r).

These are traditional examples, which had already been mentioned by Albert of Saxony in the corollaries to the *Quaestio* XXVII on Book Two of *De coelo*. But now they are related to the question of the *antipodes* which, after the geographical discoveries, made it impossible to believe in the existence of an enormous sphere of water not concentric with the sphere of the earth. To be able to stand upright, the inhabitants of any zone of the earth must put their feet and head on the straight line passing through their body and reaching the center of the world. In this way two people standing in diametrically opposed places on the surface of the earth are placed in similar positions in relation to the lowest point of the world; they are in the same situation as two hanging weights which, if allowed to drop, would converge towards that point.[20]

After briefly mentioning earlier discussions of the reasons why the inhabited lands were not covered by water (with reference to Nicholas de Lyra, Paul of Burgos and Matthias Döring), Maurolycus focused his attention on some apparently paradoxical consequences resulting from the application to the whole terrestrial globe of concepts that are usually defined with regards to smaller portions of its surface (Maurolico 1543, 17r/v). How can the construction of a leveled floor be defined? This is done by using a "level," that is, an instrument indicating a line parallel to the plane of the horizon and determining the horizontal position of a surface to which it is applied. This is unproblematic as long as we are dealing with small portions of a spherical surface: the curvature of the surface in this case is so slight that it can hardly be perceived. But if we imagine the floor as an enormously extended plane that is tangent to the surface of the earth, that floor cannot be defined as being "leveled," since a body moving on it would not remain at the same distance from the center, but this distance would diminish or increase according to whether the body moves towards or away from the point of tangency.[21] In this case, a man walking on a horizontally leveled plane would be moving upwards or downwards with regard to the center of the earth.

Another paradoxical consequence would result from imagining a very large vase full of water placed on the surface of the earth or near to its center: the curvature of the surface of the water would be different according to the position of the vase and, in consequence, the vase would contain a greater quantity of water in the area close to the center. This difference would not be noticeable on the earth, even if we were able to place the vase on top of the highest mountains or at the bottom of the deepest pits, because the distances would be too short compared to the radius of the terres-

---

[20] "Nimirum, quod utrique nos erectos arbitramur, verum est; quod vero nos illos, illique nos in caput versos putamus, falsum. Siquidem utrique recti stamus; ipsumque terrae centrum locus est infimus utrisque communis; ad quem sane duo pondera utrinque suspensa pendent, et dimissa concurrerent." Maurolico (1543, 16v).

[21] This argument was also derived from Albert of Saxony (1520, 41v.), *Quaestiones subtilissime*, in the 4th corollary to Question XXVII (already mentioned) of Book Two of *De coelo*.

trial globe. Nevertheless, these paradoxical consequences were directly deduced from the general principles given by Aristotle (Maurolico 1543, 18r/v).

The natural tendency of heavy bodies to move towards the center of the world along straight lines and their arrangement at equal distances from the center required that the nature of the resulting spherical agglomerate be homogeneous, since only in this case would the center of magnitude and the center of gravity coincide. But did the terraqueous globe formed of land and water have such a homogeneous nature? Were the two elements arranged around a single center? The observations collected during the oceanic voyages of Colombo and Vespucci were able to give positive answers to these questions and exclude the possibility of the terrestrial globe not being homogeneous, as the medieval authors had thought to explain the existence of lands not covered by water. The earth and water were not arranged around the center of the world as a spherical body made of stone or wood would have been, that is, with the center of gravity placed nearer to the heavier part: they did not have different centers (Maurolico 1543, 18v).

Does this mean that the center of gravity was no longer an important concept in the field of cosmology? Not really. Though the arrangement of land and water on the earth no longer needed to be established, the center of gravity now became an essential concept for defining in a rigorous way the motion of any heavy body. The center of gravity of a body hanging in any position was always placed on the perpendicular line extending to the center of the world and, if dropped and not stopped by any obstacle, at the end of its movement it would encounter this center.[22] This way of defining the center of gravity with regard to its possible future motion was different from Pappus' definition, where mainly the static condition of a hanging heavy body was taken into consideration. The new "definition" includes some important aspects of the method for finding the center of gravity, following Pappus in the First Proposition of Book Eight of his *Mathematical Collections*. The text of this proposition was reproduced almost *ad litteram* by Maurolycus:

> Let the body be suspended in any way so that it would be freely hanging; now from the point from which it is suspended let a straight line be drawn perpendicular to the horizontal plane, as Euclid shows in Proposition XI of Book Eleven of the "Elements." Let the same body now again be suspended in a similar way from another point, from which a new line may be drawn again perpendicular to the horizon. Both perpendicular lines should surely extend through the center of gravity, since this center is always found on the same perpendicular line, however the body may be suspended. Therefore, the point at which the perpendicular lines intersect each other will without doubt be the center of gravity that was sought.[23]

[22] "Punctum videlicet, quod utcunque ac quotiescunque suspenso corpore, semper versus universale centrum pendet ad perpendiculum; quodque, dimisso corpore, modo absint obstacula, ipsi universali centro connitur." Maurolico (1543, 18v).

[23] The original reads: "Suspendatur proposita res utcunque, ut libere pendeat; mox ab ipso suspensionis signo ad horizontis planum perpendicularis recta ducatur, quemadmodum Euclidis in 11 undecimi docet. Rursum ab alio signo similiter res ipsa appendatur; et a signo rursum perpendicularis agatur ad horizontem. Oportebit nanque utranque perpendicularium per centrum incedere gravitatis, quandoquidem tale centrum in ipsa semper ad horizontem perpendiculari, utcunque res pendeat, invenitur. Punctum igitur, in quo se vicissim perpendiculares intersecane, erit proculdubio quaesitum gravitatis centrum." Maurolico (1543, 19r/v).

6.15 Sacrobosco's *De sphera* Revised and Corrected: Cristophorus Clavius' Commentary

As mentioned earlier, this section of Maurolycus' work was almost entirely incorporated by Clavius into his *Commentary on Sacrobosco's De sphera* (Clavius 1570), within a long discussion on "whether the Earth and water formed a single globe, that is whether the convex surfaces of these elements had the same center."[24] These questions were not directly related to Sacrobosco's text and, being in a way autonomous, were discussed in a section added to the commentary. The discussion started with a close criticism of those who maintained that the spheres of the earth and water had different centers. The first argument had a long tradition: the water placed outside the center of the world would require supernatural mediation to keep in that position; but such a miracle was really not necessary and therefore should be excluded. Immediately after this explanation, Clavius referred to the geographical explorations that had revealed the existence of dry lands at the antipodes. This discovery, though it showed the falsity of the hypothesis mentioned above, could not prove anything concerning the actual position of the two centers. So what were the proofs supporting the coincidence of the centers of the two spheres? And how could their placement at the center of the world be demonstrated? This could be done first of all by observing the motion of heavy bodies, then by using astronomical arguments such as the observation of lunar eclipses, and lastly by resorting to the testimonies of sailors who had crossed the ocean and observed variations in the rising and setting of stars, similar to those observed on land. Here, we are only interested in the first argument.

Although he used most of Maurolycus' arguments based on the center of gravity of bodies, Clavius changed the order he followed and divided the proof into two parts, based on the observation on natural motion. He did so in a way that at first sight does not seem suitable to develop rigorous reasoning. Maurolycus had shown: (1) that both the earth and water had a spherical shape; (2) that their centers coincided on the basis of experience with the plumb line. Only after proving this did he criticize those who thought the center of magnitude was different from the center of gravity, and his criticism was mainly based on the new geographical discoveries, without making any reference to the motion of heavy bodies. This motion was instead taken into account in defining the center of gravity, and this argument was connected on some points with the experience of the plumb line.

In order to prove by geometrical demonstration that the centers of both earth and water could not be different, Clavius repeatedly resorted to the motion of heavy bodies. He demonstrated first of all that if all bodies have a tendency to move towards the center of the world, then they must form a single spherical body and, on

[24] Clavius (1611). This edition was included in volume 3 of Clavius' *Opera omnia*, anastatic reprint with an introduction by Eberhard Knobloch. The original reads: "An ex terra et aqua unus fiat globus, hoc est, an horum elementorum convexae superficies idem habeant centrum." Knobloch (1999, 57).

the contrary, that if they did not result in the formation of a single spherical body, they could not have a tendency to move towards the same center (Knobloch 1999, 59). He then went on to prove that the center of gravity and the center of magnitude of the body formed by both the earth and water coincided by observing that the angles formed by the plumb line on the surface of the earth and water were equal (Knobloch 1999, 59–60). In the first part of his demonstration, Clavius had taken from Maurolycus both the definition of center of gravity and the method for finding it. But these two texts seem misplaced because they are included in a geometrical demonstration that avoids any reference to the angles formed by the straight lines directed towards the center of the world.[25]

The general impression that can be gathered from this section of Clavius' *Commentary* is that different materials have been assembled without any attempt at connecting and further developing them. Such assemblage of texts could be very useful for teaching, but did not help to develop this key concept in mechanics which was destined to undergo deep transformations. This impression is confirmed by the last part of Clavius' long digression on the spherical shape of the globe formed by both the earth and water: the apparent paradoxical consequences of this doctrine are once more discussed at length, such as perpendicular buildings would not be parallel, that the motion on a plane tangent to the surface of the earth would not be horizontal but upwards or downwards, and that a vase full of water would contain a greater or smaller quantity of liquid depending on whether it were nearer to air or farther away from the center of the earth.[26]

6.16 Giuseppe Biancani's Work: Cosmography and Mechanics at the Beginning of Modern Science

Though based on more recent astronomical works, Clavius' *Commentary on De sphera* still relied on observations made by the naked eye. In the 1611 edition of his work, the German mathematician mentioned the new astronomical discoveries accomplished with the telescope but died soon afterwards without being able to include them in the discussion contained in his work. Other members of the Jesuit Order immediately realized that these novelties ought to be taken into account in any discussion concerning astronomical matters, and that they would have a great effect on the Ptolemaic model of the world. The dispute between the followers of different cosmological systems was still at the beginning and the need to reassess

[25] Knobloch (1999, 57). In the 1581 edition published in Rome, the definition of the center of gravity taken from Maurolycus is followed by the one taken from Pappus' *Mathematical Collections*, Book Eight. Since any reference to the motion of the body let loose from the point at which it was hanging is missing, Pappus' definition would have been more suitable for Clavius' argument; but its inclusion in this argument is in no way justified. Any comparison between the two texts is missing, and the new definition seems to be an unnecessary addition.

[26] Knobloch (1999, 65–66). See Maurolico (1543, 18r/v).

the nature of the motion of heavy bodies within different astronomical hypotheses was not yet felt. This new situation can already be seen in Giuseppe Biancani's *Sphera mundi*; its full title mentioned the "new findings" (*novis adinventis*) of Brahe, Kepler, Galilei, and "other astronomers." However, when discussing the position and the mobility or immobility of the earth, Biancani repeated traditional arguments (Blancanus 1620).

In Part Three, Treatise II of *On the elements*, the question of the direction of the motion of heavy and light bodies was discussed with the same arguments used by previous authors, though in a more rigorous and coherent way. Biancani, more outspoken and critical than Clavius, was utterly contemptuous of those who believed that such motion took place along parallel lines: he regarded this idea as childish and shared by ignorant people who believed that the world was like an "oven" that is, a hemisphere with an endless bottom. In such a world, heavy bodies would fall along parallel lines and men on the surface of the earth would be in an upright position only in relation to those parallel lines. For Biancani, this is contrary to experience, which shows that heavy bodies move towards the center of the world and that the upright position of men is in relation to lines passing through the same center. For him, this truth seemed to be confirmed by a simple observation and could therefore be verified by anyone standing on any point of the surface of the earth, even by those inhabiting the antipodes.

All this is based on the presupposition that the earth has a spherical shape, a thing that Biancani had not yet proved, though he thought he could reject the common experience as utterly erroneous. The traditional view of the world, which defines the structure of the elementary region of the world on the basis of the two natural motions "away from" and "towards" the center of the world, had such an influence on direct observation that it seemed unnecessary to resort to a geometrical demonstration in order to establish the true direction of the motion of heavy and light bodies. Within a naïve and primitive picture of the world, the possibility of regarding the motion of such bodies as taking place along parallel lines was excluded (Blancanus 1620, 69–70).

To correctly understand the motion of heavy bodies towards the center of the world, it was necessary to know where the center of gravity of a body was placed: only after establishing the place of that center would it be possible to draw the straight line that the body would actually follow in its movement. For Biancani, Pappus' definition of the center of gravity was the ideal picture of the body before it began to fall, and the fall was imagined as the simple motion of a point along a straight line towards another point. Once the line of direction of the fall of a heavy body had been established in this way, the movements of the other points of the body needed to be analyzed. This aspect of the question had never been considered before and it directly concerned the general concept of equilibrium.

Pappus had established that a body hanging from its center of gravity, no matter how it was positioned, would never move from its initial position because on both its sides there would be two equally balanced parts (ἰσόρροπα δύο μέρη), that is, two parts having equal and contrasting tendencies to move towards the center of the world. A body would thus be conceived as being formed, in a way, of a group of innumerable balances, all positioned with different inclinations with regards to the perpendicular line that stretched to the center of the world, but all in a state of equilibrium. Such a body, when falling, would not rotate in any way so that each one of its points would fall along a line parallel to the perpendicular (Blancanus 1620, 70). This description would make it possible to discuss in a unified way the conditions of equilibrium of all bodies, both those placed at fixed distances from the center, and those in motion towards the center or away from it: this approach was very useful for a deeper study of the problem of the balance, which after the publication in 1577 of Guidobaldo del Monte's *Mechanicorum liber* had reached a degree of complexity never achieved before.[27]

A study of this problem from a strictly mechanical point of view had actually been made by Biancani in his *Aristotelis loca mathematica*, when he discussed the second question of the *Mechanical Problems* ascribed to Aristotle. This question asks why "if the cord [more correct "support"] supporting a balance is fixed from above, when after the beam has inclined the weight is removed, [does] the balance return[s] to its original position"? (Aristotle and Hett (1936, 347–349/850a2–6). Biancani first recalled the discussion of previous authors, and then tried to answer the Aristotelian question on the basis of the concept of center of gravity.[28]

Biancani did not repeat this discussion in his *Sphera*, but preferred to include some arguments taken from the *Commentaries and Disputations on St. Thomas*

---

[27]Monte (1577, 5v–21v; 1581, 5r–25r). In these pages Guidobaldo criticized the solution of the problem of the equilibrium of the balance given by the medieval *Scientia de ponderibus*.

[28]Blancanus (1615, 155–157). In this work the Jesuit scholar had also discussed the passage on the *aporia* in Chapter XIV of Book Two of *De coelo*. On the basis of the concept of center of gravity, the motion of a heavy body towards the center of the world and its resting at this center at the end of the motion were immediately understandable: as soon as the center of gravity of the body and the center of the world coincided, the body would stop moving. But Aristotle could not have conceived of this center in this way since the concept of center of gravity was first used by Archimedes. Therefore Biancani thought that Aristotle meant the center of magnitude and hence was wrong. "Iuxta mathematicos duplex esse medium, sive centrum cuiusvis magnitudinis: aliud enim est centrum molis, aliud est centrum gravitatis. [...] Quando igitur Aristoteles ait, grave descensurum, donec ipsius medium, sive centrum, mundi centrum attingat, bene dicit, si de medio gravitatis intelligat, male autem si de medio molis, quia gravia omnia ratione centri gravitatis ponderant, neque manent, nisi ipsum maneat; quare nisi ipsum attingant centrum mundi, semper gravitabunt, et movebuntur. Verum enim vero ex antiquorum monumentis manifestum est, Archimedem, qui multo post Aristotelem floruit, primum omnium de centro gravitatis esse philosophatum, qua ratione dicendum esset, Aristotelem de centro, molis loquutum esse, et perinde non usquequaque vere." Blancanus (1615, 81). The explanation of the phenomenon described in the *aporia* was simplified and transformed by Biancani: the greater part that pushes the smaller one is regarded as being inside the solid body. "Sensus Aristotelis est, debere nos existimare, quod si quaepiam gravis magnitudo descendat ad centrum mundi, eam non permansuram, statim ac ipsius extremum centrum mundi attigent; sed eo usque descensuram, quosque ipsius medium, mundi medium, sive centrum assequutum sit; maior enim ipsius pars, in qua scilicet medium est, minorem partem propellit, donec utrinque a centro mundi aeque emineat. Omne enim grave hucusque habet propensionem, sive hucusque gravitat, v.g. si lapis illuc descenderet, non quiesceret statim ac prima ipsius pars ad mundi centrum pertingere, sed reliquae ipsius partes adhuc gravitarent, sicque ulterius primam partem impellerent, donec lapidis medium mundi medio congrueret." Blancanus (1615, 80–81).

Summa by Grabriel Vásquez, one of the leading Jesuit theologians of the time (Vásquez 1606, 464–465). These arguments were based on a strict application of the mechanical model derived from the balance, which presupposed a continuous slight shifting of the center of gravity of the earth. Any weight, however small, even the weight of a bird flying from one place to another, would alter the equilibrium of the weight existing in the terrestrial body and, in consequence, would cause a shifting and a new placement of it around the center of the world.[29] Although this shifting could not be perceived by human beings, this would not be considered a sufficient reason to deny its validity. This idea, which was a direct consequence both of the doctrine of equilibrium expounded in books on mechanics and of the customary practice of weighing by means of a balance, was the final stage of a long development of thought on the concept of center of gravity, which had begun with Maurolycus' work. At this point, Aristotle's and Sacrobosco's texts were only vague starting points and the medieval discussions had now lost any relevance. The terraqueous globe was now no longer considered from the point of view of the process by which it had been formed, but as a body already formed, having a spherical shape, and placed at the center of the world. The motions of heavy bodies proved the necessity of these assumptions and the laws of the equilibrium of weights confirmed them. But these same laws entailed the impossibility of an absolute immobility of the earth, since weights on it moved continuously; and that caused a continuous shifting of its center of gravity. A theoretical model stated the existence of a fact that could not be perceived by the senses; similar phenomena would occur more frequently in modern science.

References

Aeschylus, and Alan H. Sommerstein. 2008. *Persians. Seven against Thebes. Suppliants. Prometheus bound*. Cambridge: Harvard University Press. The Loeb Classical Library.

Albert of Saxony. 1520. *Questiones subtilissime in libros de celo et mundo*. Venetiis: heredum quondam nobilis viri domini Octaviani Scoti.

Anglicus, Bartholomaeus. 1505. *Liber de proprietatibus rerum*. Strasbourg: Georg Husner.

Aristotle, and W.K.C. Guthrie. 1939. *Aristotle in twenty-three volumes. Vol. 6. On the heavens*. Cambridge, MA: Harvard University Press. The Loeb Classical Library.

Aristotle, and W.S. Hett. 1936. *Aristotle in twenty-three volumes. Vol. 14. Minor works*. Cambridge, MA: Harvard University Press. The Loeb Classical Library.

Aristotle, and H. Rackham. 1926. *Aristotle in twenty-three volumes. Vol. 19. Nicomachean ethics*. Cambridge, MA: Harvard University Press. The Loeb Classical Library.

[29] "Terrae moles ita circa mundi centrum constituta est, ut in aequilibrio sita sit, idest, partes eius circa mundi centrum aeque ponderent, ac propterea immota consistat; quae vero in aequilibrio manent, quovis minimo ex una parte addito, vel ablato pondere, ab aequilibrii situ dimoventur, ut experientia quotidiana in lancibus, ac stateris ostendit, et rationes Mechanicorum evincunt. Cum igitur perpetuo circa Terram, res variae modo illi addantur, modo demantur (ut eum lapis in altum proiicitur, vel cum aves ab ea avolant, et ad eandem advolant, aut cum aliquid super eam saltat) necessarium esse videtur ipsa in perpetua quadam trepidatione insensibili tamen, titubare, ac vacillare." Blancanus (1620, 76).

Aristotle, A.L. Peck, and E.S. Forster. 1937. *Aristotle in twenty-three volumes. Vol. 12. Parts of animals. Movement of animals. Progression of animals*. Cambridge: Harvard University. The Loeb Classical Library.

Baldi, Bernardino. 1887. *Vite inedite di matematici italiani*, Bullettino di bibliografia e storia delle scienze matematiche e fisiche. Vol. 19. Rome: Scienze matematiche.

Blancanus, Giuseppe. 1615. *Aristotelis loca mathematica: ex universis ipsius operibus collecta et explicata*. Bononiae: apud Bartholomaeum Cochium: sumptibus Hieronymi Tamburini.

———. 1620. *Sphæra mundi, seu cosmographia demonstratiua, ac facili methodo tradita: in qua totius mundi fabrica, una cum novis, Tychonis, Kepleri, Galilæi, aliorumq[ue]Astronomorum adinventis continetur*. Bononiae: Typis Sebastiani Bonomij.

Boffito, Giuseppe. 1902. *Intorno alla quaestio de aqua et terra attribuita a Dante*. Turin: Clausen.

Caverni, Raffaello. 1895. *Storia del metodo sperimentale in Italia*. Vol. 4. Florence: Civelli.

Clavius, Christoph. 1570. *Christophori Clavii Bambergensis In Sphaeram Ioannis de Sacro Bosco commentarivs*. Romae: Helianum.

———. 1611. *Complectens commentarium in sphaeram Ioannis de Sacro Bosco*. Moguntia: Eltz.

de Aquino, Thomas. 1866. *In Aristotelis Stagiritae nonnullos libros Commentaria*. In *Sancti Thomae Aquinatis doctoris angelici ordinis praedicatorum opera omnia: ad fidem optimarum editionum; accurate recognita*, vol. 2. Parma: Fiaccadori.

Del Monte, Guidobaldo. 1577. *Guidiubaldo e Marchionibus montis mechanicorum liber*. Pisauri: Apud Hieronymum Concordiam.

———. 1588. *In duos Archimedis aequiponderantium libros paraphrasis*. Pisauri: Apud Hieronymum Concordiam.

Del Monte, Guidobaldo, and Filippo Pigafetta. 1581. *Le mechaniche dell'illustrissimo Sig. Guido Ubaldo De'Marchesi del Monte*. Venetiae: Appresso Francesco di Franceschi Sanesi.

Duhem, Pierre Maurice Marie. 1905. *Les origines de la statique*. Vol. 2. Paris: Hermann.

———. 1958. *Le système du monde: Histoire des doctrines cosmologiques de Platon à Copernic*. Vol. 9. Paris: Hermann.

Dühring, Eugen Karl. 1873. *Kritische Geschichte der allgemeinen Principien der Mechanik: Dritte wiederum erweiterte und theilweise umgearbeitete Auflage*. Leipzig: Fues.

Euripides, and D. Kovacs. 2002. Helen, Phoenician women, Orestes. Cambridge: Harvard University Press.

Grant, Edward. 1984. *In defense of the earth's centrality and immobility: Scholastic reaction to Copernicanism in the seventeenth century*. Philadelphia: American Philosophical Society.

Gutman, Oliver. 1997. On the fringes of the Corpus Aristotelicum: The pseudo-Avicenna "liber celi et mundi". *Early Science and Medicine* 2 (2): 109–128.

———. 2003. *Pseudo-Avicenna. Liber celi et mundi: A critical edition with introduction*. Leiden/ Boston: Brill.

Homerus and Alexander Pope. 1760. *The Iliad of Homer*. 6 vols, vol. 3. London.

Knobloch, Eberhard. 1999. Christophori Clavii Opera mathematica V tomis distributa ab auctore nunc denuo correcta. In *Sphaeram Ioannis de Sacro Bosco Commentarius, Christoph Clavius. Mit einem Vorwort hsrg. von Eberhard Knobloch*, vol. 3. Hildesheim: Olms Weidemann.

de Lagrange, Joseph Luis. 1811. *Méchanique analytique: Nouvelle édition revue et augmentée par l'auteur*. Paris: Mme Ve Courcier.

Mach, Ernst. 1907. *Die Mechanik in ihrer Entwicklung: Historisch-kritisch dargestellt*. 7th ed. Berlin: Akademie-Verlag.

———. [1883] 1988. *Die Mechanik in ihrer Entwicklung: Historisch-kritisch dargestellt*. Berlin: Akademie-Verlag.

Maurolico, Francesco. 1543. *Cosmographia*. Venetiis: Iunta.

Moody, Ernst A., and Marshall Clagett. 1960. *The medieval science of weights*. Madison: The University of Wisconsin Press.

Plato, and Harold N. Fowler. 1914. *Euthyphro. Apology. Crito. Phaedo. Phaedros*. Vol. 1. London: Harvard University Press. The Loeb Classical Library.

Renn, Jürgen, and Peter Damerow. 2012. *The equilibrium controversy – Guidobaldo del Monte's critical notes on the mechanics of Jordanus and Benedetti and their historical and conceptual backgrounds*. Berlin: Edition Open Access. http://www.edition-open-sources.org/sources/2/index.html.

Thorndike, Lynn. 1949. *The Sphere of Sacrobosco and its commentators*. Chicago: University of Chicago Press.

Vailati, Giovanni. 1904. *La dimostrazione del principio della leva data da Archimede nel libro primo sull'equilibrio delle figure piane. In Atti del Congresso Internazionale di Scienze Storiche*, vol. 12. Rome: Accademia dei Lincei.

Vásquez, Gabriel. 1606. *Commentariorum ac disputationum in primam sedundae S. Thomae, tomus I*. Tarvisii: Ex typografia Evangelistae Deuchini, superiorum permissu. Sumptibus Baretii Baretii Bibliopolae commorantis Venetiis.

Part III
The Expansion of Preclassical Mechanics

Chapter 7
The Social Position and Intellectual Identity of the Renaissance Mathematician-Physicist Giovanni Battista Benedetti: A Case Study in the Socio-Political History of Mechanics

Pietro D. Omodeo

Abstract The scientific work of Giovanni Battista Benedetti, one of the founding fathers of modern mechanics, is an appropriate test case for socio-political history of science. It can help us to address the tension between the social position and intellectual identity of Renaissance scholars who formed the archetype for modern scientists. This chapter deals with the manner in which socio-political coordinates informed Benedetti's science as far as its 1) demarcation, 2) content, 3) form, and 4) justification are concerned. With social coordinates, I refer to the institutional setting involving Benedetti's role as a courtier and thus to his function as a court mathematician which, in turn, was linked to the wider socio-economic interests of a Renaissance territorial state. With cultural coordinates, I refer to Benedetti's intellectual identity as a mathematician as well as his political identification with the wider interests of the Savoy ruling class. The polarity of function and identity constitutes the hermeneutic instrument for my interpretation of Benedetti's science. Such a *dialectic of position and identity* should serve as a guide for a non-reductionist socio-cultural history of science.

Keywords Socio-political history of mechanics · Social position of the Renaissance scientist · Political identity of the Renaissance scientist · Science and the court · Giovanni Battista Benedetti · Renaissance Turin

The historian of medieval science and philosophy Anneliese Maier viewed Giovanni Battista Benedetti with mixed feelings. To her Benedetti appeared as a sort of intellectual companion of Galileo Galilei, at the same time his "forerunner" in mathematical physics and a disrespectful epigone of earlier medieval *Vorläufer* or predecessors. Parisian scholastics such as Nicole Oresme and Johannes Buridane

---

P. D. Omodeo (✉)
Ca' Foscari University of Venice, ERC EarlyModernCosmology, Venice, Italy
e-mail: pietrodaniel.omodeo@unive.it

R. Feldhay et al. (eds.), *Emergence and Expansion of Preclassical Mechanics*, Boston Studies in the Philosophy and History of Science 270,
https://doi.org/10.1007/978-3-319-90345-3_7

provided Bendetti and Galileo with concepts to inquire into physics—Maier particularly had the concept of *impetus* in mind—which they did not acknowledge in their fierce attack against Aristotelian philosophy.[1] Maier shared Alexandre Koyré's view that modern mechanics was constructed around a few central concepts that essentially derived from Newton's *Principia mathematica*. Although I do not embrace the reductionist and idealistic interpretation of these historians, I mention their perplexity concerning the reception/rejection of medieval physics by Renaissance scientists to bring into focus the question of the cultural conditions for the incipient *querelle des anciens et des moderns*.[2] Nicolaus Copernicus, a sort of "unaware revolutionary,"[3] saw himself (or at least presented himself) as a Renaissance restorer of planetary theories, defended in antiquity by legendary Pythagoreans, whereas his scholastic counterpart, the Paduan Averroist Girolamo Fracastoro, presented a homocentric reform of mathematical astronomy (based on the modeling of all celestial motions through concentric spheres, in line with an Islamic/Aristotelian tradition), as a radical innovation, comparable with Amerigo Vespucci's discovery of a New World.[4] In the case of Benedetti, which I will focus on here, the rejection of the *philosophia naturalis* taught in the universities was achieved with intellectual means descending from that very philosophical tradition. Rather than viewing this fact as a paradox, it should be regarded as a sign of a profound tension in Renaissance science between the *intention* to outdo traditional authorities in order to move beyond their legacy on the one hand, and the *necessity* to rely on the shared knowledge of the time, marked by Aristotelian thought, on the other. In the following essay I will delve into Benedetti's conceptions and reconstruct their socio-cultural coordinates.

The polarity between continuity and discontinuity arguably reflects a more general tension between the material and intellectual conditions of early modern science as well as between the intentions and identities of a new genre of intellectuals, who formed the archetype for modern scientists. Benedetti's case will help us reflect on the social position and intellectual identity of such new types of scholars as well as on the way socio-cultural coordinates penetrated science, as far as (1.) its demarcation, (2.) content, (3.) form and (4.) justification are concerned. With social coordinates I refer to the institutional setting involving Benedetti's role as a courtier and thus to his function as a court mathematician which, in turn, was linked to the wider socio-economic interests of a Renaissance territorial state. With cultural coordinates I refer to Benedetti's intellectual and political identity. The former identity refers to Benedetti's philosophical interpretation of his role as a mathematician; the latter to

---

[1] Maier (1951, 304–305) established a connection between Benedetti's treatment of motion and that of Galileo.

[2] See Lehner and Wendt (2017).

[3] Copernicus' revolutionary role *malgré soi* already puzzled Thomas S. Kuhn, who called him at once "radical" and "conservative" and regarded *De revolutionibus orbium coelestium*, the book propounding the first modern heliocentric theory in mathematical astronomy, "revolution-making" rather than "revolutionary." Cf. Kuhn (1959, 135, 148).

[4] See Goddu (2010); Granada and Tessicini (2005); and Omodeo (2017b).

his identification with the wider interests of the Savoy ruling class. I do not see these cultural coordinates as *determined* by the social environment but rather as the ethical-political pendant of the socio-economic context, following Gramsci's dictum: "Material forces are the content and ideologies the form, according to a merely didactic distinction between form and content, since the material forces cannot be historically conceived without any forms and the ideologies would be individual whims separated from the material forces."[5] The polarity of function and identity constitutes the hermeneutic instrument for my interpretation of Benedetti's science. Such a *dialectic of position and identity* should serve as a guide for a non-reductionist socio-cultural history of science.

7.1 The Demarcation of Early Modern Mechanics: An *ex post* Perspective

The athletic metaphor of the "forerunner" does not have to be taken literally as a teleological claim. In reference to Benedetti it is simply meant to emphasize his outstanding originality, his capacity to advance along new intellectual paths enabling subsequent generations to explore physics further. Historians of mechanics have often extolled Benedetti's contributions to the science "of balance and motion."[6] As a matter of fact, a direct link could be established between his Archimedean treatment of heavy bodies falling through a medium and Galileo's early writings *De motu*.[7] Moreover, the young Galileo and his patron Del Monte scrutinized Benedetti's *De mechanicis*, a work that provided decisive insights for the conceptualization of what Galileo would refer to as "momento" or torque.[8] Because of these connections historians of science allotted Benedetti a central position in the Renaissance debates on mechanics. To single out and better locate his contribution, I begin with a more general treatment of the theoretical scope of early modern mechanics, in the form of an *ex post* perspective derived from a significant late seventeenth-century account. As a second step, I will integrate this tentative outline with a socio-cultural critical reworking.

Concerning the systematic peculiarities and developmental lines of mechanics in the early-modern period, it can be remarked that *prima facie* this scientific field

---

[5] See Gramsci (1975, Notebook VII, 869): "le forze materiali sono il contenuto e le ideologie la forma, distinzione di forma e contenuto meramente didascalica, perché le forze materiali non sarebbero concepibili storicamente senza forma e le ideologie sarebbero ghiribizzi individuali senza le forze materiali."

[6] Ernst Mach's *Die Mechanik in ihrer Entwicklung historisch-kritisch Dargestellt* began with this field demarcation: "Jener Teil der Physik, welcher der älteste und einfachste ist und daher auch als Grundlage für das Verständnis vieler anderer Teile de Physik betrachtet wird, beschäftigt sich mir der Untersuchung der Bewegung und des Gleichgewichts der Massen. Er führt den Namen *Mechanik*" (1912, 25).

[7] Cf. Drabkin (1960, 10–11).

[8] Cf. Renn and Damerow (2011).

does not present "demarcation" difficulties. The first theoretical and historical outline of the discipline, which was idealized but nevertheless insightful, can be found in Joseph-Louis Lagrange's *Mécanique analytique* (first ed. Paris, 1788). This work, which completed the process of subsuming mechanics under the field of mathematical analysis, can be seen as the systematical culmination of early modern mechanics.[9]

According to Lagrange's disciplinary partitions, familiar to today's reader, mechanics comprises two main areas, statics and dynamics. These are in turn subdivided depending on whether they examine solid bodies or liquids. He defined statics as "the science concerned with the balance of forces," and summarized its three fundamental laws: balance of the lever (*équilibre dans le levier*), composition of motions (*composition des mouvements*), and virtual work (*vitesses virtuel*). Lagrange integrated this theoretical schematization with a brief summary of the history of the discipline by specifying the names of the "discoverers" of the various principles, laws, and ingenious solutions to mechanical problems. In addition to the definition of general principles and laws and the solution of special problems, he considered simplification and unification to be significant to the progress of a *mathematical* discipline.

He presented the law of the lever as the most ancient law of statics. It was determined by Archimedes and reworked by Stevin and Galileo among the moderns. The reduction of the other simple machines, in particular the inclined plane, to the lever was envisaged by Pappus in ancient times but solved and brought to perfection only by modern scholars such as Stevin and others. A second law of statics, the composition of motions, was first established by Galileo and improved by several scholars: Descartes, Roberval, Mersenne, Wallis and Varignon. A third law is traced back to Galileo's concept of *momento*.[10] Galileo is also presented as the *initiator* of the "entirely modern" part of mechanics, namely dynamics as the "science of accelerating and decelerating forces and of the varied motions that they can produce."[11] It originated from the attempts to find a mathematical solution to physical problems such as the fall of bodies and the determination of projectiles' trajectories and, later on, the pendulum and centrifugal forces as studied by Huygens. Newton's discovery of universal gravitation enabled the definitive metamorphosis of mechanics into a *science nouvelle*. Eventually, the invention of calculus permitted mathematicians to express the laws of the motion of bodies in the form of equations. As far as the variation of speed of motions is concerned, the crucial improvements concerned the introduction of the idea of inertia and the composition of motions. Furthermore, the field was expanded through the study of bodily interactions during the seventeenth and eighteenth centuries by Descartes (conservation of motion) and Huygens (conservation of energy). The discipline reached perfection thanks to the work of the

[9] See Lagrange (1788, vi): "Ceux qui aiment l'Analyse, verront avec plaisir la Méchanique en devenir une nouvelle branche, et me fauront gré d'en avoir étendu ainsi le domaine."

[10] Today, the reference work on this is Galluzzi (1979).

[11] See Lagrange (1788, 158): "La dynamique est. la Science des forces accéleratrices ou retardatrices, et des mouvemens variés qu'elles peuvent produire."

likes of Jean and Daniel Bernoulli, d'Alembert, and Euler, to mention only a few of the mathematicians whom Lagrange viewed with admiration. Additionally, the principle of least action (*principe de la moindre action*) was developed in the wake of Maupertuis's quasi-metaphysical application of the principle to both light refraction and bodily shocks.[12] Here is not the place to expand on these developments of seventeenth and eighteenth-century mechanics but it is important to mention them as part of the wider picture.[13] Furthermore, the intellectual dimension of Lagrange's treatment of mechanics could be helpful as a reference point to systematically map the mental models of "classical" mechanics.

The scheme that emerges from Lagrange's historical outlook is not intricate. It can be summarized as a three-step advance of the discipline. The Renaissance systematization of ancient and medieval mechanics came first, along with the *creation* of dynamics as a specifically modern field of physical-mathematical inquiry. The second step consisted in the generalization of the science of machines into a general mathematical physics up to Newton. Third, mathematical formalism permitted the analytical treatment of mechanics and led to its complete transformation into a deductive analytical mathematical discipline. I would concisely label the three main phases of early modern mechanics as follows: first, mechanics as the science of machines, second, mechanics as the physical-mathematical science of balance and motion in general and, third, mechanics as an analytical science.

However, upon inspection Lagrange's clarifications are too general for a historical understanding of the boundaries of the discipline and its differentiation into sub-fields. His demarcation is clearly *a posteriori* relative to the tumultuous age in which Benedetti and Galileo lived and does not account for the historical development of the discipline. The definition of mechanics as the science of balance and motion and the corresponding distinction of its two main branches was itself the result of a historical process of generalization and differentiation. Moreover, as soon as one raises questions about the link of mechanics with allied sciences, methodology, the context of justification or the cultural meaning and impact, the demarcation of early modern mechanics becomes less sharp and the investigation has to take into account the *emergence* of the discipline owing to both intellectual and material circumstances. In other words, Lagrange took a snapshot of the discipline at a certain stage of its evolution. Therefore, his schemas are useful but intrinsically limited; their rigid application and uncritical reception would imply re-reading the past anachronistically. In fact, they cannot account for the concrete development of mechanics unless one assumes, first, that scientific disciplines are meta-historical hypostases and, second, that their progress is teleologically guided by inner principles. I regard none of these presuppositions as tenable. Rather, mechanics emerged from a concrete historical ground and was progressively differentiated, articulated,

[12] Lagrange (1788, 188): "Ce principe envisagé analitiquement, consiste en ce que dans le mouvement des corps qui agissent les uns sur les autres, la somme des produits des masses par les vitesses et pare les espaces parcourus, est. un *minimum*."

[13] For a thorough study of the developments of mechanics up to Lagrange, see Pulte (1989).

and organized. It is the task of the historian of science to reconstruct the paths mechanics took before it crystallized into certain forms at certain stages.

Scholars such as Ernst Mach only partly overcame the limitations inherent to a purely speculative treatment of mechanics. Mach did not question Lagrange's indication of the main problems and themes of the science of mechanics, but integrated it with historical-epistemological considerations in *Die Mechanik in ihrer Entwicklung historisch-kritisch Dargestellt* (*Science of Mechanics: A Critical and Historical Account of Its Development*) (Leipzig 1883). In particular, Mach insisted that the discipline was founded upon experience, although the social aspect of the concept of experience escaped his analysis. The importance of this social dimension was stressed by Leonardo Olschki in his three volumes on the *Geschichte der neusprachlichen wissenschaftlichen Literatur* (History of Vernacular Scientific Literature) (1919, 1922, 1927). As has already been discussed, Marxist historiography in the 1930s and 1940s pointed out the social roots of science, especially thanks to the Soviet historian of science, Boris Hessen, and the German Marxist émigré in the USA, Edgar Zilsel.[14] Hessen listed the economic realms and social spheres accounting for the *content* of mechanics in the age of Newton.[15] Alongside the Polish affiliate to the *Frankfurter Schule*, Henryk Grossman, he stated that modern mechanics resulted from a reflection on available machine technology in an age in which the emergence of the capitalist society valued the economic relevance of technology. Zilsel expanded the social discussion on the rise of modern mechanics by indicating its producers, that is, the early scientists who brought together artisanal experience, the scholastic spirit of systematization, and humanistic literacy.[16] These scholars merely hinted at the relevance of the interplay of cultural and political drives with the socio-economic foundation (*Basis*), regarding this issue as a problem of *ideology* (Hessen) or as one related to the reconstruction of the cultural background of so-called "knowledge bearers" (Zilsel). While Hessen considered the incidence of ideology to be *limited* to the *limitations* of early modern scientific conceptions and Zilsel restricted himself to the analysis of the social *function* of early scientists, the *constitutive dimension of politics and culture* and the *identity of scientists as intellectuals* remained obscure to them, owing to an overall economicist understanding of history and society. As I argued before, the integration of the structural dimension with the political-cultural can be achieved through a methodological reappraisal of Gramscian views concerning the *vital connection* of the social and the political well captured by the concept of hegemony.[17] Beginning with these premises, I will first present Benedetti's work on mechanics in general and then focus on its two constitutive elements, social position and intellectual identity.

[14] See Long (2011, ch. 1), and Omodeo (2017a, 119–150).

[15] See Hessen (2009, 41–102).

[16] See Zilsel (2000).

[17] See Chap. 3 "Socio-Political Coordinates of Early Modern Mechanics: A Preliminary Discussion" in this volume.

7.2 Benedetti's Writings on Mechanics

Benedetti developed his mechanics in a stage in which the physical-mathematical treatment of balance and motion was not sharply separated from the investigation of concrete machines and the solution of technological problems, although it showed evident bias toward generalization at the level of mathematical abstraction and the foundations of physics or "natural philosophy." We can aptly locate Benedetti's mechanics in that pre-Newtonian compartment which has been defined as the phase of *preclassical mechanics* by taking Newton's *Philosophiae naturalis principia mathematica* to be the theoretical watershed. In fact, one can assume for the sake of simplicity—that is, by abstracting from the internal tensions in Newton's work and its complex reception—that the discipline reached its classicism after this publication in 1687.[18] Benedetti's approach was marked by its connection with practice, although, as I will argue, his approach and level of formalization of mechanics reveals an accomplished division of labor between the *scientist* who *knows*, *assesses* and *projects*, and the *craftsman*, or *practitioner*, who is *experienced* and *realizes* wonderful works but lacks a *general understanding* of their principles and laws. In his role as court mathematician Benedetti embodied the former category. He was aware of the theoretical dimension of mechanics and the relevant sources; in particular, he benefitted from the re-appropriation of Archimedes' mathematical work, including his hydrostatics, the reworking of (pseudo-)Aristotle's *Mechanical Problems*, the merging of these reception lines with Jordanus Nemorarius' scholastic *scientia de ponderibus*, and the rise of innovative perspectives such as Tartaglia's ballistics, a *nova scientia* in its own right (1537), and Del Monte's determination of the center of gravity as a unifying explanatory principle of simple machines (in *Mechanicorum liber,* 1577).[19]

Benedetti's most significant contributions to mechanics are scattered in several works of his: *Resolutio omnium Euclidis problematum* (Resolution of All of Euclid's Problems) (Venice 1553), *Demonstratio proportionum motuum localium contra Aristotilem et omnes philosophos* (Demonstration of the Proportions of Local Motions against Aristotle and All Philosophers) (Venice 1554 and, second revised edition, 1555), and his major work, *Diversarum speculationum mathematicarum et physicarum liber* (Book of Various Mathematical and Physical Speculations) (Turin 1585), in particular in the third and fourth books entitled *De mechanicis* (On Mechanics), *Disputationes de quibusdam placitis Arist[otelis]* (Disputations on Some of Aristotle's Opinions), and in some of the epistles comprised in the last part

[18] Cf. Damerow et al. (2004).

[19] For an overview, see Drake and Drabkin (1969, "Introduction," 3–60). Dijksterhuis (1956) remains a valid introduction to Archimedes. On the reception of his hydrostatics, see Chalmers (2017). On Aristotelian mechanics, see van Leeuwen (2016). See also Nenci (2011). On other important aspects of the renewal of mechanics during the Renaissance, see Renn and Damerow (2010, 2011); Valleriani (2013); Becchi et al. (2013).

of the *Diversarum speculationum ... liber*.[20] The issues he dealt with included debated topics of mechanics in his time such as the determination of the equilibrium of balances, body's free fall, and cannonballs' trajectories. As for his theoretical achievements, Benedetti has been extolled for his treatment of fall acceleration as an increasing *impetus* and the conclusion that bodies of the same material fall with the same speed; he defended this view in explicit opposition to the widespread Aristotelian theory that speed is proportional to the weights of the falling bodies.[21] By considering the properties of turning wheels, he explained centrifugal tendencies in terms of an *inclinatio recte eundi* (inclination to move straightforward) in a fashion that is very close to Galileo and Descartes' later approaches. Further, Benedetti's theoretical connection of the mechanics of turning objects with the physical discussion of the Copernican motion of the Earth constituted a premise for further articulations of the problem in the *Dialogo sopra i due massimi sistemi del mondo* (1632) and *Le Monde ou Traité de la lumière* (written around 1632).[22] Benedetti's inquiry of falling bodies was the basis for Galileo's early exploration of this area, while his conceptualization of equilibrium is very close to that of torque. Moreover, Benedetti consistently applied a mathematical approach to physics. He relied on the mathematical instruments he had at his disposal, that is, geometry (not yet mathematical analysis). Just like Galileo and his school, Benedetti used the theory of proportions, derived from Euclid, as the key intellectual tool for the mathematical treatment of physical phenomena.[23]

The aforementioned contributions are substantial, and have justifiably been singled out by historians of science, but are not sufficient to comprehend the scope and the meaning of Benedetti's endeavor. In fact, his work shows great heterogeneity and the horizon of his interests reaches far beyond the boundaries of mechanics as they were established in the constitutive period for this field. Although he was convinced of the excellence of his own inquiry in mechanics as "science of machines,"[24] he did not isolate it from a wider mathematical and natural-philosophical investigation—nor did he isolate mathematics from practice and technology. Benedetti's first

[20] The standard reference work on Benedetti is Bordiga (1985). For a new general introduction to Benedetti's major work, see Omodeo and Renn (2018).

[21] See Maccagni (1967).

[22] Cf. Omodeo and Renn (2018), Chap. 4. Benedetti's importance for the development of the concept of inertia has been stressed by Koyré [1939] 1966, 47–60. As Koyré remarks (1966, 122): "Jean-Baptiste Benedetti est. très certainement le physicien italien le plus interessant du XVIe siècle; il est. aussi celui dont le rôle historique fut le plus important: en effet, son influence sur le jeune Galilée qui, dans son traité *de Motu*, le suit pas à pas, est. indéniable et profonde."

[23] Cf. Giusti (1993).

[24] In his major work, *Diversarum speculationum mathematicarum et physicarum liber,* 141, Benedetti introduced his section *De mechanicis* as follows (trans. Drake and Drabkin 1969, 166): "Now it is my desire to publish a few items that will, I believe, prove not unwelcome to those who concern themselves with mechanics, items which have never before been dealt with or have not been sufficiently well explained. I may thus either show my desire to be helpful or at least give some evidence of possessing a bit of talent and industry. And perhaps in this way alone may I leave behind me proof that I ever lived at all."

publications on mechanical issues appeared together with mathematical studies on Euclidean geometry. At the same time, he cast horoscopes—an activity requiring astronomical and computational competence—and was acknowledged for this by some of the leading astrologers of the time. The most reputed among them, Luca Gaurico, included in his famous *Tractatus astrologicus* (Venice, 1552) a nativity of Benedetti calculated by Benedetti himself.[25] In his mature years as a court mathematician in Turin, he wrote on apparently disparate subjects. They ranged from the theory and construction of sundials (e.g., *De gnomonum umbrarumque solarium usu liber*, 1574) to polemical writings on the reliability of ephemerides computation (e.g., *Lettera... intorno ad alcune nuove riprensioni, et emendationi contra alli calculatori delle effemeridi*, 1581), disputes on the quantitative proportion of water and earth in the terrestrial globe (*Consideratione .... d'intorno al discorso della grandezza terra et dell'acqua*, 1579), astrological reports and technical advice on measurement instruments.[26] His major work, *Diversarum speculationum mathematicarum et physicarum liber*, shows the unsystematic attitude of the author. Indeed, it is a miscellany of writings on the most varied topics and a document of his wide-ranging interests. The first book is a collection of mathematical exercises offering geometrical solutions to arithmetic problems, the second deals with problems of perspective relevant to architecture, the third with mechanics, the fourth with motion, natural philosophy and post-Copernican cosmology, and the fifth with Euclid's theory of proportions. The sixth book is a collection of epistles containing Benedetti's advice on countless topics: a project of calendar reform, navigation, machines, ballistics, engineering, geography, literature, mathematical philosophy and much more.

What is the common denominator of the great variety of subjects dealt with by Benedetti? What is the center around which they all gravitate? Is there one unifying principle behind the apparent disorder and heterogeneity? It should be emphasized that Benedetti first established his fame *as a mathematician*. His early treatment of motion by mathematical means was explicitly directed "against Aristotle and all philosophers" (*contra Aristotilem et omnes philosophos*). In his time "mathematics" had a wide scope. It comprised arithmetic and geometry, astronomy and astrology, and music, but also reached far beyond the boundaries of the *quadrivium* by encompassing optics, practical mechanics, architecture, and engineering. The expansion of mathematics into the fields of physics, natural philosophy, meteorology, and even metaphysics and epistemology was a crossing, a *franchissement*, of the disciplinary boundaries. Benedetti's time bears witness to several attempts to expand the boundaries of mathematics. Cardano, for one, claimed that geometry had the function of a universal logic fundamental to rational thought, and that the practical disciplines including statics, mechanics, and architecture were its subordinate fields of inquiry.[27]

[25] See Gaurico (1552, f. 76*r*).

[26] See, among other publications on this lesser-known scientific production of Benedetti, Roero (1997, 37–66) and Mammola (2014).

[27] See Cardano (1966) and Omodeo (2017b).

Benedetti's intellectual identity, however, proves much more complex than the corporate identity as a mathematician.[28] His pronounced titles vary. In a short biographical note accompanying the birth horoscope published by Gaurico he was indicated as "Phylosophus, Musicus, atque Mathematicus"; on 19 October 1589 he signed an astrological report cast for Carlo Emanuele as "Matematico e Astrologiaro"[29]; a lost manuscript collection of epistles of his was entitled *Lettere di Giovanni Battista Benedetti, Veneziano, matematico del Duca Emanuele Filiberto e Carlo Emanuele I,* but this reference is a later cataloging entry as indicated by the numbering of Carlo Emanule as "first"; contemporary admirers of his such as the Milan painter and poet, Giovanni Paolo Lomazzo, and the Danish astronomer, Tycho Brahe, called him "matematico" and "philosophus et mathematicus inprimis excellentem," respectively.[30] Probably, Brahe's designation of Benedetti as both philosopher and mathematician best captures the poles of his intellectual activity. Intriguingly enough, Benedetti generally dropped the title of "mathematician" keeping only that of "philosopher" in his publications. On the title page of his *magnum opus*, the *Diversae speculationes* (first edition, 1585), he appears as "patritius Venetus philosophus," exactly the same epithet that appears in *De gnomonum umbrarumque solarium usu liber* (1574). In the publications in the vernacular, he correspondingly appears as "filosofo del sereniss. duca di Savoia," for example, in the *Consideratione .... d'intorno al discorso della grandezza terra et dell'acqua* (1579). In this publication, his self-presentation as court philosopher is interestingly opposed to the designation of his intellectual opponent, Antonio Berga, as "filosofo nella Università di Torino," that is "university philosopher"—which is equivalent to *scholastic* philosopher. These references are telling either for Benedetti's self-perception or, to use an in-vogue expression, his *self-fashioning*.[31] In both cases, the image of court philosopher was his intended identity, whether reflected or purposely constructed (or a mixture of both). As was the case with Galileo, the Florentine courtier, the philosopher's social status and reputation was higher than that of the mathematician. This is why, among the conditions for Galileo's appointment as a courtier to the Medici, he regarded the designation "philosopher" as relevant.[32] As for the epistemological debates mirroring the disciplinary and social divides and hierarchies of the time, heated controversies began over the "certainty of mathematics." The determination of the degree of certainty of mathematics also concerned the legitimacy of using mathematics in physics.[33] In the case of Benedetti, the tension

[28] By "corporate" I refer to the *esprit de corps* of a group that considers itself a bounded entity whose interests are marked as separate from other groups. This particular meaning of corporation originated with the guild culture of the Middle Ages and it precedes the modern sense of a professional group or legal body.

[29] See Roero (1997, 57–58).

[30] Lomazzo (2006, 177): "Del Sig. Gio. Battista Benedetti Matematico"; Brahe (1916, 251–253).

[31] The expression stems from Greenblatt (1980).

[32] This theme is discussed in detail in Biagioli (1989, 1993).

[33] For an insightful treatment of the Renaissance debates on the certainty of mathematics, see Axworthy (2016, Chap. 2).

between his function as court mathematician and his identity as philosopher—and *patrizio*—underlies his science. While philosophical legitimacy was essential for the acknowledgment of the intellectual dignity of his endeavor, the practical dimension of mathematics remained fundamental for the social justification of his function as a court expert. These poles, socio-functional and identity-cultural, taken together—and only if considered together!—account for the *content* and *demarcation* of Benedetti's science, its *form* and its *legitimacy*.

7.3 The Social Position of an Early Modern Court Mathematician

For most of his life, Benedetti's social position was that of a courtier, or to be more precise, "a court mathematician." This was the setting in which Benedetti fully developed his scientific activity. After a few years in Parma with duke Ottavio Farnese, he entered the Savoy court in Turin to serve the duke Emanuele Filiberto and his successor, Carlo Emanuele, from 1567 up to his death in 1590. His commitment to the ruling family was sanctioned by the bestowal upon him of the privileges of imperial nobility in 1570. This also marked his social ascent from Venetian *patriziato*—a title he would never renounce—to higher aristocracy.

In those years, the territorial state of Savoy was going through a difficult transition toward a centralized "court society". After the restoration of the Duchy in 1559, after more than 20 years of French occupation, Turin became the new capital replacing Chambéry.[34] The dukes strove to transform it into a "modern" town, partly following the model of Florence, insofar as culture and prestige were concerned, but also the Spanish and French models, insofar as the suzerainty of the ruling family was concerned. Military force, administrative rigor, and international diplomacy were necessary for the survival of a buffer state between aggressive powers, France and Spain (as it bordered with Spanish Lombardy). Within a framework of *modernization* (which I would call a "passive revolution" as it implied the top-down importation of foreign models) taxation was reformed and the vernacular was imposed as the language of the administration. Following the Swiss example, obligatory conscription of all male subjects was imposed. Other models played a role, too, for instance Dutch military technology, which Emanuele Filiberto had come across during his campaigns as the chief commander of Charles V's imperial army in Flanders in 1556. New fortresses were built on the Alps and a citadel was erected between 1564 and 1566 in accordance with the most up-to-date precepts of military engineering.[35] Patronage attracted artists and writers as much as architects.[36] The most famous among them, Andrea Palladio, dedicated his *Quattro libri*

[34] For a comprehensive historical reconstruction of Turin in this time, see Ricuperati (1998).

[35] See Pollak (1991).

[36] See Doglio (1998).

dell'architettura (Four Books on Architecture) (Venice 1570) to the duke of Savoy and probably drafted the plan of the park of Viboccone for him, in which Benedetti is said to have constructed a fountain.[37] For reasons of prestige, as well as cultural hegemony in the field of education, the duke reopened the university, which had been closed during the foreign occupation, and attracted some of the most reputed professors of philosophy, medicine, and humanities of the time.[38]

Thus, Turin took the shape of a Renaissance capital in which the court was a hegemonic center from which powerful élites mediated between the ruler and the socio-political environment. Following the sociologist Norbert Elias, one can regard court society as a transitional social formation (*gesellschaftliche Figuration*) in European history between feudalism and industrial capitalism. A distinguishing feature of court society was the centralization of power and administration around the court. Its culmination was the absolutism of the *Ancien Régime*, prepared and imitated at a smaller scale by local efforts directed toward administrative and political centralization.[39] Although such social formations apparently depended on an individual sovereign taking all decisions (as much of the literature of the time on the *principe* and his privileges boasted),[40] it was rather a hierarchical system in which the group of experts surrounding the ruler constituted an oligarchy making the complex organization of modern states possible. Hence, the court was a political, economic and cultural hub, in which a class of experts mediated between the center of decision-making and the administrated periphery. Benedetti was one of these experts; his specific competence was mathematics. He shared the enthusiasm of his patrons (especially Emanuele Filiberto's) for mathematics and its perceived potency as an instrument for successful navigation in war and peace.[41] He was expected to supervise engineering and architecture projects, to produce instruments or machines for practical purposes, warfare, and recreation (such as fountains, sundials or nautical instruments). He was also expected to properly interact with the upper classes, respecting aristocratic etiquette and adhering to shared court values, norms and behaviors, first of all those linked to honor and prestige.[42] He had to act as the Savoy "expert" on pedagogical and technical issues linked to his profile as a mathematician and mathematical philosopher.

---

[37] Cf. Tessari (1993). On Benedetti's fountain, see Maccagni (1967, 353–354).

[38] See Catarinella and Salsotto (1998).

[39] See Norbert (2002, 73): "Durch das Bemühen um dis Struktur der höfischen Gesellschaft und damit um das Verständnis diner der letzten großen niche-bürgerlichen Figurationen des Abendlandes eröffnet man sich also mittelbar zugleich einen Zugang zum erweiterten Verständnis der eigenen berufsbürgerlich-städtischen-industriellen Gesellschaft."

[40] Doglio (2005).

[41] Emanuele Filiberto's passion for mathematics was well documented. Among others, the Venetian ambassador to Turin Giovanni Correr reported on it in 1566: "That duke is no man of letters but he loves the virtuosi. Hence, he has many of them by him; he likes to listen to their reasoning and he asks them questions. However, there is no subject that delights him more than the mathematics, as this science is not only apt but also necessary to the profession of military commander." This translation from is from Firpo (1983, 123).

[42] See Biagioli (1996, 193–238).

The primacy of courtly interests over those of science as an autonomous disciplinary endeavor can best be illustrated by considering Benedetti's correspondence. The last book of the *Diversae speculationes* is a large collection of letters, "Physica et mathematica responsa" (Answers on Physics and Mathematics). He introduced them with the following remark:

> To the reader.
>
> Nothing is more suited to virtue than being active and present through constant motion. Just like a shining star shimmering for the eyes of the spectators. It occurred to me that this or that person invited me with words or stimulated me with letters to dispute on those issues in which I am versed. In fact, I was devoted to mathematics and to highly philosophical speculations while I sojourned in the most splendid princely courts and illustrious cities, where many intelligent people displayed admirable curiosity, desire for knowledge and for discourse. I collected parts of those disputes and answers, because I judged them to be worth transmitting. I planned to reread and revise them, if I had some spare time, [having in mind two goals]: on the one hand, that science itself circulates as much as possible and grows; on the other hand, that the names of those excellent men, who stimulated me with their questions, are made immortal, as much as it is in my power, and that I can lead others to follow their example, to abandon the idle sloth (which is able to corrupt even the most talented) and start investigating, exchanging, and discussing serious subjects that could prove useful at some point and worth disseminating. For the time being, please, approach our efforts with a benevolent and judicious attitude.
>
> Regards.[43]

This quotation shows the occasional character of the topics dealt with in Benedetti's booklets, small treatises and epistles, his concern for applied knowledge, his recognition of the importance of dissemination, as well as the celebratory intention of his publication (to render the dedicatees immortal). The epistolary collection is itself the tangible witness of a network of social relations. In a sense, Benedetti acted as a cultural ambassador for the ruling family. Although it is not possible to identify all of his correspondents, one can say with certainty that most of the epistles are tied to the northern Italian environment, specifically to Turin and Venice, but some were also directed north of the Alps.[44] Many correspondents were in fact Piedmontese or Venetian patricians, courtiers, and artists linked to one of

[43] Benedetti (1585, 204, my translation): "Ad lectorem. Ut nil magis virtutis est. proprium, quam agitari, et incessabili motu prodesse. Ac veluti fulgidum sydus ante oculus spectantium commicare. Ita mihi mathematicis iisque maxime philosophicis speculationibus dedito, saepissime, ut in principium summorum aulis, et amplissimis civitatibus degenti, ubi multa semper Nobilium mira curiositate, sciendi desiderio, et conferendi cupiditate referta, versantur, ingenia, contigit, modo ab his, modo ab illis, aut verbis tentari, aut literis provocari ad disserendum, de his, in quorum studiis versamur. Quarum concertationum et responsionum, quoniam non omnino indigna existimavi, quae memoria comendarentur, partem aliquam apud me conseruivi. Ubi vero per ocium licuit, relegi, ac tandem de manu mittere decrevi. Tum ut scientia ipsa quo magis diffundetur, crescat; et quicquid valeo, sine invidia in communem utilitatem conferam. Tum ut virorum praestantissimorum, qui me suis interrogationibus excitaverunt, quantum in me erit, gratitudine ergo, nomina reddam immortalia, et eorum exemplo alios, ocio sordidiore abiecto, quod solet iurialium praecipue excelsa ingenia corrumpere, ad sciscitandum conferendum, et disserendum, de rebus seriis, et quae usui aliquando esse possint, et quandoque evulgari mereantur, alliciam. Tu interim nostris laboribus fruere, et nostram diligentiam boni, et aequi consule, et Vale."

[44] See Mamino (1989, 429–449).

these two centers; sometimes to both, as was the case with ambassadors (e.g., Domenico Morosini, Pietro Pizzamano, and Francisco Mendoza).[45]

The epistles are not organized chronologically (actually the dates are almost always omitted), but according to the importance of the addressees, who might be already dead at the time of publication. The first letter was addressed to duke Emanuele Filiberto, who had already passed on, the second to his son Carlo Emanuele, and the following four letters to the powerful nobleman Andrea Provana of Leynì. The topics are linked to Benedetti's role as court mathematician and mathematical advisor. The first of the epistles is an expert report on the reform of the calendar; the second deals with a geometrical problem (the determination of the circle circumscribing a given square) that probably emerged from the lessons imparted to the prince; letters from three to six address practical problems of navigation (such as the use of astronomical instruments at sea), and so on and so forth.

Benedetti's scientific epistles aimed not so much at exchange with other scholars as giving advice to a wide range of people beginning with the Savoy élites. There is great variety in the status and professions of the correspondents—they included mathematicians, physicians, jurists, philosophers, diplomats, experts of military art, architects, artisans, constructors of instruments and fortifications, surveyors and astrologers.[46] Benedetti was not primarily concerned with establishing a connection with the Republic of Letters. Quite aristocratically, he did not regard himself and his activity as part of a learned network but rather as the center of court interactions. This center-periphery structuring of his network is emblematic of the implementation of a rising "knowledge economy," a modern bureaucratic rationalization in Weber's sense of a hierarchical organization directed by experts.[47] In fact, the centralizing character of court society is mirrored by his acting as a sort of *technocrat* who distributed suggestions and advice from an advantaged position.

Clearly, Benedetti's social position was that of a court mathematician, a functionary acting as an expert in a wide range of realms that could fall under the compass of mathematics. Seen in this light, the mathematical formalization of practical knowledge, in particular in the realm of machine technology, engineering and architecture, corresponds to a social drive toward labor division. The division between the intellectual work of the scientific expert and that of the practitioners was essential for a progressively centralized society such as the court society. More in general, this division can be read against the background of the advance of capitalist society, in which the progressive separation of science and practice by means of labor division—later the distinction of pure and applied sciences—was the result of specialization and abstraction. As a consequence of this process, the social roots and the aims of science later escaped the self-perception of scientists them-

[45] See Cecchini and Roero (2004, 31–66).

[46] See Cecchini and Roero (2004, 31–66).

[47] For a stimulating Weberian treatment of institutional changes induced by bureaucratic rationalization (along with market commodification) in the early modern period, cf. Clark (2006), in particular the methodological considerations on 8–10. The reference work is Max Weber, *Wirtschaft und Gesellschaft: Grundriß einer verstehenden Soziologie* (1921).

selves, as the social functions became mediated by society as a whole, in a context in which the separation secures the subordination of practical labor to the intellectual.[48] In the *longue durée*, the scientific codification of practical knowledge occurring during the Renaissance was a step in a process of separation of theoretical knowledge and practical skills that would reach its acme in the age of the industrial revolution.[49]

In regards to content, the social horizon of Benedetti's scientific activity depended on the requests linked to his function as mathematician. His institutional duties immediately explain the occasional character of the themes he dealt with in his *various speculations, mathematical and physical*. The fragmentariness of his scientific work can be seen as a *formal* limitation resulting from Benedetti's courtly interactions and lack of an *academic* interest for systematization. He evidently did not acknowledge systematization as an *epistemic value*.[50] Zilsel maintained that systematization was instilled into modern science by scholastic education, which Benedetti had in fact not received. Indirectly, the *social roots* of his science can be traced back to the Hessenian fields of communication and transport (navigation, astronomy and geography), industry (machines technology, architecture, engineering), and war (military architecture and ballistics). All of these economic drives loom large over Benedetti's production. Ultimately, the social justification of his work rested on the economic and technological relevance of mathematics, which motivated princely support.

Benedetti's *position*, however, is less than half of the story. Taken in isolation, his function as a court mathematician cannot account for crucial aspects of his scientific endeavor, such as his steady engagement with philosophy and his corresponding development of a physical-mathematics at odds with Aristotelian natural philosophy. Many of the issues he dealt with *directly* descend from *superstructural* (I should say, "cultural") concerns, for instance the enhancement of mathematics at the expense of humanistic literacy and scholastic logic, or the treatment of problems specifically derived from the internal disciplinary debates in the fields of geometry, mechanics, and physics. In short, Benedetti's scientific activity was not uniquely determined by the social constraints connected with his social position but was also shaped by intellectual and political agendas. It is now time to discuss these tendencies in detail.

---

[48] For a clear exposition of these functional aspects of science in capitalist society one can still rely on the classic of Marxist sociology from Nikolai Bukharin, "Theory and Practice from the Standpoint of Dialectical Materialism." See Bukharin (1931, 9–33).

[49] Cf. Baracca et al. (1979).

[50] Cf. Daston and Galison on "epistemic virtues" in *Objectivity* (2007, 39–42). I borrow the concept to stress the rootedness of an epistemological requirement such as systematization in the social environment of the "knower." For a similar perspective, see also Kuhn (1996, Postscript-1969, 184–186).

7.4 Benedetti's Political and Intellectual Identity: *Patrizio* and Philosopher

Benedetti stemmed from the Venetian town aristocracy. As a member of the upper classes he received his education privately, outside the universities he disparaged in his early writings, and refined his culture with the frequentation of the lively intellectual circles of his hometown. His *juvenilia*, published in Venice—the *Resolutio omnium Euclidis problematum* and the *Demonstratio proportionum motuum localium contra Aristotilem et omnes philosophos*—show his sense of aristocratic superiority toward university professors, the "Aristotelians and philosophers" he deemed incapable of dealing with mathematics and physics. In the introduction to his first publication, he boasted of his autonomy from *vulgar* university teaching:

> Until now I have advanced without any mentor or teacher (under the guidance of God). I have never frequented any gymnasium or school. I have not learned what the vulgar (I mean this word without arrogance) use to estimate erudition, [such as limiting it] to the time spent at school, thus setting an end to learning when the seven years [of regular studies] are ended. As long as I live, I will continue [learning].[51]

As has been argued, the rejection of scholasticism was an important drive in the scientific culture of the Renaissance. In this specific case, Benedetti's criticism is not the artist-engineer's protest against professorial bookish culture; rather, it pays witness to his aristocratic attitude. This attitude certainly constituted an advantage for his speculations in terms of intellectual freedom and independence from authority; however, as I stressed, this also constituted a limitation as far as systematization is concerned.

He not only distanced himself from university scholarship but also from the other major cultural tendency of his time: humanistic literacy. In the *Resolutio omnium Euclidis problematum*, he showed off his distaste for rhetoric, to which he opposed the simplicity and straightforwardness of mathematics.

> Mathematics does not require much [stylistic] splendor. If some language expert tried to improve its elegance, this would have no value, because a change of the mathematical language and of the scientific terminology could easily confuse the sense [of the reasoning] and render everything obscure. Therefore, I will follow the scholarly tradition and use plain words in my demonstrations, as I disapprove of deceptive elegance. In this respect, I follow the steps of the ancients who taught the sciences and the subjects themselves using plain words. Petty teachers (indeed, charlatans and babblers) corrupted this manner of teaching. Although they do not understand the subject, their babbling obtains the highest praise by the vulgar who regard them as learned scholars. This should not be surprising, considering that the most perfect and distinguished expertise in the sciences is attained by very few, despite the fact that many people write a great deal in all kind of sciences and arts, babbling a lot and capturing the attention of the uneducated with illusions and bombastic words.[52]

---

[51] Benedetti (1553, f. 5*r*).

[52] Benedetti (1553, f. 5*v*): "Adde quod Mathematicae disciplinae, neque tantum requirunt splendorem, neque si quis peritus linguarum contendat ad elegantiam rem reducere, egregium quid effecerit, quia mutato usu Mathematicae loquendi, ipsiusque scientiae terminis, sensum facile perturbaverit, et ex nihilo nihil apprehensum obtinuerit. Quare morem scholarum sequutus, obstenta-

The counterpoint to Benedetti's rejection of rhetoric was the limitations of his communication skills. His style reveals a want of literary training. Its dryness is far from both the Italian elegance of Galileo and the philological purity of Commandino's mathematical school of Urbino. To be sure, Benedetti's lack of any classicist reverence can account for his eclectic and creative mixing of ancient, medieval, and Renaissance approaches; however mathematical humanists did not see this as a virtue. Rather, they accused him of "impurity". This explains, at least in part, the extremely negative portrait of Benedetti by the member of the Commandino school Bernardino Baldi. In his collection of short biographies of mathematicians of all epochs, *Cronica de matematici* (posthumous, 1707), he stressed the profound divergence between the *most exquisite scholars* (*più esquisiti*) and Benedetti's *impure* approach (*metodo*):

> The Venetian Giovanni Battista Benedetti occupied himself with mathematics, a field in which he served the dukes of Savoy. He wrote a book on gnomonic, which dealt with many proofs belonging to this discipline. However, he has been criticized by the most exquisite scholars for not having respected the method and the purity of explanation which mathematics requires and which was respected by the Greek masters and by their followers. He also wrote some light things, of no great import.[53]

The (mathematical) philosopher—Benedetti's self-presentation as an opponent of scholasticism, on the one hand, and of humanistic literacy, on the other, did not result in a corporate identity as practical mathematician, or engineer. In learned controversies, Benedetti *showed* the power of his mathematical approach and *demonstrated it* through philosophy. Accordingly, he presented himself as a *philosopher* and, in order to substantiate his claim, he was forced to enter the cultural arena of philosophical controversy and construct his own identity as a *mathematical* philosopher. This meant openly defending the epistemological status of mathematics. He went so far as to claim for this discipline a rank higher than Aristotelian natural philosophy, metaphysics, and ethics owing to the certainty of its demonstrations. Thus he tackled the issue *De philosophia mathematica* (On Mathematical Philosophy) in a letter to the Venetian *patrizio*, Domenico Pisani:

> I am surprised that, although you are well versed in Aristotelian philosophy, nonetheless you make a distinction between the philosopher and the mathematician in your writings, as

---

tione elegantiae explosa, verbis nudis in demonstrationibus usus sum, hac in parte veterum vestigia sequutus, qui nudis verbis scientias resque ipsas docebant, quem modum docendi, nobis devastarunt scioli vel potius circulatores, garruli, rebus ipsoque iudicio destituti, garrulitate siquidem apud vulgus, laudem summam consequuntur, et pro doctis circunferuntur, nec mirum, cum scientiarum perfecta exquisitaque perita, paucissimis detur, non obstante quod multi permulta de omnis generis et scientiis et artibus scribant, permultaque garriant, fucis suis, et ampullis imperitorum oculos perstringentes [...]."

[53] Baldi (1707, 140): "GIO[VANNI] BATTISTA Benedetti veneziano attese alle matematiche, nelle quali servì i Duchi di Savoia. Scrisse un libro di gnomonica, il quale toccò molte cose appartenenti alle dimostrazioni della detta disciplina, se non che viene ripreso da più esquisiti di non haver'osservato quel metodo, e quella purità dell'insegnare, che ricercano le matematiche, et è stato osservato dagl'ottimi Greci, e dagl'imitatori loro. Scrisse anco alcune altre cose leggiere, e di nessun momento."

> if the mathematician were not as much a philosopher as the naturalist and the metaphysician. In fact, as far as the certainty of his conclusions is concerned, he deserves the title of philosopher much more than them.[54]

This reference to the certainty of mathematical reasoning bears witness to Benedetti's acquaintance with the ongoing methodological controversy over the dignity of mathematics and its validity. He agreed with his correspondent, the Padua professor Pietro Catena, that mathematical certainty has an intelligible character.[55] As he wrote to Pisani,

> Actually, you are not the only one who makes this mistake, but this is graver in consideration of the fact that although you [Aristotelians] even label ethics as a philosophical discipline, you do not acknowledge that the divine mathematical sciences also should be adorned with the name of philosophy. In fact, if we consider this name more attentively we will clearly see that it is in itself more suited to the mathematician than to anyone else, since none of the others is more certain in his affirmations than the mathematician. And no one is more driven by the love of science in his cognition. This is evident. In fact, [the mathematician] does not rely on the senses nor accepts any presupposition that is not so true and evident to the intellect that no power whatsoever could show that it is false.[56]

During the Renaissance, mathematical philosophy was connected by definition to the reassessment of Pythagoreanism, no matter how controversial, ambiguous, and deprived of direct sources this reappraisal was. On his adherence to Pythagorean views, Benedetti mockingly wrote to the Savoy courtier, Paolo Capra: "If the souls' transmigration imagined by the father of Italian wisdom, Pythagoras, were true, I believe that your soul and mine were once the souls of hunting dogs."[57] At that time Copernicus's planetary system also counted as a Pythagorean invention.[58] Benedetti explicitly praised Pythagoras's discoveries in astronomy in the letter to the reader opening the *Diversae speculationes*, as an anticipation of his treatment of the Copernican system. Benedetti expands on cosmology in the (anti-Aristotelian) philosophical section of the work, the fourth book entitled *Disputationes de quibusdam placitis Arist[otelis]*.[59] He also defended the Pythagorean doctrine of the music

---

[54] Bendetti (1585, 298): "Miror quod cum in Aristotele sis versatus, in tuis tamen scriptis philosophum a Mathematico separes, quasi mathematicus non sit adeo philosophus, ut est. naturalis, et metaphysicus, cum multo magis quam ii philosophus sit appellandus, si ad veritatem suarum conclusionum respiciamus."

[55] Cf. De Pace (1993, 228–229).

[56] Bendetti (1585, 298): "Verum quidem est., te in huiusmodi errore solum non versari; sed gravius est., quod cum vos videatis etiam res morales sub philosophiae appellationem cadere, non animadvertatis divinas scientias mathematicas etiam philosophiae nomine ornandas esse. Quod si eiusdem nomen penitius considerare velimus, inveniemus aperte, mathematico magis illud ipsum quam cuilibet alio convenire, cum nullus ex aliis tam certo sciat id quem affirmat quam mathematicus, neque aliquis sit, qui in cognitionis, et scientiae cupiditatem magis ducantur, ut aperte patet, cum nec etiam ipsi sensui det locum, neque aliquid praesupponat, quem non sit ita verum et intellectui notum, ut nulla quaevis potentia, illud esse falsum ostendere queat."

[57] Ibid., 285: "Si vera esset animorum illa transmigratio quam sibi Italicae sapientiae Pater Pythagoras effinxerat; tuam, meanque existimarem animam canis, quandoque venatici fuisse."

[58] See "The Invention of the Pythagorean Cosmology" (Omodeo 2014a, 167–170).

[59] Bendetti (1585, f. A3*v*). On his cosmology, see Di Bono (1987, 283–300), and Omodeo (2009).

produced by the celestial bodies against Aristotelian criticism, by interpreting this doctrine not as a reference to the production of physical sounds but rather to mathematical harmony introduced into the world by the Divine Providence.[60]

In the same book of the *Diversae speculationes*, Benedetti detailed his criticism of Aristotelian natural philosophy. He first developed an alternative mathematical and atomistic theory of motion, asserting against Aristotle that void is necessary to local motion. He analyzed downward and upward motions in different media making use of the Archimedean concept of buoyancy to account for the dependence of vertical motions, in particular free fall, on the specific weight of the bodies rather than an absolute weight; he employed the (Euclidean) theory of proportions to interrelate the velocities, volumes, and surfaces of moving bodies; moreover, he clarified his basic natural conceptions on issues such as the existence of void, the definition of space and time, and the attainability of actual infinity in nature; he rejected Aristotle's theory of natural places and reflected on the connection of post-Copernican astronomy with physics, defending heterodox views such as the infinity of space and the plurality of worlds, specifically equating the Earth to the other planets of the solar system.[61] Benedetti's rejection of key elements of Aristotle's *Physics* was certainly acute and his revision of crucial concepts of natural philosophy was in line with the most original trends of the Italian Renaissance. However, his endeavor remained limited: his capacity to demolish well-established physical views and to solve specific problems was not accompanied by the implementation of a unitary vision. In this respect, scholastically trained contemporaries of his proved much more efficient in the *pars construens*, embodied by monuments of natural philosophy such as Bernardino Telesio's *De rerum natura* (1565, 1570 and 1586), Giordano Bruno's Italian dialogues (1584–1585) and Frankfurt poems (1591), or Francesco Patrizi's *Nova de universis philosophia* (1591).

The Courtier – As a young independent intellectual, Benedetti wrote in a harshly polemical tone, accusing his opponents of slavishly following the authority of their master, Aristotle, instead of relying on solid arguments based on mathematics and experience.[62] Later, as a courtier in Turin, he realized that controversies with other scholars belonging to the town élite should not escalate and he became more

---

[60] Bendetti (1585, 190–191): "Pythagoreorum opinionem de sonitu corporum coelestium non fuisse ab Aristotele sublatam."

[61] For a detailed discussion of Benedetti's *foundations of physics*, see Omodeo and Renn, *Science in Court Society* (in press), ch. 7.

[62] In the preface to the second edition of the *Demonstratio proportionum motum localium* (1555), Benedetti wrote (my translation): "They could not concede that Aristotle was mistaken in any way, because they do not regard him as a human being. Rather, they confer upon him the celestial condition of a pagan divinity. And they see even slight disagreement as a sin. Therefore [they believe that] I committed (and still commit) heresy if, according to their judgment, I do not follow the pure and authentic mind of Aristotle's doctrine in any manner." For the original text, see Benedetti's original text in Maccagni (1967, 21): "Ne vero Aristotelem ullo modo errasse concederent, cum illum non infra humanae conditionis terminum habeant, sed potius veluti coeleste quoddam numen sibi proponant, censeantque nefas esse si vel latum quidem unguem ab eo quis dissentiat, in hac potius haeresi fuisse, ac etiamnum esse, ut me germanum et genuinum sensum Aristotelicae opinionis nequaquam ex authoris mente assecutum existiment."

moderate. Benedetti's anti-Aristotelian views were thus softened by courtly etiquette. He was part of the circle of intellectuals gathering around the duke, which the professor of jurisprudence and Benedetti's correspondent, Bernardo Trotto depicted as follows:

> These learned men, played by the Prince like well-tuned musical instruments, immediately give out their specific sounds with words. And they give it their best to be clearly understood in conversations, to please the others with good arguments and to convince them of their opinions. It is like the consonance of truth. In fact, everyone says what one knows or, at least, considers to be true. Hence they discuss natural issues and at times moral ones and mathematical ones. In conclusion, one can regard him [the prince] as Apollo surrounded by the Muses by the water spring that was originated by a foot hit of the winged horse.[63]

Benedetti entered courtly conversations on the most varied issues and confronted Aristotelian views defended, among others, by the court physician Giovan Francesco Arma, or university professors such as the philosopher, Antonio Berga. In informal and polite meetings, Benedetti frankly expressed his views and objections. Although divergences of opinions emerged, education always prevailed over intellectual disagreement.

This *genteel* climate can best be captured by publications such as Arma's report to Emanuele Filiberto about a discussion on meteorological problems that took place between prince Carlo Emanuele, Benedetti, the professor of mathematics, Francesco Ottonaio, Berga and himself during the prince's meal for three consecutive days. The report, printed in 1580 under the title *Proposte tenute co'l Serenissimo Prencipe* (Issues Discussed with the Very Serene Prince), had the form of a poem. To get an idea of the tenor and the contents of such courtly conversations one could consider a quarrel over the 'solar attraction' of terrestrial elements that opposed Arma to Benedetti, Berga and Ottonaio:

> The day after [the second day], Mister BenedettiAnd Mister Berga, along with OttonaioExpressed opinions that are far from mine:That the Sun attracts everything to itself with its great brightness.As if it had hands.[64]

Arma objected that the thesis that the Sun attracts other bodies would have disruptive consequences for the *solari*, the inhabitants of the Sun. The reader apprehends that Benedetti agreed with him on this remark, but no further details are provided. Ottonaio mentioned the vapors brought upward by the Sun's action. Following the associations of ideas, the next dinner table topic concerned natural places and elemental displacements, with special reference to the motion of vapors.

---

[63] Trotto (1625, 2–3): "[...] questi huomini saputi, tocchi dal Prencipe, come instrumenti musici bene accordati, subito rendono ciascuno il suo suono con le parole et quanto meglio possono procurare d'essere intesi discorrendo, e di dar diletto con le buone ragioni, et anco di tirare gli altri al suo parere, come ad una consonanza della verità: perché ognuno dice quello ch'egli sa o crede almeno sia vero. E quindi si veggono trattare hor cose naturali, hor morali, hor mathematiche. Sì che egli quasi come uno Apolline si può dire, che sta fra le Muse, intorno al fonte, che uscì dal colpo del piede del cavallo alato."

[64] See Arma (1580, f. A2*v*): "Il Signor Benedetti l'indomani/ Col signor Berga, insiem' a l'Ottonaglio/ Forn'in pensier' a me d'assai lontani,/ Che'l Sol tirass'a sé com grand'abbagio/ Ogni cosa si com'havesse mani."

On the third day, prince Carlo Emanuele asked his learned tutors for an explanation of the origin of lightning, and why we perceive the light before the thunder. Arma's *physiological* answer was that our sight is quicker than our hearing was received with skepticism by his colleagues. However, no details of the discussion are reported except that they concerned the theory of perception, light, and optics. At the end of the 3 days, Arma wrote down all opinions for duke Emanuele Filiberto and all courtiers confirmed the reliability of the poetic report:

> All of this was presented in written formTo His Highness, reporting all speeches.Dr Berga confirmed it [the rightness of the report].Benedetti did the same.After that we discussed other issues,Hidden things and their effects.[65]

Other publications from those years mention similar courtly conversations taking place at table. For instance, the university professor of philosophy and medicine, Agostino Bucci reported in the preface to his medical book *Disputatio de principatu partium corporis* (Disputation on the Superiority of Parts of the Body) (1583) on a physiological controversy that was sparked off by his anatomical theories. Carlo Emanuele gathered his learned courtiers to discuss vexed Aristotelian-Galenic questions concerning the 'superiority' of the heart or of the brain, and the location of the soul in the body. The participants were not only physicians, as Bucci also mentions two 'mathematicians,' Ottonaio and Benedetti.[66]

The complaisant tone of courtly controversies signals the *esprit de corps* of the Savoy cultural élites prevailing over intellectual disagreements and creating a climate of polite open-mindedness. However, as soon as the reputation of the scientific *intelligentsia* was attacked from outside the court and the academic institutions, moderation gave way to the crudest polemics. In 1578, the court physician Arma was the target of a denigrating pamphlet in twenty points, composed by an anonymous author criticizing his views on the comet of 1577–1578. A harsh polemic thus started in which none of the parties renounced personal offences and insults, which went far beyond the subject matter, as is documented by an apology for Arma penned by Monsignor Della Torre, *Difesa de la πωγωνία d'il Dottore Arma* (Defense of the *Bearded Comet* by Dr. Arma). The author, who was linked to the court, published his pamphlet under the pseudonym of '*filosofo stravagante*' (extravagant philosopher).[67] Benedetti himself had to defend astrological prognostications cast for the court against the attacks of Benedetto Altavilla, a critic of the reliability of astronomical calculation from Vicenza. In his corrosive response, he singled out the mathematical shortcomings of his opponent and disqualified him. Benedetti's polemical writing appeared first in Italian in the form of a public letter to the jurisprudent Trotto, *Lettera per modo di discorso... intorno ad alcune nuove riprensioni, et emendationi, contra alli calculatori delle effemeridi* (Letter in the Form of a

---

[65] Arma (1580., f. A4*r*): "E tutto quest'in scritti fu donato/ A Sua Altezza, con tutti soi detti./ E fu dal Dottor Berga conformato./ Il che fece'l signore Benedetti./Fu poi d'altre proposte ragionato/ E de gl'occolte cose, e soi effetti."

[66] Bucci (1583, 7–8). Cf. Mammola (2013, 6–8).

[67] Omodeo (2008).

Discourse... on Recent Criticism and Corrections against Ephemerides Calculations) (Turin 1581). A Latin translation, as *Defensio ephemeridum* (A Defense of Ephemerides), was later included among the epistles of the *Diversae speculationes*.[68]

The Savoy Intellectual – The participation in courtly life required adherence to the hegemonic project of the ruling family in a period of transition from feudal society to centralized state administration. In Benedetti's case, the supervision and advice on technical matters was not only functional in regards to socio-economic needs; it also contributed to the establishment of the court as a cultural center. With respect to the cultural hegemony of the court, patronage, education and cultural diplomacy deserve particular mention. In all of these realms, Benedetti was actively engaged.

Political issues directly informed some of Benedetti's scientific writings, for instance the first of the scientific epistles of *Diversae speculationes* dealing with the calendar reform promoted by the Pope. The problem was not merely a technical one as it concerned Catholic hegemony in Europe in a time of confessional divides and clashes. Benedetti's awareness of the importance of this issue is reflected in his decision to sign his contribution on this matter, which had first appeared in 1578 as *De temporum emendatione opinio*, neither as a mathematician nor as a philosopher but as *patritius Venetus*, thus stressing his noble origin. Another instance of politically informed writings are his epistles on navigation instruments and techniques addressed to Provana, the aristocrat entrusted with the command of the Savoy ships in the battle of Lepanto.

Moreover, Benedetti contributed to courtly diplomacy by interacting with scholars and courtiers. He also assessed the quality of mathematical works dedicated to the dukes. As appears from his correspondence with Patrizi, he mediated between the famous Ferrara philosopher—and professor-to-be of Platonic philosophy at the *Sapienza* in Rome—and the dukes, on the occasion of the dedication of *Della nuova geometria libri XV* (Fifteen Books on the New Geometry) (Ferrara 1587) to Carlo Emanuele. Patrizi expressed his gratitude to Benedetti for handling on a copy to his patron (letter from Ferrara, 6 April 1587):

> Very Magnificent and Excellent Mister,
>
> I rejoice with your Lordship that you recovered from sickness quicker than believed. And I am very thankful to you for presenting my book to the very serene Prince and promising to inform me about his remarks after he has read it. If by chance the book will be forgotten, due to his many duties [negozii], I hope at least that you will remember me. If his High Serenity will give some sign that he appreciated it [my book], I will be very glad and I will be particularly grateful to your Lordship for your benevolence.[69]

[68] Omodeo (2014a, 142–149) and Omodeo (2014b).

[69] Patrizi (1975, n. XXVII, 53): "Molto Magnifico et Eccellentissimo Signore, mi rallegro con Vostra Signoria, che più tosto che non credea si è rilevata dal male, e li rendo moltissime gratie dell'haver presentato il mio libro a quel Serenissimo Prencipe, e ricevuto il favore, che Ella mi avvisi ciò che haverà detto, dopo che l'havrà letto. Et se per sorte per li molti negozii il libro andasse in oblio, spero da Lei il rimedio di un poco di ricordanza, la quale, se partorirà alcun segno che Sua Altezza Serenissima l'habbia havuto caro, mi sarà carissimo e tutto l'obbligo l'haverò a Vostra Signoria e all'amor suo verso me."

In exchange, Benedetti sent to Patrizi a copy of his discussion on the relative sizes of the elements of the earth and of the water.[70]

Patronage of the sciences and the arts played a role in the construction of cultural hegemony in the age of court society. Turin was visited by the most celebrated poets of the time, in particular Torquato Tasso in 1578–1579, and Giovan Battista Marino in 1608–1615.[71] Benedetti, for his part, contributed to the reputation of the dukes as an acknowledged scientist. The leading mathematician and astronomer of the *Collegio Romano*, Christopher Clavius, extolled his merits in a eulogy for the Savoy family, in the 1589 edition of his commentary of Euclid's *Elements*. In the dedicatory epistle, directed to Carlo Emanuele, he called Benedetti a "*scientissimus rerum Mathematicarum*" (very expert scholar in mathematics).[72] Benedetti himself dedicated his most significant publications to his patrons. Overall, the *Diversae speculationes* had an celebratory intent. It appeared in a series of elegant volumes extolling the magnificence of the court and Savoy, such as the court historian Emanuele Filiberto Pingone's *Augusta Taurinorum* (1577), a celebration of the capital, *Inclytorum Saxoniae Sabaudiaeque principum arbor gentilitia* (1581), on the rulers' lineage, and *Sindon evangelica* (1581), on Christ's shroud, recently transferred from Chambéry to Turin. The *Diversae speculationes* made the proficiency of the court intellectual, Benedetti, and his mathematical skills publicly visible. It also bore witness to the intensity of the cultural debates going on in Turin or connected with other centers. Its miscellaneous and epistolary form was suitable for displaying the variety of Benedetti's interests and for praising his patrons, supporters, friends, and colleagues as they were named one by one as dedicatees and addressees of the sections and letters.

Education was another pivotal element in Savoy cultural politics. Emanuele Filiberto did not leave too much room for ecclesiastical educators such as the Jesuits, who were in that time the major advocates for a religiously conformist formation in Catholic countries. The dukes' choice of lay tutors for his offspring, namely Benedetti and the mathematician Bartolomeo Cristini after him, is indicative of a political choice representing intellectual independence. The same desire for political autonomy applies to the dukes' university politics in a time in which the most contested field of post-Tridentine confessional and cultural struggles was academia. At an international level, the Jesuits proved the most engaged confessional group capable of implementing a religiously conformist educational system.[73] When Emanuele Filiberto reopened the Savoy university, first in Mondovì (1560) and later in Turin 1566), the Jesuits immediately opened colleges in both towns. Leading Jesuit intellectuals such as Antonio Possevino and Roberto Bellarmino were sent to

[70] Patrizi to Benedetti (Ferrara, 18 January 1588), Patrizi (1975, n. XXX, 57–58).

[71] See Doglio (1998, 621, 625).

[72] See Clavius 1589, ff. *4*r*–*5*r*.

[73] The case of the German Empire, where the Jesuits virtually obtained the monopoly of university teaching in Catholic countries, is telling. See Hellyer (2005). A thoughtful study on the political rationale behind the establishment of Jesuit education in an age of *Konfessionalisierung* and the wars of religion is found in Gui (1989).

Piedmont—a territory "dangerously" close to Calvinist Switzerland in the eyes of the Roman authorities. Initially, Emanuele Filiberto supported the Jesuits and let them occupy the chair of grammar at the expense of the renowned humanist Giovanni Battista Giraldi Cinzio of Ferrara, and almost accorded them nine chairs. However, the project did not succeed. It was met with fierce opposition by the town and the university, who thought their autonomy was being menaced.[74] Emanuele Filiberto himself changed his mind as his trust in the Jesuits fluctuated, just as his relations with Rome did. These relations were especially stained by the Cavour edict of tolerance, delivered in 1561, which guaranteed the religious freedom of his protestant subjects in the Alps. The promulgation provoked vehement protests from Rome, as well as the appointment of the intransigent Cardinal Inquisitor, Michele Ghisleri, to the diocese of Alexandria as Bishop of Mondovì, with the aim of reinforcing religious conformity. The relations between this champion of orthodoxy and the pragmatic politics of the Savoy dukes were tense, including after Ghisleri was elected pontiff as Pius V in 1566. The political issue of the independency of Savoy from excessive pontifical interference must have affected Emanuele Filiberto's decision to exclude the Jesuits from the university and the reintroduction of the teaching of humanities. Ultimately, the reasons for the defense of university autonomy in Turin are similar to those concurrently discussed at the *Sorbonne* and Padua. The loyalty of public teaching institutions to the King of France and to Venice, respectively, seemed to be better secured by marginalizing or even banning the Jesuits, as they were seen as direct representatives of papal interests.[75]

There is indirect evidence that Benedetti sided with the humanists in their criticism of the Jesuits. Patrizi, who belonged to the same Ferrara cultural environment as Giraldi Cinzio (the professor removed to make room for the Jesuits), asked Benedetti's support for the candidacy of his friend Giovanni Giacomo Orgiazzo for the position of professor of humanities in 1583.[76] The same epistle bears witness to the fact that Benedetti advised the dukes in matters concerning university policy. The sixteenth-century biographer of Emanuele Filiberto, Giovanni Tonso, even included Benedetti among those who made the University of Turin eminent (*Qui viri insignes publice docuerint*).[77] Another contemporary of Benedetti, the mathematician Cristini, mentioned him as one of those whom Emanuele Filiberto recruited in order to increase the prestige of his university.[78] Still, the role he played in the academic life of Turin remains obscure due to the lack of documentary evidence.[79]

[74] See Grendler (2002, 42–44).

[75] Cf. Omodeo (2014c).

[76] Patrizi to Benedetti (Ferrara, 21 March 1583) in Patrizi (1975, 39).

[77] See Tonso (1596, 141): "Neque vero liberalium disciplinarum omniumque artium colendarum quam susceperat cogitationem unquam deposuit: nam et publicum earum Gymnasium pro tempore in oppido Monteregali instituit: et qui viri in quacunque scientia excellerent undique conquisuit. [...] Mathematicos illustres Franciscum Othonarium, et Io. Baptistam Benedictum Venetum."

[78] Bordiga derived this information from a manuscript of Cristini's preserved in the Biblioteca Marciana in Venice. See Bordiga (1985, 596–597).

[79] The historian of Piedmontese Universities Silvio Pivano already complained in the 1920s about the lack of relevant documents. See Pivano (1928, 19–22).

Another indicator of Benedetti's attitude toward the Counter-reformation and the confessional quarrels of his time emerges from his approach to the calendar reform, in a time in which the pope's political and religious legitimacy and his authority in such matters was cast into doubt by many, especially in the reformed countries. The elimination of 10 days to make the spring equinox date correspond to its date at the time of the Council of Nicea was controversial, since it symbolically legitimized the universality of the Roman Church through Constantine. Benedetti wrote the aforementioned reform proposal on 1 April 1578 and Emanuele Filiberto sent it to Rome on 31 May. The most striking feature of Benedetti's proposal is its technical radicalism and extreme rationalism. August Ziggelaar's perspicuous description of Benedetti's proposal went as follows: "Benedetti prefers the Eastern date to be fixed according to the true motions of Sun and Moon rather than by cycles. He finds that the Prutenic tables are sufficiently exact for this purpose. Furthermore not ten days, not even fourteen, but twenty-one days should be left out in order to make the first day of January the winter solstice. The lengths of the months are to be adjusted so that they coincide with the presence of the Sun in each of the twelve zodiacal signs. Surely, these time intervals change their lengths in the course of time because of the motion of the perihelion of the Earth, but Benedetti assures us that only after 24,000 years will an adjustment be necessary. The proposal is not only interesting and original, but also very rational because, by eliminating all reference to the Moon, it makes the year entirely solar."[80] The extreme technicality of such a proposal, one can argue, was also a means to de-ideologize the issue. Therefore, it was not only rejected for its "scientific radicalism" but also for its rationalistic distance from confessional struggles. This positioning is in line with late-humanistic universalism and signals Benedetti's distance from the Counter-reformation and the militant cultural production of those years.[81] He can be seen as an organic intellectual of the Savoy ruling class who adhered to a political project, which included self-rule in matters of culture and education.

7.5 Conclusive Remarks and Prospects for Future Research

I have distinguished the social and the political-cultural coordinates of Benedetti's science and treated them separately. On the one hand, his position as a court mathematician directly determined much of the content of his writings, occasioned by the requests addressed to him as a court *expert* in technical issues pertaining to mathematics. His position also determined formal aspects of his work, in particular its occasional character and fragmentariness. On the other hand, Benedetti's identity as a philosopher directly related to his cultural ambitions and his engagement aimed to affirm *mathematical philosophy* in the intellectual arena against scholastic

[80] See Ziggelaar (1983, 211, 214).

[81] See Steinmetz (2010). For an insightful study on cultural production in the time of European confessionalization, see Mulsow (2007).

thinkers and humanistic literati. His political identity as a lay aristocrat made him an organic part of the centralizing project of the court and marked his distance from counter-reformist drives which sought to impose Roman universal interests over territorial states' autonomy. His support for a sort of party of the *politiques* resulted in treatises advising on politically relevant technical and cultural issues (e.g., navigation on the occasion of the battle of Lepanto or the calendar reform). His activity as a lay educator (e.g, his arithmetic teaching to the prince, Carlo Emanuele) is found in his pedagogical writings, some of which were published in his scientific miscellanea. In summary, both content and form, as well as the demarcation of the fields of his scientific competence as a mathematician and philosopher, depended on social settings and cultural engagement.

The analytical distinction between position and identity, between the mathematician and the philosopher, should not be taken rigidly. Retaking the initial quotation from Gramsci, it is necessary to stress the analytical usefulness of the distinction but also the limitation of a net separation of the two realms as "the material forces cannot be historically conceived without any forms and the ideologies would be individual whims separated from the material forces." In this case, the hylemorphic image, derived from Aristotelian philosophical terminology, should not be taken literally, as to imply some sort of subordination or dichotomy. Quite on the contrary, it helps stressing the interconnection, interdependency and interpenetration of the socio-economical realm (here indicated as the material pole) and the cultural (as the formal pole). At the same time, the acknowledgment of the importance of *both* poles serves to avoid renouncing structural and economic considerations, as has become common after the cultural turn of the 1980s.

Take for instance Benedetti's mathematical formalism. This can be accounted for *structurally* in functional terms of specialization and social control by means of labor division. As a court mathematician, he had to supervise engineering works carried out for the state. At the same time, mathematical formalism played a *cultural* role not only as an approach to nature but to the most varied issues that could be of interest to the court, including literature.[82] It even became a matter of style, distinguished from scholastic syllogistic and humanistic verbosity. In summary, only by taking into account both poles can one form a complete and not reductive image of Benedetti's science.

To summarize the most evident features of Benedetti's endeavors resulting from this interpenetration of elements and drives, it was courtly, secular, anti-Academic, unsystematic, occasional, elitist, learned, abstract, pleasant and useful. It was *secular*, that is, non theological, as it was linked to the interests of the ruling class and the state. It was a *useful and pleasant* science: on the one hand, it was practice-oriented but not purely empirical; on the other hand, it proved witty and fit for courtly sociability. It was *abstract and disinterested*: superior to the vulgar and tuned to aristocratic values. *Learned*: fit to be exhibited at court alongside the other arts. *Elitist*: Benedetti elevated mathematics from a practical discipline of scientist-engineers to a refined cultural activity. *Occasional*: linked to the variegate political

[82] See Omodeo (2012).

and cultural interests of the court. *Unsystematic*: fragmented, lacking the inner coherence of scholasticism. *Anti-Academic*: free from concerns about respect for university scholarly traditions. The *courtly character* of Benedetti's science was at once its strength and its weakness: it was technical and abstract without loosing contact with practice and experience—a mathematical-empirical science *in nuce*; it was (relatively) free from bookish tradition and theology but not from the contingencies of courtly life.

The fact that Benedetti never established a scientific school around himself can be seen as an indicator of the precariousness of patronized science, linked to the person of a particular ruler and not institutionalized at the level of an academic body. In the course of the seventeenth century these limitations of early court society would be solved by securing scientific continuity for patronized science through the foundation of scientific societies. These constituted an improvement over the volatility of Renaissance patronage, depending on the humors and interests of a prince, by substituting him for a corporative *persona ficta* deputed to protect, credit, and promote science (Biagioli (1995, 1418). This did not imply a diminution of the political relevance of science. As has been argued, the institution of the *Académie Royale des Sciences* as a means to patronize *all* of the sciences at once meant the conquest of a new kingdom, *la république des lettres tout entière,* for Louis XIV (Biaglioi (1995, 1438).

The meaning of Benedetti's contribution to science, however, cannot be grasped at a micro-historical level, just as the history of Savoy cannot be merely explained in terms of internal developments considered in isolation from geopolitical and *longue durée* perspectives. A better understanding of his case will require a wider reconstruction of the socio-economic settings as well as the cultural-political drives of science in the passage from the Middle Ages to modernity. This broader picture should capture the interests of the emerging urban and merchant groups, as a factor that made the value of technological advance recognizable. In turn, the acknowledgment of the social and theoretical value of practical experience led to its codification in various fields, including the *science of machines*. The *Verwissenschaftlichung* of this field—its formalization in mathematical terms and its systematization—could not have occurred without philosophy, that is, without reassessing and transforming the conceptual tools and methods of scholasticism and humanism. Thus, the study of early modern science needs a twofold investigation of its social conditions and its cultural significance, at various levels of analysis—economic, social, technological, institutional, theological, philosophical, and political.

Historians of mechanics have already studied some junctions of the socio-political context with particular care, especially the combination of mechanics with technology and practice. However, the general framework, that of the dynamic connection between economic structures and cultural-political action, requires further investigation. I would refer to this overarching issue in Gramscian terms as the problematic of the "historical bloc," that is, the historical intertwining of social-economic necessity and ethical-political freedom. During the sixteenth and early-seventeenth centuries when a *great transformation* took place toward a market society, the socio-political bloc was realized as an alliance between the production-

and-market interests of early capitalist classes and the aristocracy in the framework of emerging national states (also their preformation: the court-centered territorial states).[83] The manner in which this economic-political bloc fostered technological advance and science requires specific investigation. One should take into account state-directed investments in economy, technological projects and science, and the foundation of national academies. Jean-Baptiste Colbert's action in the time of the *Roy-Soleil* Louis XIV is an illuminating instance of scientific and economic politics occurring under the hegemony of a centralized national monarchy; while the Netherlands in the seventeenth century offers a suitable terrain for an analysis of the concomitant flowering of natural sciences and capitalistic entrepreneurship under the acknowledged leadership of the bourgeois classes of bankers and merchants.[84]

The cultural dimension of early modern science, beginning with mechanics, has to be understood as politically mediated by religion and philosophy. Famously, the natural questions brought up by the science of machines and its development into a general *physica* was at the center of philosophical and theological clashes, relative both to the scientific status of new fields of natural inquiry and their frame of comprehension. The polemics between mechanistic philosophers à la Descartes and defenders of the scholastic tradition during the seventeenth century is revealing of wider cultural struggles about models of society and epistemic values. Should the inquiry into the *book of nature* be pursued independently from the letters of the Holy Scriptures? How should experimentation and metaphysics relate to each other? Would not accepting scientific autonomy imply renouncing the centrality of religion in society? These are typical and unavoidable questions and doubts posed to early modern scientists in the time of the most severe conflict between science and religion in the history of the Latin West marked by the Roman censure of the Copernican system and the condemnation of Galileo. Several intellectuals attempted to reconcile philosophical and religious positions by developing experimental programs 'renouncing all hypotheses'—as typically occurred in state-patronized scientific academies. Continuous negotiation of the realms of jurisdiction of states, churches, and economic actors determined the various degrees of autonomy that could be accorded—or denied—to natural inquiry. At a concrete institutional level,

[83] A case in point would be mercantilism. Karl Polanyi, in his valuable reconstruction of the historical premises of the market society and free market ideology on the correlation and interaction between politics and economy in the constitution of nineteenth century liberalism, has pointed out that the tension, in early modern Great Britain, between the economic interests of those benefiting from the enclosures and the social interests of the pauperized people resulted in political regulation. This determined the pace of the socio-economic transformation that would eventually lead to the Industrial Revolution and made the process sustainable by diluting its disruptive effects. See Polanyi (2001, 35–40), whose main concern was to present an insightful case for government playing a central role in modern economic life. Cf. Polanyi (2001, 40): "England withstood without grave damage the calamity of the enclosures only because the Tudors and the earl Stuarts used the power of the Crown to slow down the process of economic improvement until it became socially bearable—employing the power of the central government to relieve the victims of the transformation, and attempting to canalize the process of change so as to make its course less devastating."

[84] For the relation between princely patrons and academies see, among others, Biagioli (1995, 1417–1453). On early capitalist Netherlands and science, cf. Cook (2007).

the balance of relative forces among the different cultural parties determined the degree of autonomy that was accorded, in various places, to education and research, to schools, universities, and academies. Hegemonic struggles concerned the values that should inform teaching centers and scientific institutions, affected the strictness of the imposition and informed curricula and research programs.

As a conclusion, I should advance, at least tentatively, an assessment of the historical position of Benedetti's mechanics in the history of modern science. The most evident feature of his science was the connection of mechanics with a wide array of fields, including cosmology, natural philosophy, and epistemology. This interweaving has escaped scholars who too narrowly dealt with Benedetti's *technical achievement* by ahistorically insulating his mechanics from its conceptual network, neglecting its interdisciplinary ties and the social motivations underlying it.[85] His integrative view on mechanics might seem blurred, as its borders are not clear-cut. This remarkable variety leads us back to the question derived from Lagrange's ex-post historical views on the developments of mechanics: the question of demarcation as a historical problem. Briefly, this is the problem of the emergence of a scientific field. With the help of Lagrange I have distinguished three phases of the historical development of early modern mechanics: from the science of machines to a physical-mathematical science of balance and motion and, finally, an analytical science. Although Benedetti's work can be located at the threshold of the emergence of a general physical-mathematical discipline, his work shows elements of all three phases. It explicitly maintains the connection with technology but clearly indicates the later development of physical generalization and mathematical formalism. I argue that it was precisely the inhomogeneity of his conceptions, the merging of social and cognitive dimensions, of technological and philosophical concerns, of disciplinary and political interests that made his work at once blurred and productive. Benedetti is representative of an age, in which the daring connection of mathematical and philosophical speculations, his crossing of social and epistemological barriers, opened up a space of possibility for a new mathematical physics. Multidirectional connections, in particular between experience and theory, technology and philosophy, were a necessary condition for the transformation of an artist-engineers' *science of machines* to the scientist-engineers' *science of balance and motion* and, later, for its elevation to a universal knowledge of the *mathematical principles* of nature.

[85] An example of technical reductionism is Drake and Drabkin's translation of *Disputationes de quibusdam placitis Arist[otelis]* which excludes some of the most compelling philosophical sections on the definition of space, infinity, and time, thus missing the connection of Benedetti's considerations on fundamental concepts and his treatment of motion through media and void. Cf. Drake and Drabkin (1969, 196–223).

References

Arma, Francesco. 1580. *Queste sono parti delle propposte tenute col Serenis. Prencipe nostro*. Turin: Appresso gli heredi del Bevilacqua.

Axworthy, Angela. 2016. *Le Mathématicien renaissant et son savoir: Le Statut des mathématiques selon Oronce Fine*. Paris: Classique Garnier.

Baldi, Bernardino. 1707. *Cronica de matematici overo epitome dell'istoria delle vite loro*. Urbino: Per Angelo Ant. Monticelli.

Baracca, Angelo, Stefano Ruffo, and Arturo Russo. 1979. *Scienza e industria (1848–1915): Gli sviluppi scientifici connessi alla seconda rivoluzione industriale*. Rome-Bari: Laterza.

Becchi, Antonio, Domenico Bertoloni Meli, and Enrico Gamba, eds. 2013. *Guidobaldo del Monte (1545–1607): Theory and practice of the mathematical disciplines from Urbino to Europe*. Berlin: Edition Open Access. http://edition-open-access.de/proceedings/4/index.html.

Benedetti, Giovanni Battista. 1553. *Resolutio omnium Euclidis problematum*. Venice: apud Bartholomaeum Caesanum.

———. 1585. *Diversarum speculationum mathematicarum et physicarum liber*. Turin: Apud haeredem Nicolai Bevilaquae.

Biagioli, Mario. 1989. The social status of Italian mathematicians (1450–1600). *History of Science* 27: 41–95.

———. 1993. *Galileo, courtier: The practice of science in the culture of absolutism*. Chicago-London: The University of Chicago Press.

———. 1995. Le prince et les savants: La civilté scientifique au 17e siécle. *Annales: Historie, Sciences Sociales* 6: 1417–1453.

———. 1996. Etiquette, interdependence, and sociability in seventeenth-century science. *Critical Inquiry* 22 (2): 193.

Bordiga, Giovanni. [1926] 1985. *Giovanni Battista Benedetti, filosofo e matematico veneziano nel secolo XVI*, 585–754. Venice: Istituto Veneto di Scienze, Lettere ed Arti.

Brahe, Tycho. 1916. In *Opera omnia*, ed. John Louis Emil Dreyer. Copenhagen: Libraria Gyldendaliana.

Bucci, Agostino. 1583. *Disputatio de principatu partium corporis*. Turin: Apud haeredes Nicolai Beuilaquae.

Bukharin, Nikolai. 1931. Theory and practice from the standpoint of dialectical materialism. In *Science at the cross roads: Papers presented to the international congress of the history of science and technology (London, 1931)*, 9–33. London: Kniga.

Cardano, Girolamo. 1966. Encomium geometriae recitatum anno 1535 in Academia Platina Mediolana. In *Opera omnia*, vol. 4, 440–445. Stuttgart-Bad Cannstatt: Fromann-Holzboog.

Catarinella, Annamaria, and Irene Salsotto. 1998. L'università e i collegi. In *Storia di Torino: Dalla dominazione francese alla ricomposizione dello Stato (1536–1630)*, ed. Giuseppe Ricuperati, vol. 3, 523–567. Turin: Einaudi.

Cecchini, Michela, and Clara Silvia Roero. 2004. I corrispondenti di Giovanni Battista Benedetti. *Physis* 41 (1): 31–66.

Chalmers, Alan. 2017. *One hundred years of pressure: Hydrostatics from Stevin to Newton*. Cham: Springer Nature.

Clark, William. 2006. *Academic charisma and the origins of the research university*. Chicago: The University of Chicago Press.

Clavius, Christophorus. 1589. *Euclidis Elementorum Lib. XV*. Rome: Bartholomaeum Grassium.

Cook, Harold John. 2007. *Matters of exchange: Commerce, medicine, and science in the Dutch golden age*. New Haven: Yale University Press.

Damerow, Peter, Gideon Freudenthal, Peter McLaughlin, and Jürgen Renn. 2004. *Exploring the limits of preclassical mechanics: A study of conceptual development in early modern science: Free fall and compounded motion in the work of Descartes*. 2nd ed. New York: Springer.

Daston, Lorraine, and Peter Galison. 2007. *Objectivity*. New York: Zone Books.

De Pace, Anna. 1993. *Le matematiche e il mondo: Ricerche su un dibattito in Italia nella seconda metà del Cinquecento*. Milan: Franco Angeli.

Di Bono, Mario. 1987. L'astronomia copernicana nell'opera di Giovan Battista Benedetti. In *Cultura, scienze e tecniche nella Venezia del Cinquecento: atti del convegno internazionale di studio Giovan Battista Benedetti e il suo tempo*, 283–300. Venice: Istituto Veneto di Scienze,Lettere ed Arti.

Dijksterhuis, Eduard Jan. 1956. *Archimedes*. Copenhagen: Munksgaard.

Doglio, Maria Luisa. 1998. Intellettuali e cultura letteraria (1562–1630). In *Storia di Torino: Dalla dominazione francese alla ricomposizione dello Stato (1536–1630)*, ed. Giuseppe Ricuperati, vol. 3, 599–653. Turin: Einaudi.

———. 2005. *Dall'Institutio al monumento, introduzione a Carlo Emanuele I di Savoia, Simulacro del vero principe*. Alessandria: Edizioni dell'Orso.

Drabkin, I.E. 1960. Introduction to *on motion*. In *Galileo Galilei, on motion, and on mechanics*. Madison: The University of Wisconsin Press.

Drake, Stillman, and I.E. Drabkin. 1969. *Mechanics in sixteenth-century Italy*. Madison-London: The University of Wisconsin Press.

Firpo, Luigi. 1983. *Relazioni di ambasciatori veneti al Senato: tratte dalle migliori edizioni disponibili e ordinate*. Vol. XI. Turin: Bottega di Erasmo.

Galluzzi, Paolo. 1979. *Momento: Studi galileiani*. Rome: Ed. Dell'Ateneo & Bizzari.

Gaurico, Luca. 1552. *Tractatus astrologicus in quo agitur de praeteritis multorum hominum accidentibus per proprias eorum genituras ad unguem examinatis*. Venice: Navo.

Giusti, Enrico. 1993. *Euclides reformatus: La teoria delle proporzioni nella scuola galileiana*. Turin: Bollati Boringhieri.

Goddu, André. 2010. *Copernicus and the Aristotelian tradition: Education, reading, and philosophy in Copernicus's path to heliocentrism*. Leiden: Brill.

Gramsci, Antonio. 1975. In *Quaderni del carcere*, ed. Valentino Gerratana. Turin: Einaudi.

Granada, Miguel Ángel, and Dario Tessicini. 2005. Copernicus and Fracastoro: The dedicatory letters to pope Paul III, the history of astronomy, and the quest for patronage. *Studies in History and Philosophy of Science* 36 (3): 431–476.

Greenblatt, Stephen. 1980. *Renaissance self-fashioning: From more to Shakespeare*. Chicago: The University of Chicago Press.

Grendler, Paul F. 2002. I tentativi dei gesuiti d'entrare nelle università italiane tra '500 e '600. In *Gesuiti e università in Europa (secoli XVI–XVIII)*, ed. Gian Paolo Brizzi and Roberto Greci, 37–51. Bologna: CLUEB.

Gui, Francesco. 1989. *I Gesuiti e la rivoluzione boema: alle origini della guerra dei trent'anni*. Milan: Franco Angeli.

Hellyer, Marcus. 2005. *Catholic physics: Jesuit natural philosophy in early modern Germany*. Notre Dame, Ind: University of Notre Dame Press.

Hessen, Boris. 2009. The social and economic roots of Newton's Principia [1931]. In *The social and economic roots of the Scientific Revolution: Texts by Boris Hessen and Henryk Grossmann*, ed. Gideon Freudenthal and Peter McLaughlin. Dordrecht: Springer.

Koyré, Alexander. [1939] 1966. *Études galiléennes*. Paris: Hermann.

———. 1966. Jean Baptiste Benedetti, critique d'Aristote. In *Études d'histoire de la pensée scientifique*, 122–146. Paris: Presses universitaires de France.

Kuhn, Thomas S. 1959. *The Copernican revolution: Planetary astronomy in the development of western thought*. New York: Random House.

———. 1996. *The structures of scientific revolutions*. 3rd ed. Chicago-London: The University of Chicago Press.

Lagrange, Joseph-Louis. (Giuseppe Ludovico)1788. *Mécanique analitique*. Paris: Chez la veuve Desaint.

Lehner, Christoph, and Helge Wendt. 2017. Mechanics in the *Querelle des Anciens et des Modernes*. *Isis* 108 (1): 26–39.

Lomazzo, Giovan Paolo. 2006. *Rime ad imitazione de i grotteschi usati da' pittori*. Rome: Vacchiarelli.

Long, Pamela O. 2011. *Artisan/practitioners and the rise of the new sciences, 1400–1600*. Corvallis: Oregon State University Press.

Maccagni, Carlo. 1967. *Le speculazioni giovanili 'de motu' di Giovanni Battista Benedetti*. Pisa: Domus Galilaeana.

Mach, Ernst. 1912. *Die Mechanik in ihrer Entwicklung historisch-kritisch Dargestellt*. Leipzig: Brockhaus.

Maier, Anneliese. 1951. Die Impetustheorie. In *Zwei Grundprobleme der scholastischen Naturphilosophie*, 113–314. Rome: Edizioni di Storia e Letteratura.

Mamino, Sergio. 1989. Scienziati ed architetti alla corte di Emanuele Filiberto di Savoia: Giovan Battista Benedetti. *Studi Piemontesi* 18: 429–449.

Mammola, Simone. 2013. Una disputa storico-filosofica nella Torino del '500: Agostino Bucci interprete di Parmenide. *Rivista di Storia dell'Università di Torino* 2/2.

———. 2014. *Il problema della grandezza della terra e dell'acqua negli scritti di Alessandro Piccolomini, Antonio Berga e G.B. Benedetti e laprogressiva dissoluzione della cosmologia delle sfere elementari nel secondo '500*. Preprint 459. Berlin: Max Plank Institute for the History of Science. https://www.mpiwg-berlin.mpg.de/sites/default/files/Preprints/P459.pdf.

Mulsow, Martin. 2007. Netzwerke gegen Netzwerke: Polemik und Wissensproduktion im politischen Antiquarianismus um 1600. In *Die unanständige Gelehrtenrepublik: Wissen, Libertinage und Kommunikation in der Frühen Neuzeit*, 143–190. Stuttgart-Weimar: J. B. Metzler.

Nenci, Elio. 2011. *Bernardino Baldi's In mechanica Aristotelis problemata exercitationes*. Berlin: Edition Open Access. http://www.edition-open-sources.org/sources/3/index.html.

Norbert, Elias. 2002. *Die höfische Gesellschaft*. Frankfurt a.M.: Suhrkamp.

Omodeo, Pietro D. 2008. La Stravagantographia di un 'filosofo stravagante'. *Bruniana & Campanelliana* 14 (1): 11–23.

———. 2009. La cosmologia infinitistica di Giovanni Battista Benedetti. *Bruniana & Campanelliana* 15 (1): 181–190.

———. 2012. Renaissance science and literature: Benedetti, Ovid and the transformations of Phaeton's myth after Copernicus. *Science & Education* 21: 1–8. Special Issue on "Science and Literature".

———. 2014a. *Copernicus in the cultural debates of the Renaissance: Reception, legacy, transformation*. Leiden: Brill.

———. 2014b. *Efemeridi e critica all'astrologia tra filosofia naturale ed etica: La contesa tra Benedetti e Altavilla nel tardo Rinascimento torinese*. Preprint 458. Berlin: The Max Planck Institute for the History of Science.

———. 2014c. Torino, 1593: Motivi dell'opposizione universitaria ai gesuiti nel contesto degli antagonismi europei del tempo. *Rivista di Storia dell'Università di Torino* 3 (1): 1–18.

———. 2017a. L'eredità di Boris Hessen: Per un approccio socio-politico alla scienza in età moderna. In *Boris Hessen, Le radici sociali ed economiche della meccanica di Newton*, ed. Gerardo Ienna, 119–150. Rome: Castelvecchi.

———. 2017b. Utilitas astronomiae in the Renaissance: The rhetoric and epistemology of astronomy. In *The structures of practical knowledge*, ed. Matteo Valleriani, 307–332. Cham: Springer Nature.

Omodeo, Pietro D., and Jürgen Renn. 2018. *Science in Court Society: Giovan Battista Benedetti's Diversarum speculationum mathematicarum et physicarum liber*. Berlin: Edition Open Access.

Patrizi, Francesco. 1975. In *Lettere ed opuscoli inediti*, ed. Danilo Aguzzi Barbagli. Florence: Istituto Nazionale di Studi sul Rinascimento.

Pivano, Silvio. 1928. Emanuele Filiberto e le Università di Mondovì e di Torino. In *Studi pubblicati dalla Regia Università di Torino nel IV centenario della nascita di Emanuele Filiberto*, 1–34. Turin: Università di Torino.

Polanyi, Karl. [1944] 2001. *The great transformation: The political and economic origins of our time*. Boston: Beacon Press.

Pollak, Martha D. 1991. *Turin 1564–1680: Urban design, military culture, and the creation of the absolutist capital*. Chicago: The University of Chicago Press.

Pulte, Helmut. 1989. *Des Prinzip des kleinsten Wirkung und die Kraftkonzeptionen der rationalen Mechanik: eine Untersuchung zur Grundlegungsproblematik bei Leonhard Euler, Pierre Louis Moreau de Moupertuis und Joseph Louis Lagrange*. Stuttgart: Steiner.

Renn, Jürgen, and Peter Damerow. 2010. *Guidobaldo del Monte's Mechanicorum Liber*. Berlin: Edition Open Access. http://www.edition-open-sources.org/sources/1/index.html.

———. 2011. *The equilibrium controversy: Guidobaldo del Monte's critical notes on the mechanics of Jordanus and Benedetti and their historical and conceptual background*. Berlin: Edition Open Access. http://www.edition-open-sources.org/sources/2/index.html.

Ricuperati, Giuseppe. 1998. *Storia di Torino. Vol. 3. Dalla dominazione francese alla ricomposizione dello Stato (1536–1630)*. Turin: Einaudi.

Roero, Clara Silvia. 1997. Giovan Battista Benedetti and the scientific environment of Turin in the 16th century. *Centaurus* 39 (1): 37–66.

Steinmetz, Dirk. 2010. *Die Gregorianische Kalenderreform von 1582: Korrektur der christlichen Zeitrechnung in der Frühen Neuzeit*. Oftersheim: Verlag Dirk Steinmetz.

Tessari, Antonio Secondo. 1993. Sul soggiorno di Andrea Palladio a Torino per le questioni militari di Emanuele Filiberto. *Studi Piemontesi* 22 (1): 9–20.

Tonso, Giovanni. 1596. *De vita Emmanuelis Philiberti Allobrogum ducis, et Subalpinorum principis, libri duo*. Turin: Apud Io. Dominicum Tarinum.

Trotto, Bernardo. 1625. *Dialoghi del Matrimonio e vita vedovile... di nuovo ristampati*. Turin: Appresso il Pizzaiglio, Stampator Ducale.

Valleriani, Matteo. 2013. *Metallurgy, ballistics, and epistemic instruments: The Nova scientia of Nicolò Tartaglia*. Berlin: Edition Open Access. http://www.edition-open-sources.org/sources/6/index.html.

van Leeuwen, Joyce. 2016. *The Aristotelian mechanics: Text and diagrams*. Cham: Springer.

Ziggelaar, August. 1983. The papal bull of 1582: Promulgating a reform of the calendar. In *Gregorian reform of the calendar: Proceedings of the Vatican conference to commemorate its 400th anniversary*, ed. George V. Coyne, Michael A. Hoskin, and Olaf Pedersen, 201–239. Rome: Pontifical Academy of the Sciences.

Zilsel, Edgar. [1942] 2000. The sociological roots of science. *Social Studies of Science* 30 (6): 935–939.

Chapter 8
A Treatise on Galileian Mechanics: *Exercitationes in Mechanicis Aristotelis* by the Jesuit Giovan Battista Zupi

Romano Gatto

Abstract This chapter will describe and comment a treatise on Galileian Mechanics, *Exercitationes in Mechanicis Aristotelis* by the Jesuit scholar Giovan Battista Zupi, revealing the sources from which the author derived his text. The treatise presents the course of lessons on mechanics that Zupi gave when he first began to teach at the Neapolitan Jesuit College. Although the ideas exposed in this work were not generally original, because the lessons are skillfully compiled using works on mechanics by followers of Archimedes, such as Bernardino Baldi and above all Galileo Galilei, it is interesting to scholars in the history of science for at least three reasons: 1) The lessons result from a skilled compilation, demonstrating Zupi's wide knowledge of mechanics, including the latest works circulating at that time. He was able to put together and harmonize chapters from the most meaningful works by different authors, thus creating an organic essay containing all the subjects concerning a modern theory of the simple machines. 2. The text is eloquent proof of the contradictions that characterized many aspects of Jesuit educational politics in the scientific field. 3) The treatise is an important documentation of the diffusion within Jesuit schools of the mechanical theories that Galileo exposed in his *Le Mecaniche*, a work he did not publish during his lifetime but that circulated at the time among the scholars featured in manuscript. This diffusion, in the case of the Neapolitan College, assumes the form of a popularization of Galilean ideas on mechanics since, at that time, the Jesuits had an almost complete monopoly on education in Naples.

Keywords Giovan Battista Zupi · Galilean mechanics · *Exercitationes in mechanicis Aristotelis*

R. Gatto (✉)
Università degli Studi della Basilicata, Potenza, Italy

R. Feldhay et al. (eds.), *Emergence and Expansion of Preclassical Mechanics*, Boston Studies in the Philosophy and History of Science 270,
https://doi.org/10.1007/978-3-319-90345-3_8

8.1 A Treatise on Mechanics in the Scientific Project of the Neapolitan Jesuits

Some years ago I published an unedited text by the Jesuit Giovan Battista Zupi, *Exercitationes in Mechanicis Aristotelis,* which shows the course of lessons on mechanics he gave when he first began to teach at the Neapolitan Jesuit College.[1] On that occasion I pointed out that, although the ideas exposed in this work were not generally original, because the lessons are skillfully compiled using works on mechanics by followers of Archimedes, such as Bernardino Baldi and above all Galileo Galilei, history of science scholars would nonetheless consider it for at least three reasons.

- First, as mentioned, these lessons result from a skilled compilation, demonstrating Zupi's wide knowledge of mechanics, including the latest works circulating at that time. He was able to put together and harmonize chapters from the most meaningful works by different authors, thus creating an organic essay containing all the subjects concerning a modern theory of the simple machines.
- Second, this text is eloquent proof of the contradictions that characterized many aspects of Jesuit educational politics in the scientific field.
- Third, it is an important documentation of the diffusion within Jesuit schools of the mechanical theories that Galileo exposed in his *Le Mecaniche*, a work he did not publish during his lifetime but that circulated at the time among the scholars featured in manuscript.[2] This diffusion, in the case of the Neapolitan College, assumes the form of a popularization of Galilean ideas on mechanics since, at that time, the Jesuits had an almost complete monopoly on education in Naples.[3]

Regarding the first point, Zupi's work will be described and commented and the sources from which the author derived his text will be revealed. It is appropriate here to reproduce what he writes at the beginning of the *Exercitationes* where, exalting the virtues of mechanics, Zupi lists the most important authors who up to that time had offered the greatest contributions:

> In fact, the nobility and the dignity of this discipline is so great that the wisest men, Aristotle, Pappus, Heron, Athenaeus, Archimedes, and among the most recent ones Alexander Piccolominus, Guidus Baldus of the marquises del Monte, Abbot Baldi and others, have introduced and illustrated it with acute lucubration.[4]

---

[1] The text of the *Exercitationes* is published in the appendix to Gatto (2006c). All the bibliographical references relate to this edition.

[2] See Galilei (2002).

[3] For more than one century, until 1650, there was no chair of mathematics established at the University of Naples.

[4] "Ea enim est huius facultatis nobilitas, ea dignitas, ut sapientissimi quoque illam praebaverint et acutissimis lucubrationibus illustraverint, ita Aristotelis, Pappus, Heron, Athenaeus, Archimedes, et ex recentioribus Alexander Piccolomineus, Guidus Baldus ex marchionibus Montis, Maurolicus, Baldus abates et alii" *Exercitationes*, 73.

When specifying how he will carry out his treatment, he later adds:

> we will follow the doctrine of Aristotle, of Archimedes, and of other experts on this matter.[5]

Although the title refers to *Quaestiones mechanicae* by pseudo-Aristotle,[6] Zupi's work has little or nothing in common with this treatise. Concerning its style and inspiration, the *Exercitationes* is essentially an Archimedean and in particular a Galileian work. But, as mentioned, Galileo's name does not appear among the authors cited or at any point in the essay, even though *Le Mecaniche* is a source from which he drew profusely. Since Zupi always cites his sources, we are inclined to believe—and here we come to the second point—that he included Galileo's name anonymously among the *aliis,* because he could not mention him explicitly.

It must be taken into consideration that the *Exercitationes* dates back to 1634 and that Zupi started to teach mathematics and ethics at the Neapolitan College in the year 1632–33. This text was therefore written in the same year or in the year following Galileo's condemnation, a period in which the Jesuits probably judged as very improper any mention of Galileo's name in their schools. But despite their ostracism of Galileo, the Jesuits did not consider it *scandalous* to teach the theories of *Le Mecaniche*. While it is true that in this treatise there is no trace of any heterodox theory, which could meet with the opposition of the Church, it is also true that after so little distance from the dramatic event of Galileo's condemnation, we would expect the Jesuits to behave in line with their principles. On the contrary, also the pragmatism that often affected their choices prevailed: if Galileo had to be excluded, then to ignore his name, that is, not to cite him, was what would be done.

So Galileo's mechanics could be taught even if he himself was not mentioned, causing some to believe that the aforesaid work was "*opera nostrorum*." It was not the first time the Jesuits behaved in this way. When Clavius published his 1589 edition of *Euclid*,[7] he had no scruples about including

> many difficult points, not well known, received from Giovanni Paolo Vernalione, with the promise that, in printing his very ingenious Commentaries about the same matter dealt by Euclid, he openly informed the world that he had received the light from him.[8]

---

[5] "sequimur autem tum Aristotelis, tum Archimedes, aliorumque in hac re peritorum doctrinam" *Exercitationes*, 73.

[6] We are referring to the Μηχανικά Προβλήματα (*Mechanical Questions*), at that time attributed to Aristotle. Today this work is almost unanimously attributed to an unidentified student of Aristotle, referred to as pseudo-Aristotle. Despite this, at that time an author had already expressed doubts about such an attribution, but Zupi was nevertheless persuaded that Aristotle was the author of the *Quaestiones mechanicae* (for this reason most Renaissance authors translated the title of the above-mentioned work from Greek into Latin).

[7] See Clavius (1589). The first edition of this work is *Apud Vincentium Accoltum*, Rome, 1574. During Clavius's lifetime, this work was published again in 1589, 1591, 1603, 1607 and 1612. The impossibility of identifying "the difficult places" by Vernalione also led to discussions. I maintain that it is the 1589 edition, see Gatto (1994, 70–73). This is also Ugo Baldini's and Pier Daniele Napolitani's opinion, see Clavius (1992). On the contrary, Rosario Moscheo maintains that it must be a later edition, see Moscheo (1993–1994).

[8] "molti luoghi difficili, non ben capiti, ricevuti da Giovanni Paolo Vernalione colla promessa, che nello stampare gli suoi ingegnosissimi Commentari su la stessa materia di Euclide, facesse testimonianza al Mondo aver avuto da lui questo lume." See Arcudi (1719, 160–161). About Giovanni Paolo Vernalione and the matter under discussion here, see Moscheo (1993–1994).

The promise was not kept because the censors of the Jesuit Society resolutely prevented Vernalione's name from being revealed because at the time he was suspected of heresy. But "the many difficult points" were nevertheless included in *Euclid* as if Clavius himself were the author. By not revealing Galileo's name, Zupi also ended up claiming authorship for the "points" of another author. However, like "the difficult points" by Vernalione, *Le Mecaniche* was an unpublished work circulating among the scholars only in manuscript form. It could therefore very easily be appropriated for such an operation.

In my opinion, the *Exercitationes* was the first step toward the realization of one of the Society's projects: to carry out Clavius' project and to finally write for the Society's colleges a text on mechanics that Clavius himself had not succeeded in producing.

We must consider that at the beginning of their school activities, the Jesuits used texts that were circulating at that time. But, driven by worries about their possible doctrinaire character, they soon decided to adopt as many works as possible written by the Society's own authors. The *Ordo servandus in addiscendis disciplinis mathematicis* by Clavius is a very precious document that casts light, not only on the pedagogic project of the Jesuits regarding the teaching of mathematics, but also on Clavius' ambitious publishing program which lay down the composition of all necessary texts for teaching all of their disciplines.[9] Clavius was absorbed in achievement of his program for the most part of his life, but since he died before its completion, there was no text for applied mathematics. The Jesuits, especially with regard to mechanics and optics, were still obliged to use books written by authors who did not belong to the Society. It therefore comes as no surprise that they tried to take possession of Francesco Maurolico's unpublished works because they were seen as the most qualified ones to complete Clavius' program.[10] They succeeded only for some of them, including the works on optics that Giovanni Giacomo Staserio published in 1611,[11] that is, in a period in which Galileo's *Sidereus Nuncius*[12] had caused great interest in this discipline.

Concerning mechanics, which in the same period attracted the attention of many mathematicians, engineers, architects and artisans, this discipline would be missing from their own textbooks for a long time to come. The absence of documents concerning the organization of mechanics courses in the Roman college has convinced some experts that the Jesuits had no interest in theoretical mechanics. This claim is denied by some documents that I found in the National Library in Naples, which attest to a great interest in this discipline in the Neapolitan College, and which also show that, starting from Staserio's teaching, mechanics became an integral part of the mathematics course at the college.

[9] See Gatto (2006a).

[10] See Gatto (1994, 81–85); Moscheo (1988).

[11] See Maurolico (1611). About the publication of this work, see Gatto (1994); Moscheo (1998). Staserio also had a manuscript by Maurolico of the *Conics* by Apollonius, but could not publish it in Naples because he could not find anyone to finance it.

[12] See Galilei (1890–1909).

8.2 The School of Mechanics at the Neapolitan College

Giovanni Giacomo Staserio was born in Bari in February 1565. On February 11, 1588, he entered the Society of the College in Messina where he studied rhetoric for 7 months and was then required to study mathematics for a year (1588–1589). He thus followed a rather anomalous curriculum because in the Jesuit schools the mathematics course belonged to the *studia superiora* which Staserio attended during the school years 1590 to 1591. The Jesuits of Messina probably soon realized that the young Staserio had a remarkable talent for mathematics. It is also probable that before entering the Society, he had already studied mathematics for a certain time, but there is no evidence of this in the catalogues of the Neapolitan province. There are two possible hypotheses: either he had special permission to study as a self-taught person, or he studied mathematics with a mathematician extraneous to the Society. Indeed, the first hypothesis does not seem likely since a novice just admitted to the Society was not generally granted this kind of permission. Only after the conclusion of the triennium of philosophical studies was this permission granted to students who had demonstrated a marked tendency for a particular discipline, as for instance in mathematics, to interrupt their normal curriculum for 1 or 2 years to apply themselves to these studies before beginning the *quadriennium* of theological studies. But Staserio's *curriculum studiorum* of the Neapolitan province catalogues does not show that he studied mathematics *extra Societate*.

The most reasonable hypothesis is the second one, that is, Staserio had attended a regular course of mathematical studies given by a teacher of mathematics for 1 year. And because there were no official courses in mathematics at the College of Messina during his time there, we must assume that he had private lessons. It is very probable that he was taught by Francesco Maurolico, who was very close to the Jesuits, and it cannot be excluded that Clavius, who was a friend of Maurolico's, intervened directly, particularly as he was concentrated at the time on creating the favourable conditions necessary to establish regular courses on mathematics among the *studia superiora* of the colleges.[13] This hypothesis can be confirmed because the following school year (1590–1591) Staserio was sent to make his novitiate at the College of Nola, whose rector was the mathematician Bartolomeo Ricci. Unfortunately there are no documents to shed light on these episodes, but there is also no doubt that the particularly confidential bond established between Maurolico and Staserio can be attested to in light of their direct habitual familiarity. Staserio knew Maurolico's nephews, from whom, after the scientist's death, he received some of Maurolico's manuscripts which he intended to publish. We have already seen that Staserio succeeded in publishing Maurolico's works on optics; he also sought financing to also publish *Conics*, but was unsuccessful. Moreover, some documents that I found among the manuscripts in the National Library of Naples demonstrate that Staserio cultivated mechanics according in particular to the

[13] See Gatto (2006a).

Archimedean approach to the centres of gravity.[14] This lends great consistence to the hypothesis that he had studied this matter with Maurolico, who was one of the few scientists of that period to stray from the traditional pseudo-Aristotelian dynamical approach to mechanics and to base his mechanics on the Archimedean theory of the centres of gravity.[15]

Staserio founded a mathematics school at the Neapolitan College where many teachers of mathematics were educated. In this school, the teaching of mechanics held an important role, as most of the documents mentioned above demonstrate. Giovanni Paolo Lembo attended this school and later became Staserio's assistant and temporary substitute before moving to the College of Saint Antão in Lisbon where he taught mechanics and hydraulics.[16] Another of Staserio's students was Francesco Antonio Camassa, who became a teacher of mathematics particularly experienced in the science of fortification, a discipline that required good knowledge of theoretical and practical mechanics.[17] Scipione Sgambati gave a course *In questiones mechanicas disputationes* at the Neapolitan College, which, contrary to what the title suggests, had little or nothing to do with the pseudo-Aristotelian mechanics inspired by the Archimedean doctrines.[18] And finally, Giovan Baptista Zupi also attended this school, whose work on mechanics we are going to examine.

In the first half of the seventeenth century in the Neapolitan College, a very remarkable mechanics school was flourishing with very highly trained teachers. One needs to bear in mind that during the same years in which the mathematics school of the Neapolitan College evolved, there were also private mathematics schools in Naples, which were much appreciated and well frequented. We are referring to the schools of Della Porta, in particular the one run by Giovanni Paolo Vernalione, and above all the school of Colantonio Stigliola, where a very good level of mechanics was taught.[19] This course was very modern and inspired by

[14] See the manuscript at the Biblioteca Nazionale di Napoli (BNN) V H 370. Staserio applied his competence in mechanics to the planning of a mill.

[15] Maurolico devoted two important works—both published posthumously—to mechanics, Maurolico (1613, 1635).

[16] A *Tractado breve de Machinas Hydraulicas* is included in a codex in the Archivio Nacional-Torre do Tombo of Lisbon (Manuscrito de livraria, 1770). In this codex the manuscripts of the lessons given by Lembo at the Jesuit college of Santo Antão in Lisbon between 1615 and 1617 are joined together. See Leitão (2001).

[17] In the school year 1631–1632 Camassa taught mathematics at the Neapolitan College. At the end of that year, at explicit request of King Philippe IV, he was sent to teach mathematics at the Estudio Reale de Santo Isidro in Madrid. Regarding his teaching, see Astrain (1916, 106).

[18] Sgambati taught mathematics at the Neapolitan College in the school year 1626–1627. The *Disputationes* are in the cc. 27–44 of his *Compendiosa tractatio aliquarum rerum mathematicarum*, ms V H 388 of the BNN which, besides the aforesaid text of mechanics, also includes *De optica partition dicendorum*, and *Sphaera*. The *Disputationes* are three in total: I *De auctoribus ac definitionibus Mechanicae*; II *De motu circulari*; III *De instrumentis Mechanicis*. About Sgambati and his *Compendiosa tractatio,* see Gatto (1994, 130–150).

[19] See Gatto (1996).

Maurolico's doctrines and was visited by many scholars.[20] For the Neapolitan Jesuits, the presence of this school was an impediment to the affirmation of their emerging mathematical school, so that they used any means possible to vanquish the competition, even recurring to that dubious weapon, the Inquisition.[21] At the same time, they tried to be adequately and culturally prepared to offer an excellent mathematical school.

8.3 Giovan Battista Zupi

Giovan Battista Zupi was born in Catanzaro on November 2, 1589. He joined the Jesuit Society in Naples on September 7, 1608 where he attended the complete course of *studia inferiora* at the college of novices of Pizzofalcone. He then attended the Maximum College where, between 1613 and 1616, he attended the triennium of philosophy and, untypically, after completing these courses did not immediately begin to study theology. Between 1616 and 1618, he probably devoted himself to the study of mathematics under the guidance of Giovanni Giacomo Staserio. Only between 1618 and 1619 did he attend the quadriennium of theological studies. Afterwards, in 1622, he was sent to teach philosophy at the college of Vibo, where he remained until 1626. At the beginning of the school year 1626–1627, he returned to Naples, but his father's death in 1628 caused the Society to transfer him to the College of Catanzaro so that he could take care of his father's affairs. He stayed in Catanzaro until 1632 when, for work reasons, he was sent to the College of Cosenza. But, having expressed the wish to return to Naples and due to a lack of mathematics teachers there, at the beginning of the school year 1632–1633 he was called back to Naples as a teacher of mathematics. He taught mathematics for 25 years until the school year 1657–1658. For some years after, he remained in the school environment as a school prefect of the *studia inferiora*, afterwards moving to the Professed House, where he died on August 26, 1667.[22]

As mentioned, Zupi authored the treatise *Exercitationes in Mechanicis Aristotelis*, a text included in a miscellaneous manuscript volume in the Biblioteca Nazionale of Naples (V H 299) titled *Ex variis Aristotelis Eruditionibus Compendiosa Tractatio per Reverendum Patrem Ioannem Bapt Suppa e Societate Iesu in Neapol° Coll° An. 1634*. This volume consists of five treatises, four of which are courses given by Zupi at the Neapolitan College: the aforesaid *Exercitationes in Mehanicis Aristotelis*

[20] About Colantonio Stigliola, see Ricci (1996, f. I), which contains an extensive bibliography about Stigliola's life and work. Stigliola wrote a very interesting treatise on mechanics (1597), which soon became very rare. An anastatic copy of this treatise was published in Gatto (1996). About the analysis of this opera, see Gatto (2006b).

[21] Between 1595 and 1596, Stigliola was charged with irreligion and taken to the church court of the Inquisition in Rome by the Jesuit Claudio Migliarese. He spent more than 2 years in prison. See Amabile (1892, documento n. 8), and Gatto (2006b).

[22] See Gatto (1994, 185–208, 211–221).

(cc. 5v–31v); *Brevis tractatus de terremotis* (73v–85r); *Compendiariae adnotationes in Sphoeram Johannis de Sacrobosco* (97r–117r); *Tractatio de morali disciplina* (127r–192r). The fifth treatise is titled *Liber problematum meterologicorum* (45r–72v) by Antonio de Bartolomeis, a teacher of philosophy at the same college from 1632 to 1635 and during the school year 1636–1637. All of these manuscripts give a very clear idea about the extent and variety of Zupi's teaching.

The eruption of Vesuvius in 1631 and the consequent disastrous seismic manifestations aroused the curiosity and interest of scientists, poets, writers and painters in Naples. Zupi, at the beginning of his teaching career at the Neapolitan College, took the opportunity to give a course on the theory of earthquakes, that is, the *Brevis tractatus de terremotis*. The presence in the aforesaid miscellaneous volume of the *Liber problematum meterologicorum* by Antonio de Bartolomeis, in which chapters XXII and XXIII are respectively titled *De terremotu* and *De igne subterraneo*, allows us to understand the different approaches and points of view of the scientist Zupi and the philosopher de Bartolomeis. While de Bartolomeis appears to present the repository of the tradition, Zupi seems intent on establishing solid scientific foundations for the phenomenon. Both authors began with the Aristotelian theory, according to which earthquakes are caused by the pressure built up by the warm and dry exhalation of underground steam. But while de Bartolomeis referred to the theory of natural places, Zupi gave a mechanistic explanation of the phenomenon, attributing it to the physical state of the steam put in violent and chaotic motion by its heating and successive rarefaction. If Zupi had clearly conferred a corpuscular structure to the steam (which appears in that violent and chaotic agitation motion), he would have given a description of that phenomenon very similar to that described by Descartes in his *Meteors*. The same effect could be caused by underground winds, or, as in the case of the eruption of Vesuvius, by a sudden underground fire in a sulphurous or bituminous zone, which, in trying to find a way out, produces a mechanical action against the overlying earth, making it shake and, at the same time, causing great landslides inside the underground cavities. In short, Zupi explains his theory with constant reference to the seismic and volcanic events at Vesuvius and other places in the Neapolitan vice-kingdom.[23]

As the treatise on mechanics and the theory of the earthquakes demonstrate, there is no doubt that Zupi was an attentive observer who was aware of the limits of Aristotelian physics. He was a scientist who was devoted to finding explanations more rational than those given by the Aristotelian theories. This is also demonstrated by the cosmological theories that he elaborated in light of the manifestation of the astronomical phenomena taking place during those last periods. In his *Compendiosae adnotations in Sphoeram Ioannis de Sacrobosco*, Zupi explained his original hypothesis concerning the crucial question of the incorruptibility of the skies. The appearance of *stellae novae* and comets in that short lapse of time and the use of the telescope had created difficulties for the theory of the solidity of skies, which appeared inadequate to explain these phenomena. Zupi rejected both the Copernican and the Ptolemaic cosmological systems: the first because it was

[23] For a wider and complete description of this treatise, see Gatto (1994).

opposed to the doctrines of the Church; the second because of its continuous reviews and adjustments. He considered the system of Tycho Brahe "*satis probabilis,*" suitable for interpreting all celestial phenomena. Zupi distinguished the universe in three skies: the first was the sky of the planets:

> and it consists of very pure and liquid air in which all the planets, supported by some intelligences, like birds turn always regular, and this is extended from the firmament to the earth, and consequently, it includes all the air.[24]

The second was the solid firmament, including all the fixed stars. The third was the empyrean according to the theological doctrine. For his theory, Zupi found theological support in doctrines expounded by Thomas, in the second letter of Saint Paul to the Corinthians and in the writings of some Church Fathers. His cosmological model was thus safe from censorship and in this way it was able to easily save the phenomena offering a suitable means for clearing any issues and controversies that could have dominated the discussion between the scholars at that time.

As with the issue of earthquakes, we cannot dwell on this aspect of Zupi's scientific profession, suffice to say that Zupi was known above all as an astronomer. A number of contemporary authors, such as Fontana, Scheiner, Riccioli and Schott, have written widely on Zupi's activities in this particular field, including his discovery of the orbital phases of Mercury and his observation, with a telescope only slightly more powerful than Galileo's, which demonstrated that the planet Mercury orbited the Sun. The crater Zupus on the Moon is also named after him.[25]

8.4 The *Exercitationes*: A Manual of Galileian Mechanics

The *Exercitationes* is a handbook of lessons structured like an actual treatise: it is divided into chapters and paragraphs with many figures, some of which are so accurately drawn that this could easily be the first draft of a work planned for publication. Even so, the work is far from the final draft stage and not entirely complete: it is sometimes extensive and overly detailed, sometimes concise, sometimes reduced to the enunciation of essential points, in some cases even finishing with "et cetera." This leads one to believe the author thought it superfluous to write down the conclusions he very clearly had in mind and would expound only orally to his students. We may conclude that the *Exercitationes* seems to be a well-planned though incomplete work, and that this incompleteness does not invalidate its scientific and pedagogic

---

[24] "et constet ex purissimo, et liquidissimo aere in quo planetae omnes tamquam avies ab intelligentiis sustentati, semper regulari girentur, huius crassities sit a firmamento usque ad terram et consequenter totum includat aera" *Compendiariae adnotationes*, cit., c. 105v. Zupi introduces this subject writing: "Quum in materia de coelo satis probavimus," that is, "as about the sky we have already much proved." Let us suppose that he explained his theory shortly before, probably in 1624 during his second year of teaching philosophy in the course of physics. This date would be justified by the fact that Bartolomeo Amico quotes Zupi's theory in Amico (1626, 274).

[25] More details can be found in Gatto (1994).

value. The treatise stands out because of Zupi's teaching abilities, his wide scholarship, his analytical treatments and the acuteness of some of the reasoning he puts forward in treating the science of simple machines. It also stands out because of his ability to explain the most difficult subjects clearly and simply, using numerical examples and the description of experimental facts to make them easier to understand.

What immediately emerges at the beginning of this treatise is that it corresponds mostly to Galileo's *Le Mecaniche*, to such an extent that it can be defined as the first example of a handbook of Galilean mechanics. As mentioned, *Le Mecaniche* was not the only work to inspire Zupi and no effort was made here to discover further sources of inspiration since, as mentioned, he always cites the authors. As we will see, Baldi is another heavily cited source, but concerning the style and the ideas of the mechanics expounded in Zupi's handbook, they correspond very closely to the above-mentioned Galilean treatise. Zupi, in perceiving the most innovative aspects of the Galilean work, adopts its principles and assumes the fundamental concept of 'moment' which he uses systematically. Zupi had a manuscript of *Le Mecaniche*, which his student Davide Imperiali had transcribed and given to him.[26] It is evident that the reading of this manuscript played a crucial role in developing Zupi's ideas on mechanics. It should be noted here that Zupi's handbook was composed in a Jesuit environment where Aristotelianism was the doctrine underlying all aspects of natural philosophy. But the *Exercitationes*, like Galileo's *Le Mecaniche*, were founded upon principles very different from those of Aristotelian mechanics. Zupi only touches on the principle of the circle, which is the basis of pseudo-Aristotelian mechanics, but he never uses it in treating his theories. This causes one to think that he mentions that principle only to attest the existence of a different approach to the mechanical science.

With regard to his teaching methods, Zupi adheres to the Society's pedagogic system. Every subject is subdivided into a certain number of paragraphs, numbered with an ordinal, and every paragraph in turn can contain other subparagraphs, often consisting of only simple enunciations or conclusions. In doing so, the author evidently intends to remark on some important results and their immediate consequences. Furthermore, he considers and solves all possible cases of a given problem, even if it involves the repetition of the adopted procedure, and finally, he enunciates the canons, or rules, for the single cases considered.

But why did Zupi never complete and publish his treatise? Lacking documents to answer this question, we can only advance a hypothesis. The reason was probably the fact that precisely in 1634, Marin Mersenne published his first edition of *Le Mecaniche* in France. Although it was a paraphrase of the Galilean text freely translated into French, it would have easily revealed that many points in Zupi's text had been almost integrally drawn from the work of the unnamable author.

---

[26] This manuscript is held at the BNN, XII D 75, cc. 1r–17r, see Gatto (1991). Imperiali himself wrote a treatise on mechanics, clearly following the Galilean style and subsequently datable to 1644. This treatise remained in manuscript form and is published in Gatto (1996). As regards Imperiali, see Gatto (1988).

8.5 The Introduction: Baldi's Influence

The *Exercitationes* is composed of an introductory chapter and three treatises. The first treatise, divided into four chapters, is devoted to the lever and the balance; the second, also divided into four chapters, treats the pulley and the block and tackle; the third treatise, in three chapters, deals with the wheel and axle, the wedge and the cochlea.

In his introduction, Zupi mentions Heron's distinction of mechanics: rational (theoretical) and surgical (practical), specifying that this difference is pursued in the aims. Theoretical mechanics investigates the principles and establishes the laws of mechanics, whereas practical mechanics applies these principles and laws to the various arts: handicraft, metallurgy, woodwork, painting, sculpture, architecture, car construction, surgical tools and so on. Zupi, citing Bernardino Baldi, affirms that the "*grave et leve*," namely the heavy and the light, are both objects of mechanics (Baldi 1621, 1). This definition leads to a clear understanding of the fact that the Jesuit teachers, like Baldi and most of the authors to treat mechanics up to that moment, had an Aristotelian conception of the philosophy of nature. Taking this into consideration, as we will see, Zupi was one of the first scientists to gather important innovations from Galilean mechanics and we can conclude from this that he was only half modern. Unlike Maurolico and Galileo, Zupi, following Baldi and the other Renaissance scientists, was an Archimedean but not an anti-Aristotelian scientist. Indeed, he has to be counted as one of the scientists who considered the Ἐπιπέδων ἰσορροπιῶν (*On the Equilibrium of Planes*) to be an interpretation of the Μηχανικά Προβλήματα (*Mechanical Questions*), thus finding a way to reconcile the pseudo-Aristotelian dynamic approach with the Archimedean static one. Maurolico's and Galileo's approaches were different because these scientists resorted to the principles of the Ἐπιπέδων ἰσορροπιῶν to develop some models as an alternative to pseudo-Aristotelian mechanics. If we consider that the ideological and cultural foundations of the Jesuit Society were deeply immersed in Aristotelianism, then we must conclude that Zupi was a progressive scientist. It is true that assuming the works of Maurolico, Baldi and Guidobaldo and Galileo's *Le Mecaniche* to be models did not implicate any doctrinaire revision, but it is also true that to stray from the Aristotelian mechanical tradition meant breaking with the unitarian epistemological model of the Jesuit schools. Therefore, concerning natural philosophy, Zupi is Aristotelian, while concerning the mechanics of simple machines, he is a follower of Baldi and, first and foremost, of Galileo. From Baldi, Zupi takes on the Aristotelian distinction between heavy and light bodies, the definition of centre of gravity and the distinction between natural, violent and mixed motion. Like Baldi, he too distinguishes between *heavy because of nature* and *heavy because of violence,* defining the first as:

> what because of its own propensity goes to the centre of the world[27]

[27] "quod insita propensione in centro mundi fertur" *Exercitationes*, 73. This definition is taken verbatim from Baldi (1621, 1).

and the second as:

> what is propelled because of an external force exerted on it.[28]

The definition of *leve*, that is, a light body, is the same as that defined by Baldi:

> what because of nature goes to the centre.

With regard to the centre of gravity of a body, Baldi says that the internal point of a body is where the whole weight of the body can be considered to be concentrated, namely:

> The point inside the heavy body, such as that by which the body is suspended, will always incline downward, that is, directly toward the centre of the universe.[29]

He then expounds three possible motions of a body: natural, violent and mixed. The first motion, due only to the action of gravity, is why heavy bodies move downward along a vertical line joining the body to the centre of earth, and why light ones move upward along the same line. The second motion, due to an external cause, compels the body to deviate from the vertical direction. The third motion, resulting from the combined action of gravity and violence, is a motion whose trajectory is composed of a rectilinear and a circular line. In this case, Zupi again follows Baldi's text and does not realize that the conclusion is erroneous because the trajectory of a mixed motion is parabolic. He also ignores Tartaglia who, although he had not understood the real parabolic nature of such a motion, affirmed that at no point could it be a rectilinear motion.

8.6 The Presence of *Le Mecaniche*

There are other parts in the *Exercitationes* where the influence of Baldi's text can be perceived but, starting from the treatment of the lever and the balance, Zupi's main reference text is Galileo's *Le Mecaniche*. This aspect is immediately clear if the introductory paragraph of the *Exercitationes*, *De utilitatibus mechanicorium instrumentorum* is compared with that of *Le Mecaniche*, *Delle utilità che si traggono dalla scienza meccanica e dai suoi strumenti*. It will become clear that the first text not only has nearly the same title as the second but, with regard to its subject, is nothing but a summary of the Latin translation of the introduction to *Le Mecaniche*.

Exercitationes
There are some people who maintain that they can deceive nature by means of mechanical instruments, and that they are able with a force, namely a power, as small as possible, to surpass every weight, and having conceived such an

---

[28] "quod per vim extrinsecam impressam pellitur" *Exercitationes*, 74. Baldi writes: "quod impressio extrinsecus pondere ab impellente pellitur," Baldi (1621, 1).

[29] "Illum punctum in corpore gravi, quod corpore undecumque suspenso semper deorsum, id est directe versus centrum universi, pellet" Baldi (1621, 1).

expectation they declare that they will do many and astonishing things. But disappointed by such a great and vain expectation, they realize that they wasted, as some say, their efforts and oil. Therefore we establish that it is not possible that with a machine, whatever it is, a power smaller than a weight can move this weight in the same time with the same speed and for the same space in which it is moved by a power equal to or smaller than the weight.[30]

Le Mecaniche
I have seen the generality of mechanics in deceiving themselves into going on to apply machines to many operations impossible for their own nature; by the success whereof they have been disappointed with their result, and likewise others have been frustrated in their hopes of success, conceived as a guarantee, given by the promises of them. I think I have found that the principal cause of these mistakes lies in the belief and constant opinion these mechanics had, and still have, that they are able with a small force to move and raise great weights; in a certain manner with their machines they deceive nature, whose instinct, or rather its most very positive constitution, is that no resistance can be overcome, but a force more powerful than that.[31]

It is obvious that Zupi had understood perfectly the innovation in Galileo's text whereby it is impossible "to deceive nature" in any way, and that mechanics is not a magic art but a science governed by strict laws. In the science of weights, power and resistance observe a "compensation principle." As enunciated by Galileo, to ignore this is the reason for the failure of such a task.

Zupi reiterates Galileo's passage containing the aforesaid principle, translating it almost literally.

Exercitationes
In order to say it more clearly, four things must be considered: Firstly, the weight to be transferred from place to place, secondly, the force, that is, the power that should move it, thirdly the space, that is, the distance between the one and the other term of this motion, namely the dexterity and delay of motion.[32]

[30] Sunt qui putent mechanicis instruentis naturae institutum fallere se posse, atque quavis minima vi seu potentia posse se quodlibet pondus superare, qua spe concepta, multa et mira se facturos praedicant. Sed tantum inani spe decepti, operam, et oleum, ut aiunt se perdidisse animadvertunt. Statuimus igitur fieri non posse, quibuscumque machinis ut potentia pondere minor, eodem tempore ac velocitate pondus per idem spatium moveat quo potentia ponderi aequalis ut eo minor.

[31] Ho visto ingannarsi l'universale de mecanici, nel volere a molte operazioni, di sua natura impossibili, applicare machine dalla riuscita delle quali, ed essi sono restati ingannati, ed altri parimente sono rimasti defraudati della speranza, che sopra le promesse di quelli avevano conceputa. Dei quali inganni parmi di aver compreso essere principalmente cagione la credenza, che i detti artefici hanno avuta ed hanno continuamente, di potere con poca forza muovere ed alzare grandissimi pesi, ingannando, in un certo modo, con le loro machine la natura; instinto della quale, anzi fermissima constituzione, è che niuna resistenza possa essere superata da forza, che di quella non sia più potente.

[32] Quod ut clarius illigitur, quattuor sunt considerando: primo pondus movendum ut ex uno loco in altrum transferendum, 2° vis, seu potentia quam tale pondus debet movere, 3° intervallum seu distantia ad teminum a quo et ad quem talis motus fieri debet seu velocitas et tarditas motus.

Le Mecaniche
Going on therefore according to the nature of these studies, let us first propose four things to be considered. Firstly, the weight to be transferred from place to place; and secondly, the force and or power that has to move it; thirdly, the distance between the one and the other term of the motion; fourthly, the time in which that motion has to be made: And the time becomes the same thing as the dexterity and speed of the motion; so determining one motion to be faster than another, which in lesser time covers an equal distance.[33]

Beyond Zupi's final part, the text, which should have been completed with "fourthly, the time, namely the dexterity... ," evidently remained in the quill of this author and is clearly the literal translation of the corresponding text by Galileo. What I would like to emphasize here is that Zupi not only quotes Galileo's "compensation principle" in the same terms as in *Le Mecaniche* but, in order to clarify it, also adopts the same example used by Galileo in his work. He explains that a weight of 100 libras can be carried by a power of 10 libras, provided it is divided into 10 equal weights of 10 libras, and each of them is carried out separately, repeating the run 10 times. This process would take ten times longer than the time necessary to transfer the entire 100 libras at once, provided that the given force remains unaltered. By means of a machine, the entire weight of 100 libras can be transferred at once, but the motion of the machine is accomplished in a time interval according to the ratio between the resistance and the employed power. In the given example, the ratio is 100:10 and therefore the time needed by the machine to move 100 libras is ten times longer than the time the machine would need to move a resistance equal to the power (10:10).

To explain the usefulness of the machines, Zupi again faithfully follows in Galileo's footsteps. In translating and summarizing the text of *Le Mecaniche,* he writes that essentially there are three benefits deriving from the use of mechanical tools. The first benefit is that, when using these tools, a smaller force is able to move in its entirety a large weight at one time. The second benefit is that some tools are suitable for working in places where it would otherwise be difficult, if not impossible, to operate, for instance, the pulley enables a bucket of water to be hoisted from a well, or a pump can empty the bilges of ships. The third benefit derives from the fact that some machines exploit natural or animal forces, as for instance, the wheel of a mill set in motion by the current of a river or by the power of a horse.[34]

Like Galileo, Zupi dwells in particular on the first benefit in order to explain the sense of his affirmation that by using a machine, a smaller force is able to move in its entirety and at the same time a great weight. This seems to contradict what he had established earlier, namely, that "nature cannot be deceived" in any way as it is

---

[33] Ci si fanno avanti quattro cose da considerarsi: la prima è il peso da trasferirsi di luogo a luogo; la seconda è la forza o potenza, che deve muoverlo; terza è la distanza tra l'uno e l'altro termine del moto; quarta è il tempo, nel quale tal mutazione deve esser fatta; il qual tempo, torna nell'istessa cosa con la prestezza e velocità del moto, determinandosi, quel moto essere di un altro più veloce che in minor tempo passa eguale distanza.

[34] All of these examples are also taken from Galilei (2002).

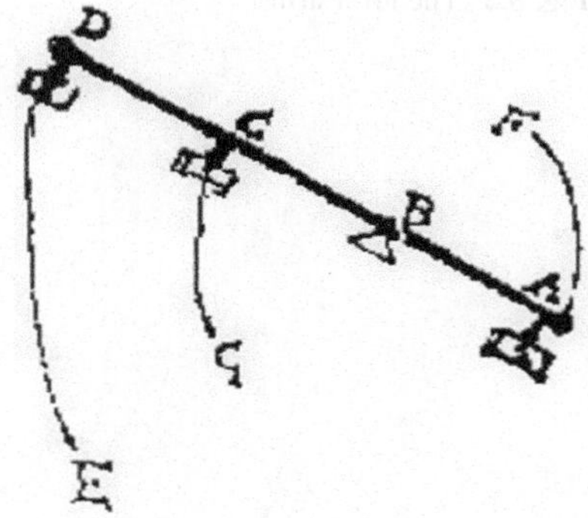

Fig. 8.1 The lever's working

impossible to move a resistance using a power smaller than it. To clarify his assertion, Zupi adds the following example, which differs from that in Galileo's *Le Mecaniche,* but only in the parts on how the fulcrum divides the lever (2 to 1 for Zupi; 5 to 1 for Galileo).

A lever AD is given whose fulcrum is set at point B such that BD = 2AB (Fig. 8.1). A two libra-weight set at the extreme A will be lifted, covering the arc AF up to point F, by an equal power placed at point C, such that BC = AB. This power also covers an arc CG equal to the arc AF, which covers resistance. But we can get the same result by employing a smaller power and applying it at a point more distant than point C, as for instance, point D. Here the two libra-resistance placed at A will be lifted by the power of only one libra, but the arc DE that this power covers will be double the arc AF covered from the resistance placed at A. And since the arcs AF and DE are contemporaneously covered in the same time, the greater arc DE will be covered with a speed greater than the smaller arc AF.

Zupi does not explain why this happens. Indeed, he has not yet demonstrated the principle of the lever and so is unable to justify his assertion.[35] Zupi showed that if the power needed to lift a given resistance is halved, the line of the arc ideally drawn by the resistance during its motion duplicates. In this way, Zupi aims at justifying Galileo's "compensation principle," according to which there is a compensation among weight, space, time and velocity, so that whatever is gained in force, the same amount is lost in velocity and time.

Zupi is only interested here in clarifying the concept of distance because, as he writes:

> There are some who are greatly wrong in measuring the distance that is between the fulcrum of the lever, that is, the centre of the balance, and the weight, that is, the power, and consequently they are wrong when building machines.[36]

[35] On the contrary, Galileo introduces this example only after having demonstrated the principle of the lever.

[36] "sunt etiam qui in mensuranda distantia, quam intercedit inter fulcimentum vectis, seu centrum librae et pondus, seu potentia, maxime errant, unde in fabricandis machinis falluntur" *Exercitationes*, 76.

Fig. 8.2 The lever arm

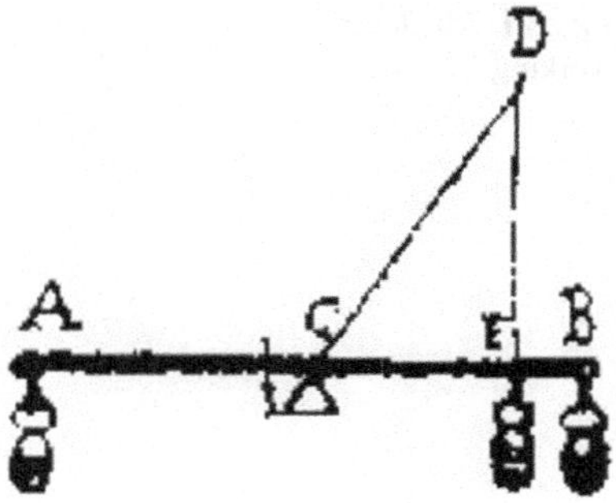

In order to explain his assumption, he considers a balance AB with equal arms on which equal weights are suspended from the extremes A and B (Fig. 8.2):

> then it is certain that these weights have their moment and the centre of gravity in the middle point C, suspended from which they will balance each other.[37]

But if the extreme B is moved to D:

> the weight A will come down, and the moment will be not at B, but at F[38]

because the distance must not be measured from point C to point D, but from point C to point F where the perpendicular from D intersects CB, and as CF < AC, the moment of the weight at A will be greater than the moment of the weight at D and therefore the balance will incline to the side A.

In this case, Zupi also refers to the concept of the moment that he has not yet defined. He therefore appropriates some parts of Galileo's *Le Mecaniche* but he does not follow its logical and rigorous method. Galileo in fact defines first the moment and then, in a special chapter, clarifies some particular aspects, such as the distance connected to this concept.

8.7 Treatise I (Lever and Balance)

At the root of Zupi's theory of the simple machines is the lever, geometrically considered as a line of invariable length characterized by a point of support, the fulcrum, whose position as regards the moving power and the resistance to be moved determines its kind. Zupi explains that there are two different classes of lever: one in which the fulcrum is between the power and the resistance and another in which the resistance is between the fulcrum and the power.

Before dealing with the lever, Zupi briefly dwells on the principles of pseudo-Aristotelian mechanics. In this case, he again seems to be a faithful follower of

---

[37] "tunc certum est, haec pondera habere suum moment et centrum gravitatis in medio C a quo puncto suspense, aequiponderabunt" *Exercitationes*, 76.

[38] "descendet pondus A, neque momentum erit in B sed in F" *Exercitationes*, 76.

Baldi, and like him, shows that the four properties of the circle, which are the foundation of pseudo-Aristotelian mechanics, are not "miraculous" but simply properties inherent in the geometric nature of the circle. But it should be noted here that the presence of these principles in the *Exercitationes* is an end in itself: the pseudo-Aristotelian theory is present throughout the treatise.

Zupi introduces the theory of the lever giving 7 axioms:

1. equal weights, suspended from equal rays or from equal arms, are balanced and are sustained by equal powers
2. a weight, suspended either higher or lower, has always the same weight
3. given equal weights, that which is suspended from the longest arm, is the heaviest one
4. the centre of gravity of a body is only one
5. this centre of gravity, with respect to its body, is always in the same place
6. this centre of gravity carries the bodies downward
7. with regard to the quantity, the body is gravity and moment. In fact there are often things that are equal about the body, but are unequal about their gravity, etc.[39]

The first and the third axiom, put together, reproduce the first axiom of Book 1 of Archimedes's *Equiponderantis*. The second axiom establishes the invariance of the weight with the height of its point of suspension. Axioms 4, 5 and 6 concern certain properties of the centre of gravity of a body. Finally, in the last axiom, Zupi distinguishes between gravity and moment establishing that equal bodies can have different gravity and moment. Indeed, after gravity the sentence stops with "et cetera," but we can argue that Zupi intended to write "moment." In fact, immediately afterwards, he gives the same definition of moment given by Galileo in *Le Mecaniche*:

> moment is gravity of the bodies composed of the natural propensity of the body to go downward and of the position and the location of the same body, for which a body that hangs from the longest arm of the balance is said to have a greater moment.[40]

---

[39] "Primo: aequalia pondera ab aequalibus radiis, seu brachiis suspensa, aequiponderarent et potentia aequali sustineri; 2° pondus sive sublimius, sive humilius suspensum aequiponderare; 3° aequalium ponderum, quod maiori brachio suspensum est, magis ponderare; 4° unius corporis, unum tantum esse centrum gravitatis; 5° hoc centrum gravitatis respectu sui corporis semper in eodem esse situ; 6° suum hoc gravitates centrum pondera deorsum ferri; 7° aliud esse in re quanta corpus gravitatem et momentum. Saepe enim aliqua sunt aequalia in ratione corporis, et inaequalia sunt gravitatem, etc." *Exercitationes*, 79.

[40] "Momentum est gravitas corporis compositum ex naturali propensione rei gravis ad motum deorsum, et ex positione, et collocatione eiusdem rei; qua re, corpus, a longiore librae brachio pendet, maius habere moment dicitur." Galileo's definition is: "Moment is the propensity of descending, caused not so much by the gravity of the mobile body, as by the disposure which divers grave bodies have in relation to one another; by means of such a moment, we often see a body less grave to balance another of greater gravity: as in a steelyard, a great weight is raised by a very small counterpoise, not through the excess of gravity, but through the remoteness from the point hereby the beam is upheld, which conjoined to the gravity of the lesser weight adds there unto moment, and impetus of descending, wherewith the moment of the other greater gravity may be exceeded. Moment then is that impetus of descending, compounded of gravity, position, and the like, whereby that a propensity may be occasioned" *Exercitationes*, 79.

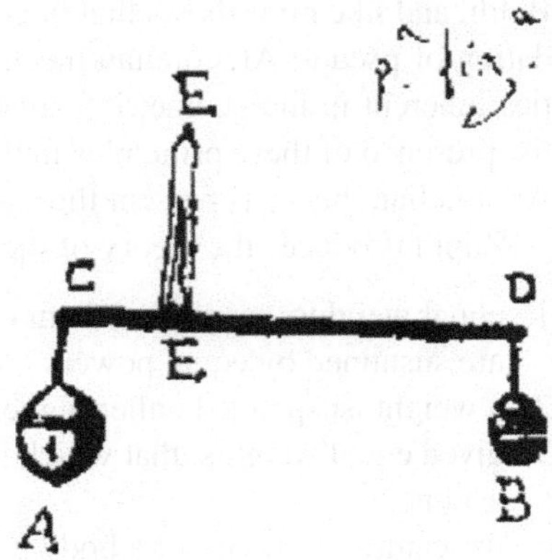

Fig. 8.3 The first class lever

At this point Zupi introduces the law of the lever, but does not demonstrate it. In fact, he writes:

> First of all I say that the power, sustaining a weight hanging from a lever, with this weight has the same ratio that the part of the lever between the fulcrum and the weight has with the part of the lever between the fulcrum and the power.[41]

Therefore, with reference to Fig. 8.3, we have:

$$A:B = CE:ED \tag{1}$$

So, if AB = 3 BC then to sustain the weight at C it will be enough to apply at A a power equal to one third of that weight. Therefore, as regards this class of lever, Zupi concludes that the closer point C, which is supporting the weight, is to the fulcrum, the smaller the power to apply at point A to sustain it, so that the power at A is always smaller than the resistance placed at point C inside AB. But if the fulcrum is between the resistance and the power, namely the lever is of the second class, then:

> the ratio between the power and the weight is the same as that between the distance of the weight from the fulcrum and the length of all of the lever[42]

that is, referring to Fig. 8.4:

$$A:C = BC:AB \tag{2}$$

[41] "Dico primo potentia sustinens pondus vecti appensum eandem ad ipsum proportionem habet quam habet pars vecti inter fulcimentum et pondus ad partem vectis inter fulcimentum et potentiam" *Exercitationes*, 79.

[42] "eadem erit proportio potentiae ad pondus quam est. inter distantiam ponderis a fulcimento, ad totum vectem" *Exercitationes*, 81.

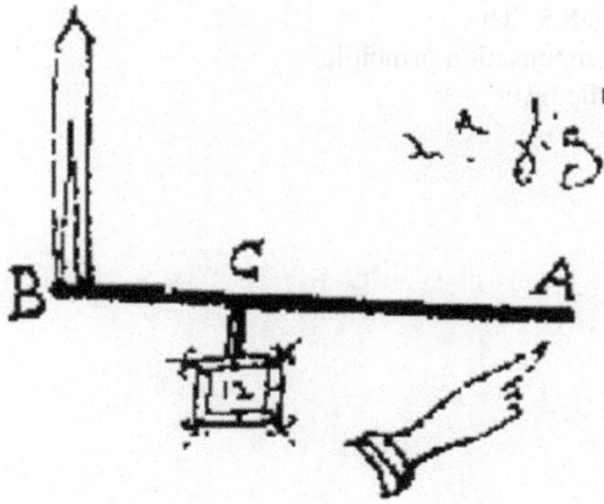

Fig. 8.4 The second class lever

Again, Zupi does not demonstrate the law (2).[43]

This class of lever gives Zupi occasion to mention *Problem XXVIII* of the *Quaestiones mechanicae*, which, at that time, caused much interest among the scholars:

> Why when two men together carry a weight with a long stick or something similar, they are not equally burdened when the weight is not at the middle point, but the carrier nearer to the weight is mostly burdened.

Introducing the question, Zupi points out that (2) allows us to deduce:

> what part of the weight two powers can sustain if another power is set at the place of the fulcrum.[44]

Indeed, the question deserves to be analyzed with theoretical reasoning but once again Zupi uses an example which, although it does not have any demonstrative value, is nevertheless very apt to make the scholars understand how the forces are distributed in this case.

Undoubtedly, Zupi's behavior is justified only if we consider his pedagogic concerns and in particular the need to choose an example with a possible paradigmatic value, like a general rule to apply whenever a similar case occurs. A weight of 12 pounds is suspended from point C of a lever AB so that BC = 1/4AB and consequently CA = 3/4AB (Fig. 8.4). Then the power at A sustains 1/4 of the whole weight, that is, only 4 pounds. If in the place of the fulcrum B another power sustaining the same weight is set, we will deduce that this one sustains 8 pounds.

[43] In reality he wants only to show that (1) and (2) allow the various problems of the two different classes of lever to be solved: to determine the position of the fulcrum such that a given power can lift a given resistance; to determine the power or the resistance once assigned the place of the fulcrum and the resistance or the power and so on. Zupi does not solve these problems arithmetically, that is, by determining the unknown term of the proportion (1) or (2), but practically, dividing materially the lever in a convenient number of parts and placing, according to the cases, fulcrum, power and resistance in such a way as to satisfy (1) or (2).

[44] "quantum ponderis substineant duae potentiae si loco fulcimenti statuatur alia potentia" *Exercitationes*, 81.

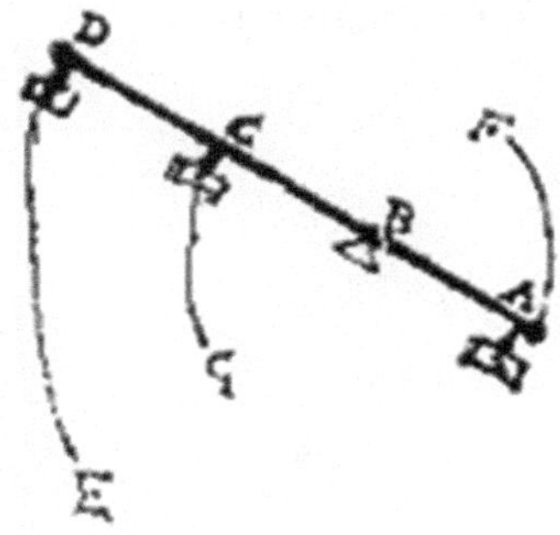

Fig. 8.5 The "compensation principle" in the lever

We must believe that Zupi's resolution was inspired by Baldi, who had successfully solved *Problem XXVIII* starting from the basis that each of the two support points on the transporters' shoulders, A and B, was the center of fulcrum and power. More precisely, if A is considered the fulcrum and the power is applied at B, then it will occur that.

$$A:C = BC:AB$$

that is, the (2); on the contrary, if B plays the role of fulcrum and the power is applied at A, then it will occur that.

$$B:C = AC:AB$$

Thus, as in Fig. 8.4, point C is closer to the extreme B. Since $AC/AB<BC/AB$, $A < B$ and the transporter at B will sustain a weight greater than that at A. It is evident that for his example Zupi had considered only the first part of Baldi's demonstration, namely, the case of fulcrum at A, but it is also clear that he had contemplated this case only to simplify the question.

Zupi concludes this topic by applying Galileo's "compensation principle" to the lever. He explains that with a lever it is possible to lift a weight greater than the employed power, but not in the same time, that is, at the same velocity by which the power can lift a weight equal to itself. The lever AB, whose fulcrum is at B and whose distance DB is twice the distance BA (Fig. 8.5) is able to lift a weight placed at D through a power applied to A, but in the same period of time in which A covers the arc AF, D covers the arc DE which is twice as long as AF. On this occasion, Zupi does not demonstrate his assertion. He restricts himself concluding:

> From this we can clearly deduce that what we acquire with a lever we will lose in velocity.[45]

[45] "Ex quo aperte colligitur quantum medio vecte facilitates acquiritur tantum derogari de velocitate," that is, the same as what Galileo himself affirms in Galileo (2002, 61, ll. 601–602): "comprendesi, come negli altri strumenti, in questo ancora, quanto si guadagna in forza, tanto perdersi di velocità," that is: "we understand that, as in the other instruments, also in this one, the more it acquires in strength, the more it loses in velocity" *Exercitationes*, 82.

8.8 The Balance

The above-mentioned considerations concern a weight hanging from one of the two extremes of a lever. Zupi completes the matter by also considering the case of a weight supported above or below the arm of the lever or of the balance, that is, an equal-arm lever. As a model, he assumes Baldi's treatment the theory of Guidobaldo,[46] who had considered all possible cases inherent in this instrument. But, unlike Baldi and Guidobaldo, Zupi systematically uses the concept of moment producing a very simple theory completed with some important aspects of Galileo's theory of balance, such as the correct consideration of the distance between the fulcrum and the weight. Zupi considers all the three possible cases occurring when the weight is supported by a lever: the center of gravity of the weight is above the lever, it is below it, or it coincides with one of its points. In the first case

> when the lever is inclined as regards the horizon, such as in the case of the lever IBG and of the lever FBH, then we must measure the distance between the fulcrum and the point of the lever where the perpendicular to the horizon falls, like point O in the lever FBH and point E in the lever IBG, in which the perpendiculars DO and DE fall.[47]

When the weight is lifted (see HBF in Fig. 8.6) the perpendicular from its center of gravity to the horizon (*linea directionis*) is the line DO, and it is as if the weight hung from point O, which is distant from the fulcrum B, is less than point C. On the contrary, when the weight is lowered (see GBI) the perpendicular from its center of

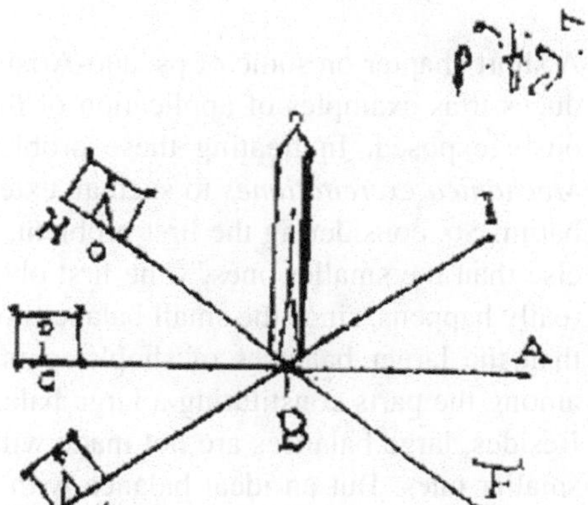

Fig. 8.6 Weight supported above a lever

[46] See Dal Monte (1577).

[47] "cum scilicet vectis horizonti est inclinatus ut in vecte IBG etiam FBH, tunc enim adhuc mensuranda est distantia inter fulcimentum et punctum vectis in quod cadit perpendicularis ad horizonte ducta, ut in vecte FBH est punctum O et in vecte IBG est punctum E in quae puncta perpendiculares, DO, et DE cadunt" *Exercitationes*, 83. This is what Galileo also maintained in the chapter *Alcune considerazioni sopra le cose dette* in Galileo (2002, 53–54).

gravity to the horizon is the line DE, and it is as if the weight hung from point E, which is more distant from the fulcrum than point C.

Zupi could therefore conclude that

> while the weight is lifted, the direction line on which the center of gravity is drawn nearer to the fulcrum, and [the weight] acquires smaller moment, and consequently the ratio between the lines OB and BA is smaller than that between the lines CA and BA.[48]

Then, in such a position, the weight would be sustained by a power less than that necessary when the lever lies in the horizontal position. Precisely the opposite occurs when the weight is lowered because in this case the moment of the resistance increases, and consequently a greater power will occur to counterbalance the lever.

The same criterion is valid if the center of gravity of the weight is below the lever: the moment of the resistance increases or decreases according to the point of intersection of the perpendicular from the center of gravity of the weight, with the horizontal being more or less distant to the fulcrum of the lever with respect to the equilibrium horizontal position. Finally, if the center of gravity of the weight is a point of the lever, Baldi shows that

> whether it is lowered, or lifted up, the same power will always be required to sustain it.[49]

And the same occurs with the lever of the second class.

8.9 The Resolution of some of Pseudo-Aristotle's Problems

A short chapter on some of pseudo-Aristotle's problems then follows. Zupi introduces it as examples of application of the general theory of the machines previously exposed. In treating these problems, he constantly refers to Baldi's *In Mechanica exercitationes* to such an extent that he sometimes cites it almost verbatim. So, considering the first problem, "why the largest balances are more precise than the smaller ones,"[50] he first observes that this seems to contradict what really happens, since the small balances of jewelers are more sensitive and precise than the larger balances of shopkeepers. But first he explains that the attrition among the parts constituting a large balance is greater than that of a smaller one. Besides, large balances are not made with the same detailed care and skill as the smaller ones. But an ideal balance with no attrition is much more sensitive and precise the longer its arms are. In fact, the wider the arcs described from the ends of the arms, the more precise are the measures valued on them, and the longer arms

48 "cum deprimitur pondus linea directionis magis recedit a fulcimento, maiusque momentum habet et consequenter maiorem proportionem habet linea OB ad BA, quam CB ad BA, igitur et maiori indigebit potentia" *Exercitationes*, 83–84.

49 "sive elevetur vectis, sive deprimatur, semper pondus ab eadem potentia sustinebatur" *Exercitationes*, 85.

50 "cur maiores librae esactiores sint minoribus" *Exercitationes*, 85.

will describe wider arcs. He also explained that the frauds perpetrated with the balance have two possible causes: the imperfect equality of the length of the two arms or the imperfect equality of the weights of the two arms. In both cases, it happens that equal weights placed at the ends of the two arms do not have equal moments, so to put the goods at the end of the longest or heaviest arm to achieve equilibrium, it will be necessary to put a weight greater than that of the goods at the end of the other arm.

The second pseudo-Aristotelian problem asks why a balance with the *spartum* at the top, when removed from its state of horizontal equilibrium, returns to its original position once the cause of such a removal stops. But if the *spartum* is below the balance, however, it stays in the acquired position. Zupi solves this question in the same way as Baldi, but because he systematically uses the concept of moment, his theoretical results are simpler than Baldi's. Like Baldi, he begins by criticizing the pseudo-Aristotelian resolution, writing:

> In this problem nearly everyone is wrong in drawing the figure, so that it is very hard to achieve the meaning of the author.[51]

And although Figs. 8.7 and 8.8 are in his text, Zupi in his reasoning actually refers to Figs. 8.9 and 8.10 in Baldi's treatise, which he does not reproduce. In fact, he explains that the balance AH, in horizontal equilibrium, has its center of gravity at C and its *spartum* at D (Fig. 8.13).

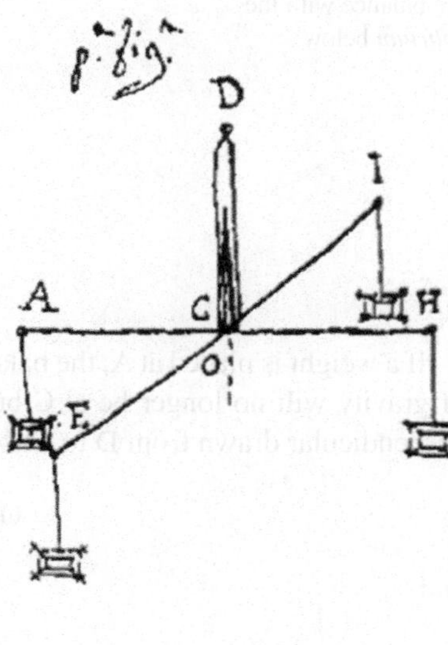

Fig. 8.7 Balance with the *spartum* at top

[51]"In hac questione fere omnes in schemate lineando errant ita ut difficile admodum sit authoris sensum assequi" (*Exercitationes,* 87) where Baldi writes "Pessime omnes schema hoc linearunt, ita ut difficillimum sit autoris inde sensum assequi," see Baldi (1621, 19).

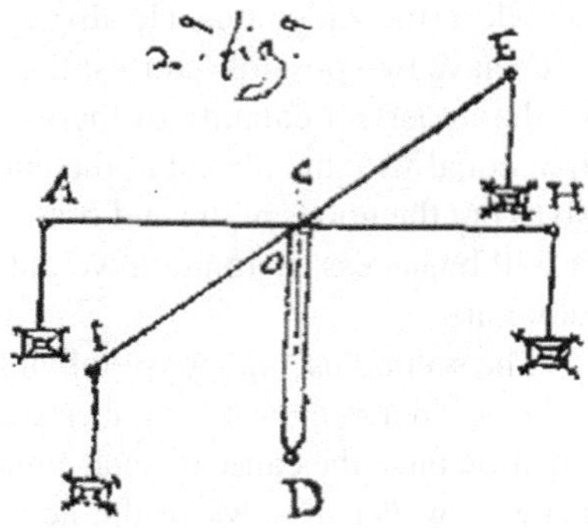

Fig. 8.8 Balance with the *spartum* below

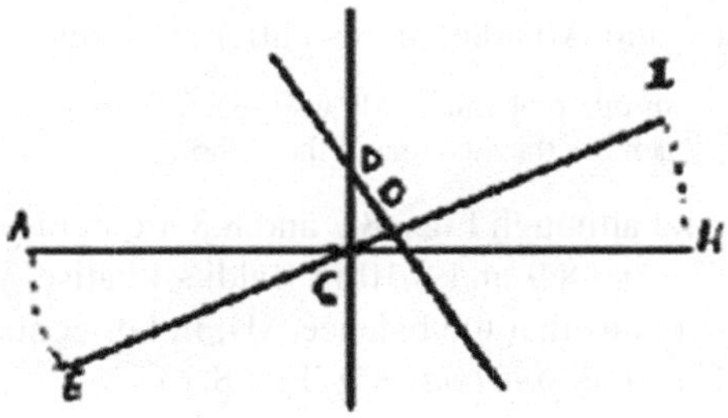

Fig. 8.9 Baldi's figure of the balance with the *spartum* at top

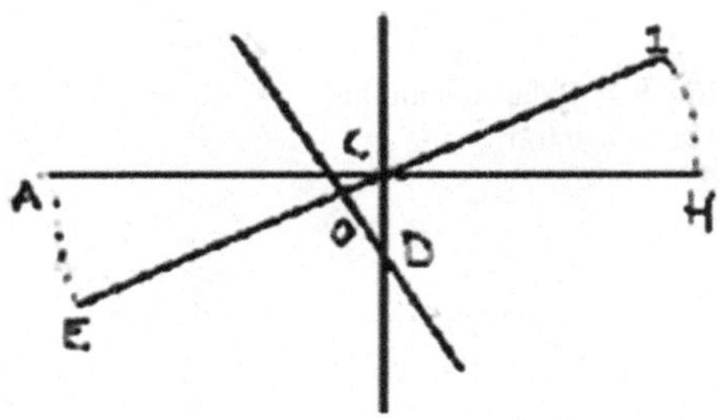

Fig. 8.10 Baldi's figure of the balance with the *spartum* below

If a weight is placed at A, the balance will assume the position ECI and the center of gravity will no longer be at C but will be moved to O, O being the foot of the perpendicular drawn from D to EI. And since.

$$OE = EC + CO$$

and

$$IO = IC - CO = EC - CO$$

the arm of the balance where the weight is will be longer than the other one. Once the weight at E is removed, point O, which inclines downward, will join point C and the balance will assume its horizontal position again.

The same thing cannot happen in the case of a balance whose *spartum* is below because, if it is removed from its horizontal position because of a weight placed at A, its center of gravity will move to point O and downward to point C and the balance will assume the position EOI (Fig. 8.10). Once removed from E, the weight that had been placed at A, point O, will not return to C, but on the contrary, will also continue to go downward, producing a rotation of the arms of 180° around O, thus turning into a balance with the *spartum* placed upward.

Like Baldi, Zupi also considers the case of the *spartum* coinciding with the center of gravity of the balance. He says that in this case

> if they are perfect, they will stay in the same place, either they are lower down, or they are lifted up.[52]

In fact, the perpendicular through the *spartum* always divides the balance into two equal parts so that there is no reason to remove the arm of the balance from its current position.

Zupi's entire argumentation is founded on the assumption that the center of gravity must always be on the perpendicular to the arm of the balance passing through its suspension point. Starting from this assumption and reproducing all the examples shown by Baldi, Zupi explains why it is almost impossible that a pole with a pointed base stays upright by itself. The difficulty is due to the fact that, even the most skillful artisan cannot prevent the presence of a little imperfection, which would cause the center of gravity of the pole not to be on its perpendicular axle.

The same thing happens to the spinning top, whose center of gravity, because of the defects of its material and the inevitable imperfections of its construction, is always out of the perpendicular to the horizon through its tip. Therefore, until the top is spinning it stays upright, because its center of gravity also turns around the same axle of rotation and the movement does not allow it to incline and fall to any side. In fact, while inclining to the right it will be pushed left, and vice-versa.

The discussion of *Problems V* and *VI,* concerning the rudder and the sail of the ship respectively, closes this chapter. The first asks how it is possible that the force of only one man can move a very heavy ship by means of a small rudder set in the stern; the second wants to know why, if the sails and the wind of two ships are equal, the one with the highest mast, that is, with a higher topsail, moves faster. These are two of the four problems that pseudo-Aristotle devoted to navigation and that Baldi solved very brilliantly with a rich and well-organized treatment, which can be actually be considered as a small essay on the art of navigation. It is not difficult to recognize that in solving the aforesaid problems, Zupi was an attentive reader of *In Mechanicas exercitationes*. He reassesses Baldi's reasoning in light of the concept of moment being able to gather and clearly illustrate its most meaningful results. Following Baldi, Zupi begins by confuting pseudo-Aristotle's solution according to the fact that the rudder is a lever whose fulcrum is the hinge connecting it to the ship, the sea is the resistance, while the force of the helmsman is the moving power. But Zupi, having in mind Baldi's demonstration, affirms that:

[52] "si perfectae sint, in eodem situ manere, sive deprimantur, sive elevantur" *Exercitationes*, 88.

this fact does not really explain how it can happen that, by a small movement of the rudder, such a great movement in the bow is generated.[53]

In order to explain why the small movement of the rudder is able to give the ship such a great boost, it is necessary to consider not the sea as resistance but the fulcrum. The resistance is the ship and the power is the force of the helmsman.

As regards *Problem VI*, Zupi also limits himself to an enunciation of Baldi's conclusion, without giving any justification.

The reason is that the spar, namely the tree of the ship, is like a lever whose fulcrum is at the bottom of the ship, the weight is at that part where the tree comes out of the hull, the moving power is the wind that swells the sails. So, the longer the spar, the longer the lever, and the ship is moved more easily by the wind. Here it is necessary to retain that if the ship is straight, the stern of the same ship should be lifted by the wind and the bow should be submerged into the water, because, really, the water being a liquid, the ship is not always straight and is pushed forward by the wind.[54]

This question deserved a wider and more detailed discussion since it introduced a new type of lever, namely the angular lever, which caused great debate at the time. Zupi himself had largely treated every aspect of this matter in letters to Davide Imperiali, which can be considered exemplary lessons. So the few lines now devoted to the angular lever in the *Exercitationes* cannot be interpreted as a demonstration of the scant interest in this particular kind of tool. I am convinced that here, as in other points of his treatise, Zupi simply intends to take note of a matter that he should subsequently develop, or that should be derived from other authors such as Baldi or Galileo.[55] Zupi could also have intended here to introduce the parts of the question that he had already treated in his correspondence with Imperiali. This conviction is based on what Zupi himself writes in concluding this matter:

the issue could also be explained in another way (*Exercitationes*, 90).

This brings to mind the treatment of this subject in the aforesaid correspondence.[56]

[53] "Verum hoc non esplicat quo modo fieri posse ut parvo existente motu gubernaculi tam magnus in ipsa prora oritur motus" *Exercitationes*, 89.

[54] "Ratio est, quia malus, sive navis arbor, est veluti vectis cuius fulcimentum est in fundo navis, pondus vero est in ea parte ex qua malus e carina, exit potentia movens est ventus in velum inpingens. Igitur quo sublimior est antenda eo longior evadit vectis, atque facilius movetur navis ab eodem vento. Hic notat quod si navis rexisteret a vento deberet puppies illius elevari, et prora in aquas demergi, quia vero propter maris liquiditatem, navis non resistit semper a vento impellitur ad partes anteriores" *Exercitationes*, 89.

[55] To confirm the aforesaid theory, Zupi adds an argument drawn by Baldi, that is, because the sails are more aloft they are more exposed to the wind "such as the towers, that the higher they are, the more they are hit from the winds" "sicut turres, quo sublimiores sunt eo magis a ventis infestantur," see *Exercitationes*, 90.

[56] One of the most important subjects Zupi treated in this correspondence is how, in a lever, in computing the moments the arms of the power and of the resistance have to be considered. And this especially in consideration of the angular lever, a matter that Zupi explains in detail to his scholars. This correspondence is published in an appendix to Gatto (1996).

8.10 The Second Treatise

The second treatise consists of four chapters devoted to the pulley and to the block and tackle. Zupi begins with the fixed pulley, that is, the simplest case of this kind of machine, and goes on to build up step by step the general theory of the block and tackle, considering the different ways to increase the number of the pulleys used. He generally follows the text of Galileo's *Le Mecaniche*, but also introduces some of his own innovative elements.

At the base of this theory lies the demonstration that the pulley is a lever. The fixed pulley in particular is a lever of first class, whose fulcrum is the center I of the same pulley and the power and the resistance are applied at the ends of the rope wound around it, which is the same as considering them to be applied at the extremes A and B of the diameter AB parallel to the horizon (Fig. 8.11).

Since the power and the resistance are applied at the same distance from the fulcrum, this kind of pulley is simply a balance with equal arms. And nothing changes when the power lifts the weight, causing the pulley to rotate around its

Fig. 8.11 Fixed pulley

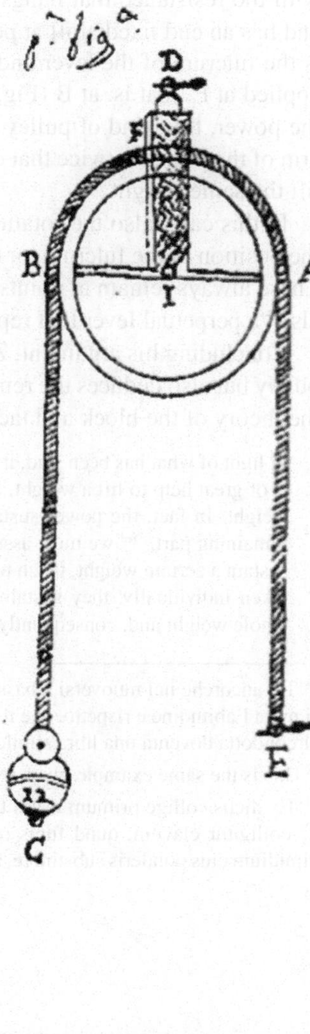

center, because the points of application of the power and of the resistance are always A and B. The fixed pulley can be considered as a "perpetual balance" which, rotating around its center, continually reproduces itself. The idea of the perpetual balance belongs to Galileo who in *Le Mecaniche* at this point writes:

> And, as moving downward the force F turns around the pulley ABC, the weight and the force do not change yet the behaviour and the position as it concerns the two distances AD, DC; yet, in turning the pulley, it becomes a balance similar to the AC, but a perpetual balance.[57]

Reproducing the same arguments developed by Galileo, Zupi explains that since the moments of the forces involved are equal, the fixed pulley is not advantageous but very convenient in a many everyday activities, for example, in drawing water from a well. When a bucket of water is lifted from the well using only the arms, not only the weight of the aforesaid bucket is lifted, but also that of the arms of the person lifting. In lifting the same bucket of water with a fixed pulley, the benefit gained is that the weight of the body of the lifting man can be added to the strength of the person's arms (Fig. 8.11).[58]

The mobile pulley is a lever of second class, so called because it is lifted together with the resistance that hangs from its center. The rope winds it from underneath and has an end fixed aloft at point C so that the line of the rope CA, that is point A, is the fulcrum of the lever and the other end is sustained from the power, which is applied at E, that is, at B (Fig. 8.12). As the resistance is between the fulcrum and the power, this kind of pulley is therefore a lever of second class, and because the arm of the power is twice that of the resistance, a power equal to half the weight will lift the same weight.

In this case, also the rotation of the pulley does not change anything as regards the position of the fulcrum, or the application points of the resistance and the power. These always remain at points A, I and B respectively, so that this kind of pulley is also "a perpetual lever that reproduces itself."

Concluding his argument, Zupi underlines not only the advantage of this type of pulley but also deduces the remarkable property that is a fundamental assumption of the theory of the block and tackle:

> In light of what has been said, it is possible to understand, 1st that this second type of pulley is of great help to lift a weight. 2nd, that the nail, which the rope is tied to, sustains half the weight. In fact, the power sustains half the weight so therefore the nail will sustain the remaining part. 3rd we must assume that if there are two powers that with the same pulley sustain a certain weight, taken together they must be equal to the weight. The reason is that taken individually, they sustain half the weight, and therefore both together sustain the whole weight and, consequently both together are equal to that.[59]

---

[57] "Ed ancorché nel muoversi a basso la forza F si venga a girare intorno la girella ABC, non però si muta l'abitudine e rispetto, che il peso e la forza hanno alle due distanze AD, DC; anzi la girella circondotta doventa una libra simile alla AC, ma perpetuata" Galilei (2002, 62, ll. 630–633).

[58] This is the same example given by Galileo (2002).

[59] "Ex dictis collige primum usum trocleae hoc 2° modo maxime iuvare ad ponderis elevationem, 2° colligitur clavum, quod funis religatur dimidium ponderis sustinere. Ponitur enim potential dimidium eius ponderis substinere, igitur clavus reliquam partem sibstinebit. 3° si duo erunt poten-

Fig. 8.12 Mobile pulley

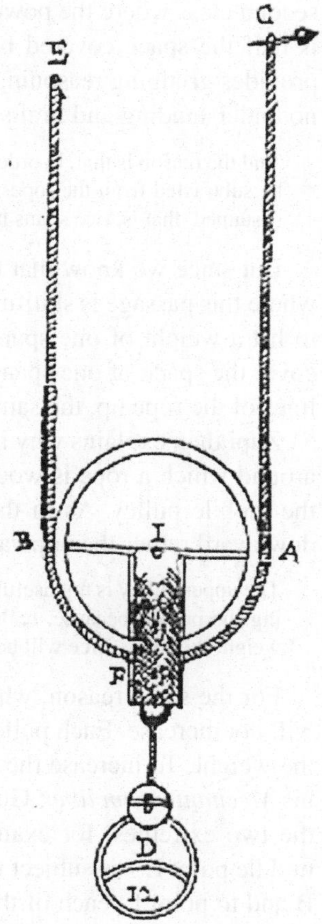

With this reasoning, inspired by the analogous theory formulated by Galileo in *Le Mecaniche*, Zupi establishes the fundamental principle of the theory of the block and tackle, namely, that every vertical line of rope passing around a pulley sustains a part of the total weight being lifted.

Before dealing with this theory, Zupi remarks that also in the case of the pulley, the "compensation principle" established by Galileo, continues to hold:

> a diminution of the power cannot be achieved by the speed not being changed.[60]

In the case of the fixed pulley, which is reducible to a balance with equal arms where the power is equal to the resistance, the space covered by the power is equal to that covered by the resistance. In the case of the mobile pulley, that is, a lever of

---

tiae beneficio huius trocleae pondus aliquod substinentes, quod utraque simul esse deberet ponderi aequalis. Ratio est quia singulae dimidium ponderis substinent, atque utraque simul totum pondus substinet; et consequenter utraque esset illius aequalis" *Exercitations*, 92.

[60] "non posse haberi potentiae diminutionem remanente eadem velocitate" *Exercitations*, 92.

second class where the power is half the resistance, the space covered by the power is half the space covered by the resistance. In explaining this second case, Zupi provides grudging reasoning so that he sometimes appears to be in the dark or with no understanding and limits himself to writing:

> And the reason is that, in order for the weight to raise the space of one span, two spans must be subtracted from the ropes of the second figure [see Figure 12] by which the weight is sustained, that is, two spans must be subtracted from the ropes with the power.[61]

But since we know that the source of this treatment is Galileo's *Le Mecaniche*, where this passage is skillfully explained, we succeed in understanding that in order to lift a weight of one span, both lengths BE and AC of the rope (Fig. 8.12) will cover the space of one span, but as it is the power applied to E that pulls these two lines of the rope up, the same power will uncoil a length of rope of two spans.

Zupi then explains why it is opportune to add to the mobile pulley a fixed pulley around which a rope is wound, which is coiled up after having been wound under the mobile pulley. As in the case of the fixed pulley, the power has been pulled downward rather than upward (Fig. 8.13). Zupi immediately remarks that.

> The upper pulley is not useful for diminishing the forces and is only advantageous in applying the power, because, really, if the upper pulley is taken away, in order to sustain the weight the same force will be required.[62]

For the same reason, when two or more pulleys are placed in series, the power will not increase. Each pulley is in fact a balance where the power must be equal to the weight. To increase the power, the pulley must be placed in a different way. In his *Mechanicorum liber,* Guidobaldo showed that if the lever AB—of which one of the two extremes, for example A, is the fulcrum, sustaining the weight G at its middle point L—is subject to the action of two equal powers applied to the extreme B and to point L, each of them will sustain the third part of the weight. Zupi demonstrated the same proposition in terms of moment. The power at L being equal to that at B and AB = 2AL (Fig. 8.14), the moment at B is the double that at L. Therefore the power at B will sustain double the weight sustained by the same power, but without the help of the lever, whereas the power at L will sustain the part of the weight equal to itself. So, if we divide the weight into three equal parts, two of them will be sustained by the power at B and one by the power at L. At this point Zupi remarks that since AB is a lever of second class, the power at B will sustain two of the three parts of the weight, not by itself, but "*adiuta a fulcimento* at A," that is, with the help of the fulcrum at A. In fact, on the fulcrum another power acts, which, together with the weight at G, has the same ratio that the power at B has together

[61] "Et ratio est quia, ut pondus ascendat per spatium palmare, debet ex duobus funibus secundae figurae, quibus pondus substinetur detrahi duo palmi quod fit cum potentia duos palmos funi trahit" *Exercitations*, 93.

[62] "trocleam superiorem esse inutilem ad diminutionem virium et solum maxime iuvare ad applicandam potentiam, quod quidem et superiori troclea detracta eadem vis requiritur ad pondus A substinendum" *Exercitations*, 94.

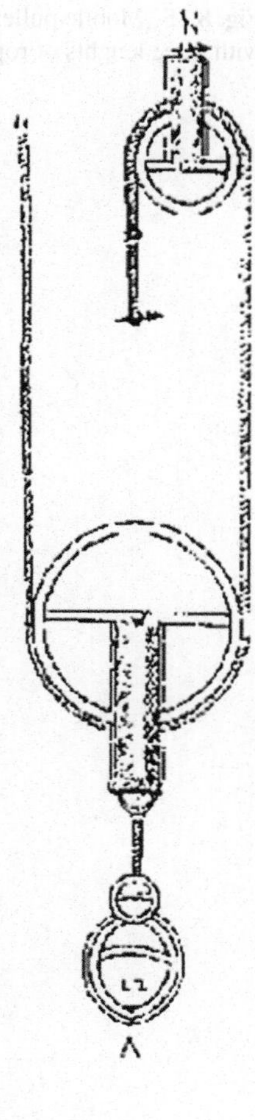

Fig. 8.13 Mobile pulley with the support of a fixed pulley

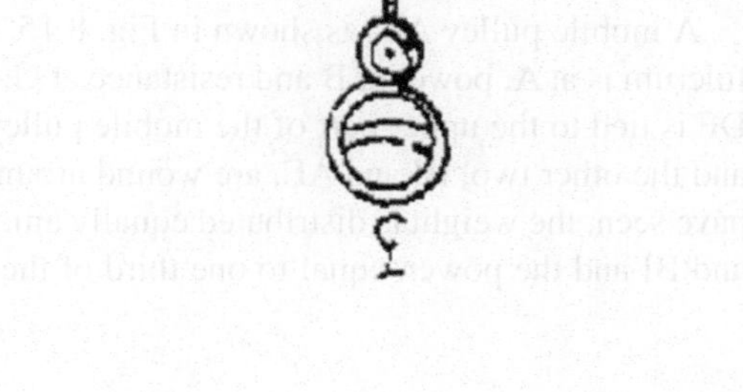

Fig. 8.14 The mobile ulley as a lever

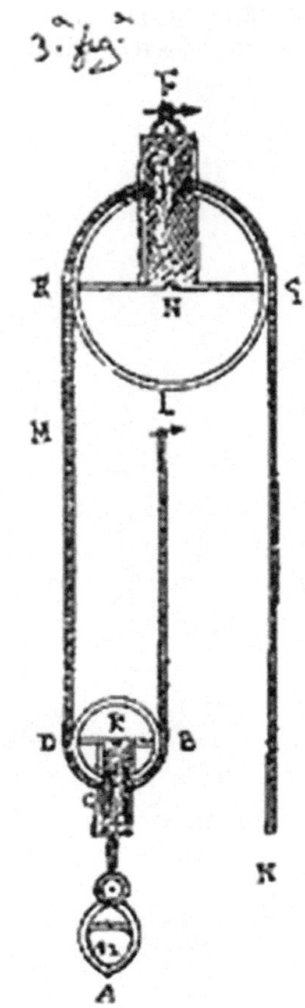

Fig. 8.15 Mobile pulley with three lenghts of rope

with the weight. The remaining two thirds of the weight are equally sustained by the power at B and by the fulcrum at A.

In conclusion, each of the two equal powers, applied to B and to L, will sustain the third part of the weight suspended from L. So, if a weight G is suspended at the middle points E and F of two levers AB and CD and their fulcrum is sustained at point B and D respectively, the first by a power applied to A and the other by a power applied to C, then each of these two powers will sustain one third of the weight G. This result is fundamentally important for the theory of the block and tackle since all the laws concerning the multiplication of the power by means of the block and tackle arise from it.

A mobile pulley AB, as shown in Fig. 8.15, is a lever of the second class whose fulcrum is at A, power at B and resistance at G. It has three lengths of rope, the first DF is tied to the upper part of the mobile pulley sustaining the weight at its center, and the other two, BI and AE, are wound around the same pulley. Therefore, as we have seen, the weight is distributed equally among the three lengths of rope AE, DF and BI and the power, equal to one third of the weight applied at I, will be enough

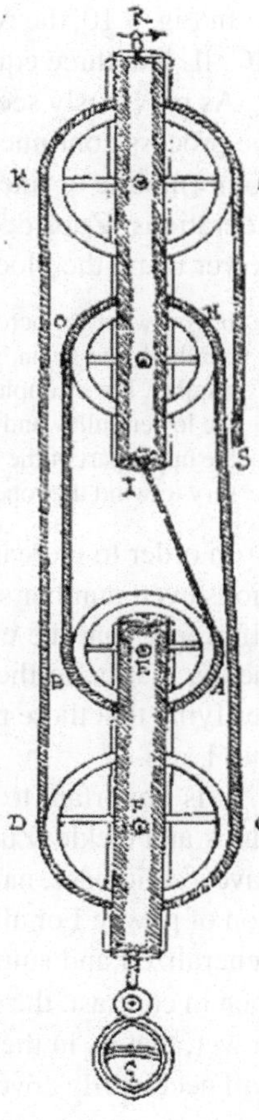

Fig. 8.16 Block and tackle with four lenghts of rope

to lift it. Zupi clarifies once more that the presence of the upper pulley does not increase the power but facilitates the application of the same power.

If a second pulley is added to the lower part of the pulley sustaining the weight, and the end I of the rope is tied to the upper part of the second fixed pulley (Fig. 8.16), we attain a block and tackle that functions in the same way as the two joined levers in Fig. 8.14, BA and DC, respectively, with the fulcrum at B and at A, the resistance at their centers E and F and the powers applied to A and to C. The weight will therefore be equally distributed among the four lengths of rope KD, OB, IA, NC, and the power equal to one fourth of the weight, applied to the free end of the rope S, will be enough to lift it. But if in the same block and tackle the end I of the rope is not tied to the lower part of the second fixed pulley but to the upper part of the second mobile pulley (Fig. 8.16), a power one fifth of the weight will be able to lift it.

In Fig. 8.10, the two diameters KF and ON function like two levers, the ropes FS, NC, IL like three equal powers and the ropes OB, KD like two fulcrums.[63]

As previously seen, it suffices that the power at S is one fifth of the weight. And the process continues by increasing the number of the pulleys and by changing the point (mobile or fixed pulley) to which one end of the rope is tied. From these considerations, Zupi deduces the following general rule to increase the effect of the power using the block and tackle:

> If you want to increase the power according to an even number, namely six, eight, one hundred, and so on, you must add to the upper pulley as many pulleys as half of the chosen number. For example, if the power has to be increased according to six, put three pulleys in the lower pulley and tie one of the end of the rope to the upper pulley; indeed the pulleys of the upper part of the block and tackle have to be multiplied as many times as they are necessary to wind the rope around the three lower pulleys.[64]

In order to increase the effect of the power according to an odd number, the previous even number should be considered and the above rule applied with the single difference that the end of the rope must be tied to the lower part of the block and tackle, that is, to the upper part of the last mobile pulley. There is no difficulty in verifying that these rules are the same as those enunciated by Galileo, but translated into Latin.

It is important to note here that at the end of his treatment of different kinds of block and tackle, Zupi repeats that Galileo's "compensation principle" continues to have a solid base, namely, that in preserving the same speed we do not have a reduction of power. For all cases, Zupi gives the same reasoning as Galileo, which can be generalized and summarized as follows: If the power is the nth part of the weight, then in contrast, the space of the moved weight is nth times the space of the moving power, that is, in the same time the weight covers the space of one palm, the power will necessarily cover the space of n palms because, if the weight ascends one palm, this palm must be subtracted from the n ropes from which it equally hangs.

8.11 The Treatise III

The third and final treatise is devoted to the other machines: the wheel and axle, the wedge, the screw and the cochlea. The inclined plane, which is one of the most important matters in Galileo's *Le Mecaniche*, is not treated.

---

[63] "duae diametri KF, ON se habent ut vectes AB CD decimae figurae [here Figure 14] et funes FS, NC IL ut tres potentiae aequales, et funae OB, KD ut duo fulcimenta" *Exercitations*, 99.

[64] "si velis potentiam augere secundum numerum parem scilicet ut sex, ut octo, ut centum, etc., in inferiori troclea tot urbiculos adde, quot dimidium numeri assumpti conflant. Exempli gratia, si sit augenda potenti aut sex, colloca tres urbiculos in inferiori troclea, et alterum funis extremum in superiori troclea, sit relegatum; in superiori vero troclea tot multiplicentur orbiculi quot ad funem per tres inferiores circumducendum necessarii videbuntur" *Exercitations*, 101.

Fig. 8.17 Wheel and axle

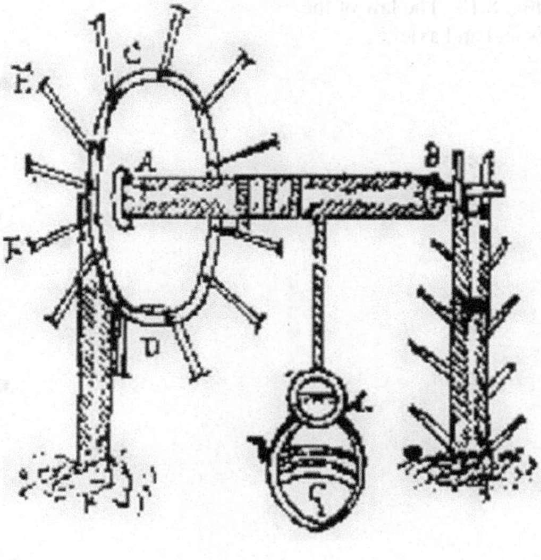

The description of the wheel and axle is the same as in Book VIII of Pappus's *Mathematical* collections:[65] there is a wheel CD whose axle AB has a rope wound around it, which the weight to be lifted G is tied to. A number of bars, such as E and F, to which the power is applied, are fixed to the external circumference of this wheel (Fig. 8.17).

By causing the wheel to turn, the axle will also necessarily turn, winding the rope that lifts the weight. Zupi shows that this tool refers to the lever BAD whose fulcrum is at point A between the power applied to D and the resistance hanging from B (Fig. 8.18), so that the ratio between the power and the weight is equal to the ratio between the semidiameter of the axle and that of the wheel and the bar. Because there are many bars around the wheel, this tool can easily be used as a perpetual lever. Zupi remarks that the usefulness of this tool lies in the fact that

> by means of it we produce with only one operation what we would normally produce with many more and repeated operations.[66]

Zupi also explains that this tool can be used in two different ways, with the axle placed either horizontally (the winch) or vertically (the capstan). The different positions do not change the nature of the tool, which in any case observe the same law, that is, of the lever of first class. Zupi also remarks that for this device the "compensation principle," according to which all that is gained in power is lost in speed, is valid because

> the space of the weight moved is to the space of the moving power as the power is to the weight.[67]

---

[65] See Pappus Alexandrinus (1588).

[66] "quod sine illo per plures et replicatas operationes facere deberemus, eo mediante unica operatione perficimus" *Exercitationes*, 104.

[67] "spatium ponderi moti ad spatium potentiae moventis, sicut se habet potentia ad pondus" *Exercitationes*, 104.

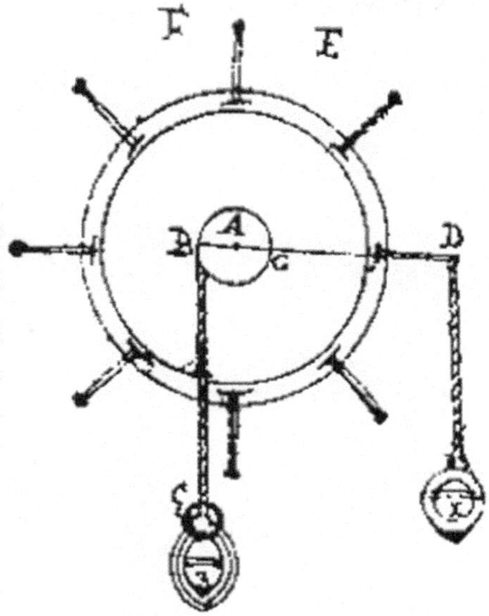

Fig. 8.18 The law of the wheel and axle

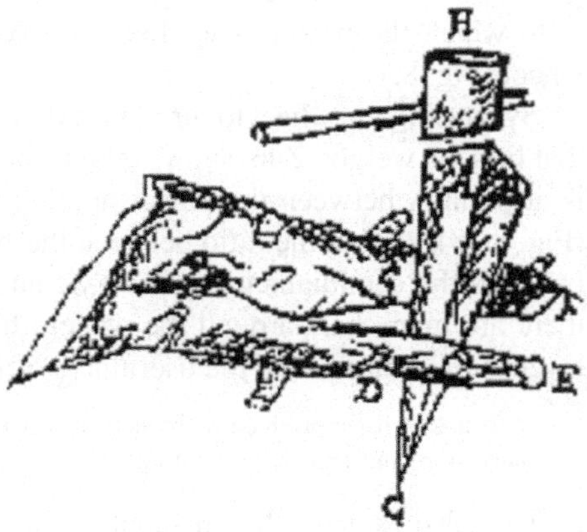

Fig. 8.19 Wedge

In moving to the wedge, Zupi begins by referring to the different interpretations of the nature of this device. Aristotle maintained that it consisted of two levers of the first class, AGC and BDC (Fig. 8.19), having their fulcrum at G and D, respectively; the resistance is at the common point C and the power is at H.

Guidobaldo considered AGC and BDC as two levers of second class with the fulcrum at their common point C; the power at H is the resistance at point G and point D respectively of the wood. Because of the strike inflicted by H, these points are separate so that the strength of the wedge would be dependent on the combined effect of the two levers and on the strike. Zupi's analysis of this treatment is rather

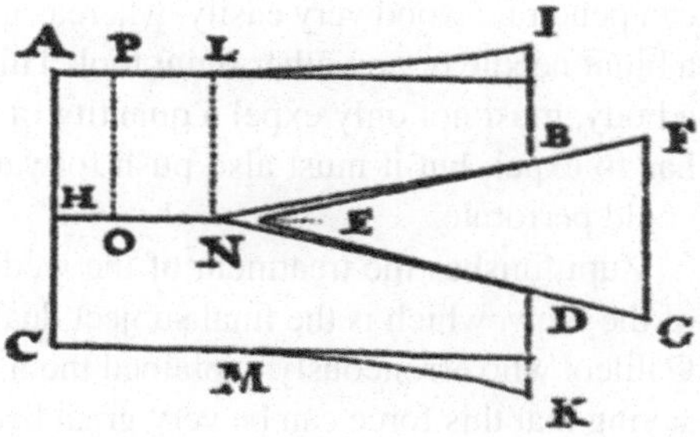

Fig. 8.20 Wedge according to Baldi's explanation

poor and he makes no mention of Guidobaldo's formulation of a second hypothesis according to which the wedge is composed of two inclined planes joined together, and that a wedge with a smaller angle has a greater ability to penetrate because it is easier to move forward on a plane with a smaller inclination than on one with a greater inclination. Zupi criticized both Aristotle and Guidobaldo and openly declared that he shared Baldi's point of view. Following Baldi, he explained that with regards to Fig. 8.20, at the beginning of the scission, parts B and D of the body ABCD are pushed toward I and K respectively, and the two separated parts AIBH and CKDH form two levers, the first with its fulcrum at L and the second at M. For both levers, the dilating power is applied to BD and the resistance is the material at the point of separation N.

Consider the perpendicular LN to the line HE, the greater the ratio between BN and NL, the easier the material will be separated. The value of this ratio changes continuously because by advancing the scission the fulcrum moves forward. So when the wedge arrives at O, the fulcrum is moved to P and the ratio between BO and OP increases with regard to the ratio between BN and NL. It is therefore possible to conclude that as the wedge penetrates the wood, the resistance decreases and the wedge itself advances more easily. Zupi concluded his brief explanation by writing:

> Bernardino Baldi teaches us that this happens because wedges are linked to the nature of the surface. Indeed, nails, awls, and daggers are linked to the nature of the line of the surface. And so strength is given to the line of penetration and to those bodies that are linked to them. I like this thought because I can clarify and confirm it also by reasons of physical character.[68]

Although the wedge is a solid, because of its tapering form, it is closely linked to the nature of the surface and acquires the ability of penetration that is peculiar to flat and thin bodies. Needles, which are very thin, are linked to the nature of the line and for this reason they have a great ability to penetrate. A needle or a very sharp tool

[68] "Bernardinus Baldi docet hoc provenire quia cunei maxime ad superficiei naturam accedunt: clavi, vero subulae pugiones accedunt ad naturam lineae superficiei vero et lineae penetrandi vim conceditur et iis corporibus quam ad illos accedunt. Haec sententia mihi placet quare eam explico et rationibus etiam physicis hoc modo confirmo" *Exercitationes*, 105.

can penetrate wood very easily, whereas it is very difficult to do the same thing with a blunt needle or any other blunt tool. This is because a blunt object, in penetrating a body, must not only expel a quantity of material greater than what a sharp object has to expel, but it must also push forward a part of the material that a sharp tool could perforate.

Zupi finishes the treatment of the wedge with a brief consideration of the force of the blow, which is the final subject dealt with in Galileo's *Le Mecaniche*. Unlike Galileo, who erroneously explained the nature of this force, Zupi limited himself to saying that this force can be very great because it results from the natural gravity of the striking hammer and the gravity deriving from the striking power. It is therefore not surprising that an axe, no matter how heavy it is, does not even scrape a piece of wood if it is simply placed upon it. If the axe, however, is brought down violently on to the wood, then it can easily chop it.[69]

The *Exercitationes* finishes by mentioning very briefly Archimedes' cochlea, as defined by Guidobaldo, as a wedge wound around a cylinder. Zupi gives no explanation of how this tool works, but affirms that the force of the cochlea increases either with a greater number of helixes, or by turning it with longer bars or using handlebars. He concludes by remarking on the validity of the "compensation principle":

> all that is acquired in strength, is lost in time and in speed.[70]

The third treatise is very concise with the treatment of the wedge and the cochlea presenting no analytical parts. Everything is dealt with in qualitative terms and very concisely.

The final part of Zupi's text is very sparse, being little more than a draft of notes. The inclined plane, as noted, is not treated although this device is treated in Guidobaldo's presentation of the theory of the wedge. This omission leaves the question unanswered of whether or not Zupi addressed the inclined plane in his lessons. The brevity of Zupi's treatment of the wedge, the force of a blow and the cochlea, should not to be considered as scant regard for those matters. As in other parts of the *Exercitationes*, Zupi refers directly to other authors whose doctrine he assumes entirely, so that he finds it superfluous to reproduce them in his text. Really, the final pages of the *Exercitationes*, Zupi's analysis is reduced to little more than notes, so that the text seems almost incomprehensible to those who do not have good knowledge of the works on mechanics of that period.

[69] This is the object of pseudo-Aristotle's *Problem XVIII*: "Why if a heavy axe is put on a wooden stump and a great weight is on it, the axe does not chop the wood appreciably, but if the axe is lifted up and then the wood is struck, this is broken in half even if the weight of the striking axe is very small with respect to a weight that sets on [the stump] exerts exerting pressure on it."

[70] "quantum virium acquirimus, tantundem amicti temporis, et velocitatis" *Exercitationes*, 106.

References

Amabile, Luigi. 1892. *Il Santo Officio dell'Inquisizione in Napoli*. Vol. 2. Lapi: Città di Castello.

Amico, Bartolomeo. 1626. *De coelo et de mundo*. Neapoli: Apud Secundinum Roncaliolium.

Arcudi, Alessandro Tommaso. 1719. *Galatina letterata*. Genova: Nella stamperia di Giovan-Battista Celle.

Astrain, Antonio. 1916. *Historia de la Compañia de Jésus en la asistencia de la España*. Madrid: Administracíon de Razón y Fé.

Baldi, Bernardino. 1621. *In Mechanica Aristotelis problemata exercitationes; adiuncta succinta narratione de autoris vita et scriptis*. Moguntiae: Typis et sumptibus viduae Johannis Albini.

Clavius, Christoph. 1574. *Euclidis Elementorum libri XV*. Romae: Apud Vincentium Accoltum.

———. 1589. *Euclidis Elementorum libri XV*. Romae, Apud Bartholomaeum Grassium.

———. 1992. *Corrispondenza. Edizione critica a cura di U. Baldini e P.D. Napolitani*. Vol. 1. Pisa: Dipartimento di matematica dell'Univeresità di Pisa, sezione di didattica e storia della matematica.

Dal Monte, Guidobaldo. 1577. *Mechanicorum liber*. Pisauri: Apud Hyeronimum Concordiam.

Galilei, Galileo. 1890–1909. *Sidereus Nuncius*. Vol 3 [1892]. In *Le opere di Galileo Galilei, Edizione nazionale*, ed. Antonio Favaro. Florence: Barbéra.

———. 2002. *Le Mecaniche: Edizione critica e saggio introduttivo di Romano Gatto*. Florence: Olschki.

Gatto, Romano. 1988. Un matematico sconosciuto del primo seicento napoletano: Davide Imperiali (con un'appendice di lettere e di manoscritti inediti). *Bollettino di storia delle scienze matematiche* 8 (1): 71–135.

———. 1991. Il codice napoletano delle Mecaniche di Galileo: una nuova attribuzione. *Giornale critico della filosofia italiana* 6 (11): 369–378.

———. 1994. *Tra scienza e immaginazione. Le matematiche presso il collegio gesuitico napoletano (1552–1670 ca.)*. Florence: Olschki.

———. 1996. *La meccanica a Napoli ai tempi di Galileo. In appendice "De Gli Elementi Mechanici" di Colantonio Stigliola riproduzione anastatica e le inedite "Meccaniche mie" di Davide Imperiali*. Neapoli: La Città del Sole.

———. 2006a. Christoph Clavius "ordo servandus in addiscendis disciplinis mathematicis" and the teaching of mathematics in Jesuit colleges at the beginning of the modern era. *Science and Education* 15: 235–258.

———. 2006b. Gli elementi meccanici di Colantonio Stigliola: Un trattato archimedeo di meccanica del XVI secolo. *Bollettino di storia delle scienze matematiche* 26: 289–331.

———. 2006c. *Un trattato inedito del '600 di meccanica galileiana: Le "Exercitationes in Mechanicis Aristotelis" del gesuita Giovan Battista Zupi*. Neapoli: La Città del Sole.

Leitão, Henrique. 2001. Galileo's telescopic observations in Portugal. In *Largo campo di filosofare*, ed. J. Montesinos and Carlos Solís, 903–911. La Orotava: Fundación Canaria Orotava de historia de la ciencia.

Maurolico, Francesco. 1611. *Photismi de lumine et umbra, Diaphanarum partes seu libri tres*. Neapoli: Tarquino Longo.

———. 1613. *Problemata Meccanica cum appendice, et ad Magnetem, et ad Pixidem Nauticam pertinentia*. Messanae: Ex typographia Petri Brae.

———. 1635. *Admirandi Archimedis Syracusani monumenta omnia mathematica, quae extant, quorum catalogum inversa pagina demonstrate, ex traditione doctissimi viri D. Francisci Maurolico, nobili Siculi*. Panormi: apud D. Cyllenium Hesprium.

Mersenne, Marin. 1634. *Les mechaniques de Galilée, mathématicien et ingénieur du Duc de Florence. Avec plusieurs additions rares, et nouvelles, utiles aux architectes, ingénieurs, fonteniers, philosophes et artisans. Traduites de l'italien par le L.P.M.M.* Paris: Guenon.

Moscheo, Rosario. 1988. *Francesco Maurolico tra rinascimento e scienza galileiana. Materiali e ricerche*. Messina: Società messinese di Storia Patria.

———. 1998. *I Gesuiti e le matematiche nel secolo XVI. Maurolico, Clavio e l'esperienza siciliana*. Messina: Società Messinese di Storia Patria.

———. 1993–1994. Matematica, filologia e codici in una lettera inedita della fine del XVI secolo. *Helion* 33–34: 159–241.

Pappus Alexandrinus. 1588. *Mathematicae collectiones a Federico Commandino urbinate in Latinum converase, et commentariis illustratae*. Pisauri: Apud Hyeronimum Concordiam.

Ricci, Saverio. 1996. *Nicola Antonio Stigliola. Enciclopedista e linceo. Con l'edizione del trattato Delle Apparenze Celesti a cura e con un saggio di Andrea Cuna*. Rome: Accademia nazionale dei Lincei.

Stigliola, Nicola Antonio. 1597. *De gli elementi Mechanici*. Neapoli: Nella stamperia di Porta Regale.

Chapter 9
Heaviness, Lightness and Impetus in the Seventeenth Century: A Jesuit Perspective

Rivka Feldhay and Ayelet Even-Ezra

Abstract This paper examines the mechanical project of a Jesuit scientist-engineer Paolo Casati (1617–1707), as presented in his *Mechanicorum libri octo* (1684). The core of the project consists in an attempt to physicalize the mathematical science of machines while using Archimedes's lever principle and his theory of buoyancy. Casati, however, aimed at enriching this science with principles originating in the Aristotelian science of motion. The fusion of Archimedean and Aristotelian elements—including the concept of impetus as the sole cause of motion—allowed him to deal with the motion of machines in terms of a hydrostatic theory of motion, neutralizing the Aristotelian notion of "positive lightness," but without erasing it. This he did while engaging himself with experiments on "positive lightness" done at the Academia del Cimento (1657–1667). While this analysis throws light on the transformation of Aristotelian science in Jesuit schools of the seventeenth century, it also frees some Jesuits, at least, from a far too simplistic historiographical category of Galileo's "*nemici*" (foes) and inserts them instead into the field of debates relevant for understanding the emergence of "the new science."

Keywords Lightness · Levity · Heaviness · Gravity · Impetus · Paolo Casati · Jesuit science · Academia del Cimento · Archimedean tradition

---

R. Feldhay (✉)
Cohn Institute for the History and Philosophy of Science and Ideas & Minerva Humanities Center, University of Tel Aviv, Tel Aviv-Yafo, Israel
e-mail: feldhay@post.tau.ac.il

A. Even-Ezra
History Department, The Hebrew University, Jerusalem, Israel
e-mail: ayelet.even-ezra@mail.huji.ac.il

R. Feldhay et al. (eds.), *Emergence and Expansion of Preclassical Mechanics*, Boston Studies in the Philosophy and History of Science 270,
https://doi.org/10.1007/978-3-319-90345-3_9

9.1 Introduction

In the years 1658–1660, the *Accademia del Cimento* performed a series of experiments intended to refute the doctrine of positive lightness (*levitas*). The *Cimento*, considered by historians to be the first academy wholly dedicated to scientific experiments, was set up in Florence in 1657 by Prince Leopold de Medici (1617–1675) with the full support and protection of his brother, Grand Duke Ferdinand II (1610–1670). The academy was dissolved about 10 years later (1667), leaving behind one publication, the well-known *Saggi di Naturali Esperienze fatte nell'Accademia del Cimento* (1667).[1] The experiments performed by the academy have left their traces, however, in more than one historical source. A description of one of the experiments related to "positive lightness" can be found in a letter from January 11, 1658, written by Giovanni Alfonso Borelli (1608–1679)—holder of the chair of mathematics in Pisa at the time and prominent among his fellow academicians—to Vincenzo Viviani (1622–1703)—Galileo's last devotee and member of the academy himself. In his letter, Borelli remarked that the experiment had been performed before the Grand Duke, Prince Leopold, and numerous "peripatetics" at Pisa, where the Court was then staying.[2] Another reference and description are found in the unpublished diary of the academy, included in the Galileo collection of the "Biblioteca Nazionale Centrale" in Florence,[3] and the full report was then printed in the first edition of the *Saggi*.[4] In addition, the long fourth chapter of Borelli's *De motionibus naturalibus a gravitate pendentibus* (1670) was dedicated to the refutation of the doctrine of "positive lightness," as is clearly indicated by its title "*Positivam levitatem in rerum natura non dari*" and though this specific experiment does not appear there, others performed in the academy do.[5] At least one more experiment in the same series had a similar career (Middleton 1971, 221–224). Further experiments may have been performed by the academicians for reasons other than just the refutation of the doctrine of "positive lightness." Nevertheless, they could and were interpreted as experiments capable of refuting this doctrine as well. One such experiment dealt with the motion of smoke in a void (Middleton 1971, 147). It was performed four times by the academy in June–July 1660. One of those performances at the academy was ascribed to Borelli. Borreli was so preoccupied with the doctrine of "positive lightness" that he found it necessary to remark of the experiment that it was relevant for a refutation of this doctrine, and not just for proving the existence of the void.

---

[1] For a modern translation into English, see Middleton (1971, 92–255); on the Academy, see Beretta et al. (2009); Boschiero (2003, 2007); Galluzzi (1981, 2001).

[2] G followed by a number is one of the 347 volumes of the Galilean manuscripts at the Biblioteca Nazionale Centrale in Florence (FBNC). Here G. 283, 35r–36r, cited by Middleton (1971, 225).

[3] G. 262, 83r; G. 268, 117r–18r in Middleton (1971, 224, n. 252).

[4] For an English translation, see Middleton (1971, 221–225).

[5] In proposition xc, we find a version of the first part of this experiment. Borelli refers explicitly to experiments performed in the Accademia, see Borrelli (1670, 128 and 170).

Much less known is the fact that a few of the experiments reported in the *Saggi* have also been described in the first book of the *Mechanica* (1684) written by the Jesuit mathematician Paolo Casati (1617–1707).[6] Casati was quite a well-known figure in the Society of Jesus. Of noble origins, having the looks, manners and rhetorical talents of a typical Italian "gentiluomo," he was chosen by the superiors of his Order for a delicate mission at the court of Queen Christina of Sweden–concerning the conversion of the queen to Catholicism, to which he set himself in 1652. Between 1644 and 1649 Casati had taught philosophy at the Jesuit college in Bologna. After returning from Sweden, he was called to the Collegio Romano where he stayed until 1658. In 1650–1651 and 1653–1655 he taught mathematics, and in 1655–1658, scholastic theology. From that period, among other works, he left two unsigned manuscripts (*De machinis hydraulicis* and *Varia hydraulica*) that probably represent the first lectures on hydraulics given at a Jesuit college (Baldini 2000, 229/n.33). Casati then moved to the province of Venice, where he assumed a series of prominent roles in the Jesuit administration, occasionally serving as professor of mathematics. Among his various other occupations he was also involved in some large hydraulic projects at the Veneto (Baldini 2000, 113/n.5). Thus, he can best be defined as one of the scientist-engineers,[7] after the model of Galileo, whose social and intellectual role in the field of science in Italy became prominent in the second half of the seventeenth century. In addition, Casati was one of the most authoritative writers in the tradition of Jesuit mechanics. On his death in 1707 he had published a series of works on various subjects. Undoubtedly, his *Mechanica* was the most comprehensive among them (Casati 1684).[8] The structure of the text, comprising more than eight hundred pages in eight books, may be described as follows: The first two books are dedicated to the theoretical principles of Archimedean statics (such as an explanation of centers of gravity) interwoven into an Aristotelian theory of matter and motion (especially the notions of heaviness and lightness and the concept of impetus that had become part of the Aristotelian legacy during the fourteenth century). The first book includes the teaching of methods of calculation (for example: Book 1, Chap. V: "By what calculation the center of gravity of bodies is found"), but also the application of the principles to some traditional physico-mathematical problems such as the motion of the center of gravity of the earth (Book 1 Chap. XII: "Whether the earth moves by the motion of trepidation"). The second book is wholly dedicated to a theory of motion as far as it is related to machines, including a theory of impetus as the cause of motion, but also the calculation of the "forces of motions." Here too application to well-known, traditional problems is effectively treated, for example, Book 2 Chap. VII: "Is it

[6] For a good account of Casati's life and work, see Gavagna (1999, 3–15).

[7] On the category of engineer-scientists, see Valleriani (2010, 207–211).

[8] Pierre Duhem dedicated a chapter to Jesuit treatises in his *Les Origines de la Statique* Duhem (1905, XVI). Even though Casati's book was published in the 1680s, in the letter to the reader the author remarked that it was based on materials for the courses he had given at the Collegio Romano during the 1650s. The inclusion of the academicians' experiment may testify that Casati updated the text later on.

better to increase the size of a machine or to combine?" All these are presented in the form of questions to be answered, or problems to be solved, rather than in the form of a geometrical treatise with definitions, postulates and theorems. Finally, each of the last six books is dedicated to the description and problems related to one of the simple machines: the balance, the lever, the wheel and axle, the pulley, the wedge and the "Cochlea"–namely Archimedes' screw. Judging by its form, Casati's *Mechanica* seems to be a very traditional text. Its subtitle, mentioning the principle of the lever as the one and only principle to which all machines may be reduced,[9] follows Guidobaldo del Monte's century-old project (Del Monte 1577; Drake and Drabkin 1969). Its style—chapters organized around a concept, a method and most often a problem—may be considered as belonging to the ancient tradition of problems dating back to the time of Aristotle, as well as to the writings of some classical mathematicians.[10] For all these reasons, a contemporary scholar may have to undertake a rather deep reading of the contents of the materials to be able to identify the book as a specific product of the "preclassical" Jesuit tradition of teaching mechanics.

In the introduction to the second book Casati unfolds the rationale of his project, which was to give a physical explanation of the forces of machines by reference to the laws of motion that govern them and also to prove these forces mathematically, as promised in the subtitle.

> Now we should more closely approach the science of machines itself, which we call mechanics. This is subsumed to geometry, because it does not abstractly consider, as geometry does, the pure quantity of bodies and the extension of mass, but [only] insofar as [quantity and extension are] connected to heaviness or lightness [...]. Although that part of statics as well, which we name hydrostatics, equally engages itself in considering the heaviness of bodies, it nevertheless takes upon itself a different study, in that it principally explores motion fitting the nature of individual bodies in relation to the diversity of liquids into which they run. Mechanics, however, explores the propensity for motion or rest innate in bodies only to the extent that it is able to apply force to their perceived ability by means of a suitable contrivance of instruments. Wherefore, in order to succeed in building machines capable of moving loads by a sure method, we must have the causes of mechanical motion known before we proceed to the machines themselves.[11]

---

[9]The full title of the book reads as follows: "Eight books of Mechanics in which by one and the same principle of the lever, the forces are physically explained & geometrically proven."

[10]On the tradition of writing problems, see Blair (1999, 171–204).

[11]Casati (1684, Introduction to Book II, 125). The original reads as follows: "Nunc propius ad ipsam machinalem scientiam accedendum, quam Mechanicam dicimus. Haec Geometriae subjicitur, neque enim, ut illa, puram corporum quantitatem molisque extensionem abstracte considerat, sed quatenus gravitati illigatam aut levitati; [...] Quamvis autem ea quoque Statices pars, quam Hydrostaticen indigitamus, se pariter in corporum gravitate consideranda exerceat, aliam tamen sibi contemplationem assumit; motum siquidem corporum singulorum naturae congruentem, pro humorum in quos incurrunt, diversitate, potissimum speculatur. Mechanice vero eatenus solum ingenitam corporibus propensionem in motum aut quietem explorat, ut earum facultati perspectae vim possit opportuna instrumentorum machinatione inferre. Quapropter ut certa methodo machinas oneribus movendis pares construere valeamus, motus machinalis causas ante cognitas habere necesse est quam machinas ipsas aggrediamur" Casati (1684, Introduction to Book II, 125).

This passage presents us with one clue to Casati's project. The subject matter of the science of machines—in contrast to mechanics as a practical art— concerns forces understood in terms of weights in motion studied geometrically. However, unlike many of his predecessors, Casati openly declared that both for the sake of theoretical knowledge—namely *scientia*—and in order to succeed in the practical construction of machines in a methodical way, one must first understand natural laws governing weights in motion. These are obviously anchored in the concepts of heaviness and lightness and in the concept of impetus as the cause of motion.

But the letter to the reader discloses a complementary aspect of Casati's project. Not only does he claim the necessity to know the laws of motion in order to understand and prove the forces of machines, he also suggests that he may investigate the physical cause of the wonderful motions of machines. One may identify here the beginning of the trend to create a feedback loop between science and technology that came into fruition in the following centuries.[12]

Our aim in this paper is to reconstruct the conceptual framework of the *Mechanica* through an analysis of two key concepts: 'heaviness' and 'lightness' in their context, and through a more superficial exploration of the way the concept of 'impetus' functions within this text. This analysis is expected to clarify the foundations upon which Casati hoped to establish his science of machines. Contrary to the preliminary impression, Casati's text is far from being a pale epigone of Guidobaldo's *Mechanica.* Unlike Guidobaldo, who mostly preferred to leave the science of motion to philosophers while concentrating on the mathematical aspects of Archimedes' statics, Casati seems to have aimed at some kind of *physicalization* (Schuster 2013) of the mathematical science of machines that developed in the sixteenth century, while fusing it with principles originating in the Aristotelian science of motion, as taught in the Jesuit environment.[13] Casati's book thus represents the culmination of a tradition that emerged in the seventeenth century among Jesuit mathematicians and philosophers who took part in the conversation about the science of nature in Europe in the seventeenth century (Anstey and Schuster 2005, 1). The Italian Jesuits were particularly challenged by the work of Galileo and his followers, who strove to improve and perfect an Archimedean approach to nature but were unable to put aside Aristotelian concepts, ideals and methods.[14] As will become clear later on, Casati, while working within an Archimedean framework, was painfully aware of the need to abandon major elements of the Aristotelian discourse, but was constrained by what he deemed to be the unsatisfying alternatives and by the society's official policy of hostility to novelties. Thus, his authorial voice often sounds undecided and hesitant. Nevertheless, his aim is clear: to perfect a

[12] See Chap. 1, by Renn et al. in this volume.

[13] On the *physicalisation* of the Aristotelian "mixed mathematical sciences," see Schuster (2013, 24). On the science of motion developed by Jesuits in the second half of the seventeenth century, see Chap. 10 by Feldhay and Elazar in this volume.

[14] For the interpretation of Descartes, Galileo and Beeckman in terms of *preclassical* mechanics, see Damerow et al. (2004).

mathematical science of mechanics in which statics is enriched by a causal science of motion.

Our argument proceeds in three steps:

1. After briefly sketching the historical context of the discussion of heaviness and lightness, we shall reconstruct Casati's position, emphasizing two points: (a) his adoption of the distinction between *gravitas* and *gravitatio*, common among sixteenth-century Aristotelians, which allows him to incorporate the concept of Archimedean weight into his discourse without seeing this as an anti-Aristotelian move, (b) his efforts to neutralize the notion of *positive levity* without, however, actually denying it. Here we will show how Casati makes use of experiments performed at the *Accademia del Cimento* to legitimize his approach.
2. After showing how Casati's notion of heaviness provides him with tools for dealing with phenomena of motion in accordance with Archimedes' laws of hydrostatics, we shall briefly discuss his concept of impetus as the cause of motion.
3. Lastly, we shall attempt to extricate Casati's work from the context in which Jesuit mathematicians have been usually inserted by traditional historiography, namely that of simple opponents of the new science. Instead, we shall delineate the context to which he really belonged, namely the field of debates around the *Accademia del Cimento*. This field, we claim—following the work of Galluzzi, Boschiero and others[15]—was much more diverse than Galileo's dichotomous image of *amici* and *nemici*. Casati maintained his connections with the Tuscan mathematicians and philosophers long after the *Accademia* was dispersed. In particular, he corresponded with Magliabechi (Gavagna 1999)—the Medici's librarian who kept encouraging him to print his book, which he finally did in 1684.

9.2 Aristotle's and Archimedes' Concepts of Heaviness: Contradictory or Complementary? Some Milestones on the Way toward an Archimedean-Aristotelian Approach

The notions of absolute heaviness and lightness were necessary elements in a coherent Aristotelian theory of matter and motion, as may be seen in the fourth book of his work *On the Heavens* (*De caelo*), the *locus classicus* of Aristotelian cosmology, where Aristotle defends his theory against the atomists. For Aristotle, the idea of a cosmos implied an order embodied in the form of a continuous material sphere—that is, a plenum with no space void of matter—with a clear center and boundary. Both center and boundary were fully charged with physical, and not just geometrical meaning. To be 'heavy' meant, primarily, to move naturally

[15] See note 3.

toward the center in a straight line, and to be 'light' meant to move naturally toward the boundary in a straight line.[16] Thus heaviness and lightness were physically and not just geometrically related to different kinds of motions. Such motions, again, were distinguished by their quality, and not simply by their (geometric) direction. Contrary motions, in this framework, were capable of inhibiting each other, unlike motions on different geometrical planes—the horizontal and the perpendicular, for example—that could, in principle, be combined in an Aristotelian world. In this sense motions up and down, of heavy and light bodies, were constituted as contrary motions.

One may rightly say that the meaning of absolute heaviness and lightness encapsulated Aristotle's controversy with the atomists. According to Aristotle's description of that controversy in the first two chapters of *On the heavens*, the atomists believed that the cosmos consisted of random aggregations and collisions between indivisible atoms of matter differing from each other in size and shape, and separated by tiny vacuous spaces. Neither qualitative differences between elements, nor qualitative characterization of motions were part of the atomists' cosmos. Within this framework, heaviness and lightness were relative terms, whereby wood, being heavy in air but light in water, moved toward the center in air but toward the boundary in water. Aristotle was aware of these empirical phenomena, as he began his account of his predecessors' opinions with the distinction between an absolute and relative meaning of the terms 'heavy' and 'light' (Aristotle and Guthrie 1939, 308a6–33). He also admitted that only two elements—earth and fire, but not water and air—were absolutely heavy and light. Nevertheless, he insisted that fire is always light and moves upward, while earth and all earthly things move downwards or toward the center and he went on explaining the reasons why "some bodies always and naturally move upwards and others downwards, and others both upwards and downwards" (Aristotle and Guthrie 1939, 310a17–19). This explanation epitomizes the basic principles of his theory of matter and motion (Aristotle and Guthrie 1939, Book IV, Chaps. 3–6).

Archimedes was neither peripatetic nor atomist. In his scheme, bodies differed quantitatively only in their specific heaviness. Since he did not offer any kind of explicit cosmological framework, though his text on buoyancy did at least presuppose one, his mathematical account of certain phenomena—specific heaviness, for example—could be described as neutral. However, Eutocius of Askelon (b. 480 A.D.), the single commentator of Archimedes to have survived from antiquity, tends to stress the incompatibilities between Aristotle's and Archimedes' respective concepts of heaviness.[17] Eutocius' commentary was extremely influential in the sixteenth century. Together with Archimedes' original texts, it had been translated into Latin by Jacobus de Sancto Cassiano Cremonensis in the 1450s, and entered the first printed edition of the works of Archimedes in 1544.

[16] "[...] in calling things heavy and light we mean that they have a capacity for a certain natural motion," Aristotle and Guthrie (1939, 307b32–4).

[17] See Mugler (1951, 202, fn. 11).

Nevertheless, the relative neutrality of the concept allowed it to be integrated into different worldviews, exemplified by Buonamici, who chose to integrate an Archimedean concept of specific gravity into his basically Aristotelian *De motu* although he rejected Archimedes' theory of buoyancy (Buonamici 1591, Chaps. 28–30). Galileo, by contrast, resented Buonamici's move,[18] which he understood as an act of appropriation of Archimedes into an incompatible conceptual framework.

During the 1580s there were quite a few mathematicians—among them Guidobaldo del Monte, Bernardino Baldi and the Jesuit Christopher Clavius—who saw Aristotle and Archimedes as successive links in one tradition. In the first stage, this group was dedicated to the broadening of the Archimedean project in quadratures and the method of exhaustion, as well as to the application of Archimedes' method of centers of gravity to homogenous and non-homogeneous solids (Baldini 1992a, 244–248). An impressive body of works—both Jesuit and non-Jesuit—was dedicated to the development of Archimedean techniques in the area of *centro-baryca*, among them Galileo's *Theoremata circa centrum gravitatis solidorum* (1587–88), Marino Guetaldi's *Promotus Archimedis* (1603) Luca Valerio's *De centro gravitatis solidorum libri tres* (1604) and later on the Jesuit Paulus Guldin's *De centro gravitatis trium specierum quantitatis continuae* in three volumes (1635).[19] The Jesuits played a major role in this discourse and Archimedes acquired a respectful status in their curriculum. Some preferred a purely geometrical approach to Archimedes. Others, however, did not. The first two professors of mathematics at the Collegio Romano—Balthazar Torres and later on Christoph Clavius—established the connection between the Archimedean tradition as it was revived by Federico Commandino (1506–1575) in northern Urbino, and the Aristotelian tradition of the *Quaestiones Mechanicae* developed by Francesco Maurolico (1494–1575) in Sicily (Scaduto 1949, 126–141). Clavius' correspondence testifies to a continuous exchange of letters with the masters and revivers of both traditions in Italy.[20]

It is clear therefore that around the Jesuit intellectual center of the Collegio Romano a group of mathematicians dedicated to Archimedes' work did not exclude the possibility of developing, or even broadening, an Archimedean approach by associating it with the mechanical tradition stemming from the earlier pseudo-Aristotelian *Quaestiones mechanicae* (Aristotle 1585) and occasionally with the problems of motion discussed in Aristotle's *Physics*. The crystallization of such discourse has left its traces in many Jesuit texts upon which we cannot dwell here.

---

[18] In Galileo's words: "…it appears to me that Signor Buonamico imposed on Archimedes, deducing from what he said more than he set forth or that can be deduced from his propositions, inasmuch as Archimedes neither denied nor admitted positive lightness …" Drake (1981, 66).

[19] Many can be added to this list, among them: Clavius' edition of the *Elements* (1589), printed by Bartolomeo Grassi, presents Archimedes' work as the last stage of the work of Euclid thus articulating a purely geometrical interpretation of Archimedes, Dollo (1992, 201). A. F de Gottignies, V. Giordani and D. Quarteroni were also inspired by Archimedes, Baldini (1992a, 273).

[20] Although some of those letters have been lost (Clavius–Commandino), others—those written to Guidobaldo del Monte and Bernardino Baldi—have survived and been published, Philips (1939, 193–222). Both are cited in Baldini (1992a, 246).

Suffice to mention Clavius' citation of the story of Archimedes' ship in his "*Prolegomena*" to the commentary of Euclid's *Elements*, the praise of Aristotle's and Archimedes' machines in Antonio Possevino's *Bibliotheca Selecta*,[21] Grienberger's praise of mechanics as the most physical of all mathematical sciences and his *laudatio* of Archimedes for his inventions as well as his demonstrations and his emphasis on the relevance of mechanics for the study of astronomy and cosmography.[22] Finally, one may refer to Maria Bettini's attribution of the name Archimedes to Grienberger, by which he meant to emphasize the combination of theoretical and practical abilities embodied in the person of his teacher, thus rejecting the image of Archimedes as a pure geometrician (Gorman 2003, 4). Such discourse, we shall see, did indeed open up many new possibilities but was no less beset with internal and external difficulties.

A glimpse into the possibilities opened up for Jesuit scholars—not only mathematicians but also natural philosophers—by the incorporation of Archimedes into their program of studies may be afforded by looking into the unpublished notes (*reportationes*) of physics courses given at the Collegio Romano in the 1580s (Wallace 1977, 308, n. 16; 1981, 114–120; 1984, 109). We shall limit ourselves to one example, quoted both by Wallace and by C. Dollo (Dollo 1992, 214–215) that indicates terminological modifications in the Aristotelians' vocabulary of "heaviness" and "lightness." This enabled a certain distinction between qualitative and quantitative elements, leading to the accommodation of Archimedean theoretical elements into an Aristotelian discourse on motion and machines.

In his disputations on the treatise of the elements of Aristotle's *On the heavens*, M. Vitelleschi distinguished between two aspects of the concept of heaviness and lightness: *gravitas* and *levitas* as qualities referring to an internal inclination to move up and down; *gravitatio*—the motion itself—whose modality expressed the quantitative ratio between the heavy body and the medium, and between the force acting on the body in motion (*potentia*) and the power resisting that motion. Thus he says:

> Laying aside the solutions of others, I reply that heaviness and lightness may be taken in two ways. The first is with regard to the *actus primus*, namely the quality itself moving up or down. The second [is] with regard to the *actus secundus*, namely the movement itself [...] At the end of the movement, therefore, the heaviness or lightness understood in the first mode I mentioned above is not increasing, but the heaviness in the second mode, the *gravitatio*, is increasing. For this is nothing other than the excess of the motive force on the resistance of the medium.[23]

---

[21] Cited in Gorman (1998, 33).

[22] Anon. *Praefatio* 1591 attributed to C. Grienberger by Michael J. Gorman (2003, 40).

[23] "Omissis aliorum solutionibus, respondeo gravitatem et levitatem sumi posse dupliciter. Primum pro actu primo, scilicet pro qualitate ipsa movente sursum aut deorsum. Secundo, pro actu secundo, scilicet pro motu ipso [. . .] In fine ergo motus non augetur gravitas aut levitas primo modo accepta ut supra dixi, augetur tamen gravitas secundo modo, hoc est gravitatio. Haec enim nihil est aliud quam excessus virtutis motivae super resistentiam medii." Dollo (1992, 215) quoting M. Vitelleschi, *Libros de Coelo Disputationes, Tractatiotertia De Elementis*, cc. 124v–125r, in Dollo (1990, 184–5).

The strategy of Vitteleschi, then, as that of many others after him was a kind of duplication of the scientific terminology, which would enable philosophers to deal with Aristotle's text in the traditional way but would also allow the field of Aristotelian physics to be broadened by embracing Archimedes' work as part of the same tradition. In spite of constant tensions between mathematicians and philosophers in the Jesuit milieu, it seems that an accommodation of Archimedes within the Aristotelian framework, to which all Jesuits were obligated, had occurred by the end of the sixteenth century and was developed during the seventeenth.[24]

The last years of the sixteenth century, however, saw the emergence of a few works in which Archimedean concepts and methods were used for criticizing and revising Aristotle's doctrine of motion, not least by Giambattista Benedetti (1530–1590) and Galileo Galileo (1564–1642). In his *Resolutio* from 1553 (Benedetti 1553; Drake and Drabkin 1969, 130),[25] Benedetti asserted that bodies of the same density would fall with equal speed regardless of (absolute) weight, and his argument depended on the application of Archimedes theory of buoyancy to the problem of falling bodies. This has become known as the buoyancy theory of fall. It stirred up much attention when it was later accompanied by Benedetti's anti-Aristotelian rhetoric. In his last work in 1585, Benedetti also dealt with the existence of the void and objected to Aristotle's doctrine of absolute heaviness and lightness (Benedetti 1585). Though it is not clear whether Galileo had actually read Benedetti's works before writing his *De motu antiquiora* (c. 1590),[26] he had certainly heard of the buoyancy theory of fall, which he used as a central principle in his old theory of motion. He also shared with Benedetti an anti-Aristotelian tone that often distorted Aristotle's thought. One of the main targets of this attack was the notion of positive lightness. *De motu*, however, contained much more than that. It started with an Archimedean explanation of the physical order of the elements in the cosmos, identifying heaviness with density and arguing that the denser elements needing less space were naturally concentrated around the center of the universe. It went on to deny the existence of more than one kind of matter, and affirmed one kind of natural motion, namely motion downwards: a concept of heaviness was all that was needed to explain the basic nature of matter, and any upward movement had to be explained in analogy to the rising up of the balance's arm. Bodies ascend because they are pushed up by the descent of heavier bodies. Galileo then explicitly derogated Aristotle's doctrine of the relation between weight and velocity in free fall and offered his own doctrine of the differences of specific weight between mobile and medium as the cause of natural motion. Hence the equal speed of all bodies of the

[24] See Baldini (1992a, 1992b, 75–78, 157–8); Dear (1995, Chap. 2); Dollo (1990); Elazar (2011); Feldhay (2006); Palmerino (2003); Wallace (1977, 1981, 1984).

[25] Benedetti (1553, 17–18): "I say that if there are two bodies of the same shape and the same material, whether equal [in size] to each other or unequal, they will, in the same medium, move over an equal distance in an equal time." See also the introduction to Drake and Drabkin (1969, 32–41).

[26] On Galileo's possible reading of Benedetti and his relation to Benedetti's legacy, see Renn and Damerow (2012, Chap. 3.10).

same density, as had already been stated by Benedetti about 40 years earlier (Drabkin and Drake 1960, 13–41). Although Galileo was heavily influenced by Archimedean mathematics, he was no less acquainted with the tradition of the *Quaestiones Mechanicae*, knowledge he acquired while teaching mathematics in Pisa and writing his *De motu antiquiora*. But once he had written his own *Le mechaniche* (Gatto 2002), he was ready to insert a correction to his own buoyancy theory of fall. The theory was already corrected in his *Discorso intorno alle cose che stanno in su lácqua o che in quella si muovono* of 1612 (Favaro 1968, IV, 63–140), where the resistance of the medium was understood not simply in terms of its specific heaviness but also in relation to the moment of velocity of the medium (Shea 1968, 149–153).

On the one hand, then, Galileo—like Guidobaldo del Monte and Stevin before him—thought that Aristotle's *Mechanical Questions* provided the physical basis for Archimedes' mathematical theories. On the other hand, he took it as his major role to criticize, even ridicule, Aristotle's physics, and to replace natural philosophy taught in the universities of the time with his new kind of apodictic science, which he characterized as mathematical physics.

Galileo's loud and well-communicated dispute of 1612 with a number of Tuscan Aristotelians over the problem of floating bodies brought to the fore the gap between Aristotelian and Archimedean approaches to matter and motion. Galileo denied the Aristotelians' doctrine of the relation between coldness, density and weight, and insisted on the need to understand upward and downward motions quantitatively. In addition, Galileo presented the Aristotelian and Archimedean frameworks as being mutually incompatible. For the Jesuits, who were deeply invested in the study of Archimedes, believing it was compatible with, or at least not necessarily opposed to their Aristotelian commitments, Galileo's presentation did not make it any easier to accommodate his buoyancy theory. Nevertheless, in 1614 a Jesuit novice, a former Florentine student of Galileo, expressed his wish to present in public a mathematical problem through which he would demonstrate Galileo's theory. The problem had been written by the professor of mathematics, Christopher Grienberger (1561–1636) and was performed at the Collegio Romano in 1614 (Bardi 1614).

The problem of *heaviness* and *lightness* touched upon issues relevant not only to the theory of motion, but also to theories of quality that were to become prominent in the second half of the seventeenth century, such as air pressure and the existence of the vacuum, popularized by Torricelli's barometric experiments of 1644. These issues were related to the question of whether air had weight in its natural place. From the point of view of the Jesuits, the development of Galileo's mechanistic discourse around these questions would become the greatest challenge to their Aristotelian commitments after the suppression of the Copernican debate. The decade 1657–1667 saw the work of the *Accademia del Cimento* also focusing around these questions.

The historians Paulo Galluzzi (1981) and Luciano Boschiero (2003, 2007) have argued that, contrary to the rhetoric of neutrality predominant in the published work of the *Accademia del Cimento*, the abundant unpublished diaries and letters indicate an intense preoccupation with major natural-philosophical issues. This was reflected in the contents of the experiments performed, focusing as they did on the problem of the qualitative or quantitative substantial nature of coldness and heat, on the question whether a vacuous space could be created, on air pressure and on the weight of air. In addition, Boschiero has shown that the field of knowledge claims about these issues indicated at least four major positions (Boschiero 2003):

1. Aristotelians insisting on the impossibility of the vacuum but ready to accommodate the notion of the pressure of air;
2. Gassendists believing both in atmospheric pressure and in the possibility of creating a large vacuous space;
3. Cartesians admitting air pressure but unable to deal with it in terms of weight, and in addition considering the world as a plenum and denying the existence of natural or artificial vacuum;
4. Galileanists in the academy who, following Torricelli's strategy, preferred to avoid open discussion of the void, refrained from building a pneumatic pump and focused on proving their mechanistic principles against qualitative arguments of the Aristotelians and on developing explanations of dynamic processes in static terms.

Furthermore, Boschiero has clearly shown that there was an intense debate within the academy itself between a majority group of Galileanists and a minority group of Aristotelians, both over the natural philosophical principles that lay behind a mechanistic and Aristotelian worldview and over the role of experiment and mathematics in establishing those principles.

Against this background, Paolo Casati—a Jesuit mathematician and enthusiastic follower of Archimedes—should not automatically be catalogued as an old-fashioned Aristotelian. Our reading of parts of his *Mechanica* is an attempt to delineate anew the relation between his discourse and that of the experimenters of the academy. This is part of a larger argument: beyond antagonisms and competition, the Galileanists of the academy and many of the prominent Jesuit mathematicians shared a common field of discourse in which they were invested and which they took care to reproduce.

9.3 Heaviness and Lightness in the *Mechanica*

The goal of Casati's monumental work is to understand the forces (*vires*) of machines by means of which we resist the innate propensity of bodies to motion or to rest.[27] Such forces are constituted through two main concepts: heaviness, which Casati claims he cannot leave aside in order to know what it is that should be conquered by art, and the causes of motion of machines, for in order to construe machines capable of moving weights in a certain, methodical way, it is necessary to

[27] "Machinarum vires, quibus innatæ corporum in motum aut quietem propensioni obsistimus" Casati (1684, Liber Primus, 1).

know the causes of the motion of machines before we turn to the machines themselves. Casati dedicates the first two of the eight books of his work to an exploration of heaviness and the causal explanation of motion, assuming that after having established such fundaments, the rest of what he has to teach about machines will flow easily so that the force of each and every machine may be explained without any difficulty. Thus he says: "These basics having been cast further, it will not be burdensome to build, and to set out the properties of individual machines, be they simple or composite, to the extent that when the things discussed in this second book are properly understood there will remain hardly any difficulty in the rest of the work."[28]

The first chapter introduces the idea of "the center of heavy and light things." Diversity, order and place are the basic concepts with which he chooses to open. Since there are many and diverse bodies in the universe, there is a need for an order to guarantee that these bodies find themselves in the right place. This order may be kept by each body receiving from nature a *facultas*, an innate inclination by which it remains in its proper place or moves itself toward its proper place if previously removed from it by some other force. However, since there is an equal distribution of equals, all these bodies may be arranged in no other way than as a sphere, the only body that has a perfect distribution of its parts. A sphere has a single point from which all radii extend, and it is in relation to this point—called "center"—that all bodies are measured: every movement is seen in terms of moving toward the center or away from it. Having familiarized us with this system, Casati turns to an explanation of heaviness and lightness: moving toward the center is called *gravitare*, and away from it, *levitare*; bodies that lie closer to the center are held to be heavier *secundum speciem*, and those further away, lighter (Casati 1684, 2).

A few points are worth noting at this stage. The verbs for being heavy and light are introduced before the cognate adjectives (*gravis*, *levis*), and these are introduced in the comparative form (*gravior*, *levior*). Concepts such as order or place are referred to naturally, but words pertaining to heaviness and lightness are introduced as if deriving from a more specialized discourse. For example, moving toward the center "is called" (*dicitur*) *gravitare*; bodies [...] "are held" (*censentur*) to be heavier (Casati 1684, 2). This is the point where Casati begins explaining the difference between heaviness or lightness *secundum speciem* as opposed to absolute heaviness and lightness, instructing the reader how to compare two bodies, measuring their heaviness *secundum speciem*. This emphasis on the distinction between the world of words and that of phenomena alerts the reader to the subtle strategy working in this text that aims to create a whole network of alternative vocabulary for studying in depth the physical-mathematical meaning of heaviness and lightness, without, however, dismissing the old, qualitative meanings of these terms. Casati, however,

[28] "His porrò jactis fundamentis operosum non erit in ædificare, & machinarum singularum vires, sivè simplices illæ sint, sivè compositæ, exponere: adeò ut iis ritè intellectis, quæ hoc secundo libro disputabuntur, vix quicquam in reliquo opere supersit difficultatis" Casati (1684, Liber secundus, 125).

does not make this explicit. He simply introduces his new, more elaborate vocabulary as a heuristic device for studying phenomena related to heaviness.

After the short exposition of the cosmological context of his teaching, the lesson moves on to another rule, introducing another couple of concepts: "Generally speaking, bodies that are heavier *in specie* are those that are denser, while bodies that are lighter *in specie* are those that are rarer" (Casati 1684, 2). Casati confirms this rule by two examples from experience. The explanation of this correlation between heaviness-lightness and density-rarity turns the system drawn earlier into a more complex model by combining the geometrical form of the sphere with the physical distribution of matter or substance in a space:

> For just as radii extending further from the center to the sphere toward the surface move away from each other, it is reasonable that bodies having more matter and substance in a smaller volume would be placed together in a narrower space; but those that are contained in larger dimensions would occupy greater spaces, where the radii are more distant from each other. So that clearly by this reason an equal distribution of substance occurs in the whole sphere.[29]

This is the reason why the same body ascends after being rarified, like water that turns into vapor: the same amount of matter in larger spatial dimensions desires to move away from the center. It will rest there until, being pressed into smaller volume, it will descend. This center, in respect of which bodies move or rest, is called the center of heavy and light bodies and—Casati brings us back to the real (Aristotelian) world—this is believed to be the center of the universe, or at least of this elemental sphere.[30] He also adds that if it were not so, we should think that a separated part of the moon would return to the moon, being the center of its movement. Needless to say, this is a significant cosmological remark by which Casati refers to a non-Aristotelian cosmology that he implicitly—but not emphatically—rejects.

The opening chapter of Casati's book anticipates the wider significance of what will follow. The broad cosmological picture of the universe presented there signals Casati's view that an adequate explanation of the forces of machines must be firmly anchored in the principles of natural philosophy, and in fact, in the principles of natural motion. This introduction appears to aim at establishing mechanics as causal knowledge, or *scientia*. In a pleasant, non-aggressive way Casati implies that the

[29] The original reads: "Cùm enim radij à sphæræ centro ad superficiem ducti longiùs à se invicem recedant, æquum fuit, ut quæ plus habent materiæ atque substantiæ sub minori mole, in angustiore spatio collocarentur; ea verò, quæ sub majoribus dimensionibus continentur, ampliora spatia occuparent, ubi radij magis distant: ut videlicet hac ratione æqua substantiæ distributio fieret in totâ sphærâ" Casati (1684, 2–3).

[30] "Quare centrum hoc, quod motus, vel quies corporum respicit, dicitur Centrum gravium, & levium; atque idem creditur esse cum centro universi: vel saltem (ne parùm utili nos disputatione torqueamus) centrum eorum, quæ in hac sphærâ elementari gravia, aut levia dicuntur," Casati (1684, 3): "Therefore this center to which the motion or rest of bodies relates is called the center of heavy and light things, and it is believed to be identical to the center of the universe, or at least (to avoid being dragged into useless disputation) the center of the things called heavy or light in this elemental sphere."

status of a "mixed mathematical science"—or "practical mathematics"—is no longer compatible with the praxis of Jesuit mathematicians and their aspirations. It is for this reason that this first chapter in a book purporting to study centers of heaviness does not offer a mathematical [or operative] definition of the concept. Rather, it starts with an explanation of heaviness and lightness as an expression of the faculty endowed by nature to all material inanimate bodies, natural and artificial alike, to strive to return to their place among other bodies of the same specific heaviness when they are displaced.

A comparison with Galileo's *De motu antiquiora*, which might have been the source of inspiration for Casati's opening chapter, shows the great similarity existing between these two texts. The idea that any discussion of motion should start by modifying Aristotle's presentation of the cosmos, introducing the geometrical form of a sphere as a kind of justification for the distribution of matter in the cosmos, is Galilean, and Casati repeats it. However, Galileo's strategy in justifying this picture is very different from Casati's. The latter, as we have seen, wished to add a clear geometrical dimension to the Aristotelian discourse on heaviness as a quality expressing the body's propensity to move to its natural place. Galileo had used the same picture as a convenient point of departure for attacking Aristotle's theory of matter:

> For if it is true, as ancient philosophers believed, that there is a single kind of matter in all bodies, and those bodies are heavier which enclose more particles of that matter in a narrower space (as those same philosophers asserted, for which they were perhaps wrongly criticized by Aristotle in *De caelo* Book 4), then surely it was logical that bodies containing more matter in a narrower space should also occupy narrower places, such as are those that are nearer the center. (Drabkin and Drake1960, 15)

Unlike Casati, Galileo starts from the notion of matter: there is only one matter, and it is the number of particles contained in a (specific) space that determines its being heavier or lighter. He explicitly takes the side of Aristotle's opponents regarding matter-theory. Casati makes no such explicit, direct claims. As a Jesuit, he must adhere to the tenets of the Aristotelian anti-atomistic theory of matter. However, even though Casati seems to stick to Aristotle, in fact he adds something that allows him, later on, to develop—or rather copy—a kind of Galilean hydrostatic theory of motion. Thus, concepts of one matter and particles are slipped in as a matter of course, in a vague, non-provocative form. "Denser" bodies are actually those with "more matter and substance in a narrower volume";[31] The logic of the sphere—presupposed above—enables and entails an equal distribution of substance. In other words, Casati actually shows that there is always a relation between specific heaviness and density, by means of which he establishes the connection of both concepts to the concept of natural place. This is an argument that allows him to adopt Galileo's reductive explanation of motion in media in terms of the difference between the body's and the medium's specific heaviness, while removing other considerations—such as consideration of the form of the body—from his explanations.

---

[31] "quæ plus habent materiæ atque substantiæ sub minori mole" Casati (1684, 4).

The price Galileo had to pay for his simple model, however, was the neglect of many other factors which complicated the analysis of motion of bodies in different kinds of media. This sometimes made it extremely difficult—or even impossible—to observe in nature, or during an experiment, the effects predicted by Galileo's theory. Casati was not alone in quickly discerning the weaknesses of Galileo's *simplistic* or *reductive* models when tested experimentally, as the members of the *Accademia del Cimento* had tried to do. This is clear from the way he criticized some of the *Cimento's* experiments and used such critique for promoting his own Archimedean-Aristotelian approach.

Chapter two is a digression from the main course of Casati's argument, mostly leaving the main issue of heaviness and motion and focusing on the problem of positive lightness through discussion of a few experiments. As Casati says later, since we need machines mainly for conquering heaviness, and not lightness (holding air in water for example), the mechanical discussion from the next chapter onwards will concentrate on heaviness alone. Nevertheless, through the digression on positive lightness he tries to achieve two goals: a) to show his dedication to a mathematical-experimental discourse, and b) to refute or at least criticize more radical interpretations of some experiments that tend to undermine the delicate balance he is trying to maintain between Archimedean and Aristotelian elements and their fusion in his text.

After a short repetition of his cosmological principles of order and natural place Casati starts developing his position. Others[32] say that bodies ascend only as much as they are pushed by descending bodies, so that heaviness remains the only existing internal force. But, says Casati, I can do the same trick:

> For were you to say that the things called light rise as far as they are pushed forward by heavier things descending to a lower place, I would be equally free to remove all positive heaviness, lightness alone being admitted. I will solve all things in the same way by saying that things considered heavy are those that are less light, and that they descend only to the extent that beaten they are pushed down by lighter things ascending in place, not that they are driven downward by an internal faculty.[33]

Thus positive heaviness is no more compelling than positive lightness if only one agent is sought. At this point, Casati presents to the reader two equally possible "mirrored" explanations for motions up and down and their causes: internal faculty or external push. He also explicitly claims that it will be quite reasonable to accept each of them ("*utramque admittere rationi maxime consentaneum fuerit*"). Leaving the reader with these two competing explanations, the text turns to a series of experiments that show how difficult it is to resolve such a dilemma.

---

[32] See, for example, Borelli (1670, proposition 41, 93ff). The argument is that it is superfluous to assume two principles of motion when one will suffice to explain everything.

[33] "Nam si ea, quæ levia dicuntur, eatenus dicas ascendere, quatenus à gravioribus in inferiorem locum descendentibus propelluntur; mihi æquè liberum erit tollere omnem positivam gravitatem, solâ levitate admissâ; & omnia pariter solvam dicendo ea gravia censeri, quæ minùs levia sunt, atque ideò tantùm descendere, quòd extrinsecùs à levioribus ascendentibus loco pulsa detrudantur, non quòd ab internâ facultate deorsum impellantur" Casati (1684, 4).

Casati describes six different experiments that deal with measuring the weight and motion of different bodies in different media, and aims at convincing his interlocutors or students that the attempt to negate positive lightness is hard to defend. Sometimes, however, he admits that his own thesis—positive lightness exists—cannot be proved by experiment either. For example, the first experiment he describes deals with a heated glass ball with a curved back pipe full of rarified air, pulled down into water by a piece of lead inserted into the ball and there weighed. The hot air is allowed to cool down, and the glass ball is again weighed in water. The experiment is designed to compare the weight of hot and cold air in water, following his basic assumptions that rarefied (hot) air has lightness while condensed (cold) air has heaviness. His conclusion from the experiment is that it is not easy to convince that bodies strive with their *nutus* (inner inclinations) not only downwards but also upwards.[34] Casati's sensitivity not only to the role of experiment in the investigation but also to its discursive importance is untraditional. Rather, his text manifests a sophisticated attitude to the art of designing experiments and to the ways instruments and machines can be used to demonstrate natural principles. The text also discloses his attention to technical detail.

The fact that Casati—like many other contemporary Jesuit mathematicians—became a legitimate speaker in the preclassical mechanical discourse of the seventeenth century may best be shown by comparing a description of an experiment done at the *Cimento* and reported in the *Saggi* (CCVII) to Casati's reference to, repetition and critique of the same experiment in his *Mechanica* (Casati 1684, 23ff).

The experiment is designed in two stages. In the first phase, a wooden cylinder is positioned perpendicularly on a flat horizontal plane. The surface of the cylinder's base and the surface of the horizontal plane are lined with metal plates "well flattened and polished." A little bank is built around the cylinder into which quicksilver [mercury] is poured. Thus, with no air penetrating between the plates, the two bases strongly adhere to each other. The cylinder is then tied to one side of an equal-armed balance, and a weight (L) equal to that of the cylinder is hung on the other side. Obviously, the equal weight is unable to overcome the force of adhesion. More and more weight is then added to the balancing arm of the balance until the cylinder is lifted up. The additional lifting force (M)—regarded as the force capable of overcoming the resistance to detachment—is measured and found to be around three pounds. In the second phase of the experiment, the wooden cylinder is situated within another cylinder, the bases of both again tightly held together using the same method. The second vessel is then filled with quicksilver. According to the *Saggi* text, the cylinder does not move at first. But once its base is freed by hand, the wooden cylinder forcefully rises up; a weight of five pounds (Q) laid on top of the cylinder to keep it down is taken as the measure of the force pushing the cylinder up when it is freed from adhesion (Fig. 9.1).

From the outset, the academicians are heading to prove that nature knows only one kind of force—namely heaviness—arguing that heaviness can cause both

[34] Original reads: "… non leviter suadere potest corpora suis nutibus non deorsum tantum sed etiam sursum conari" Casati (1684, 5).

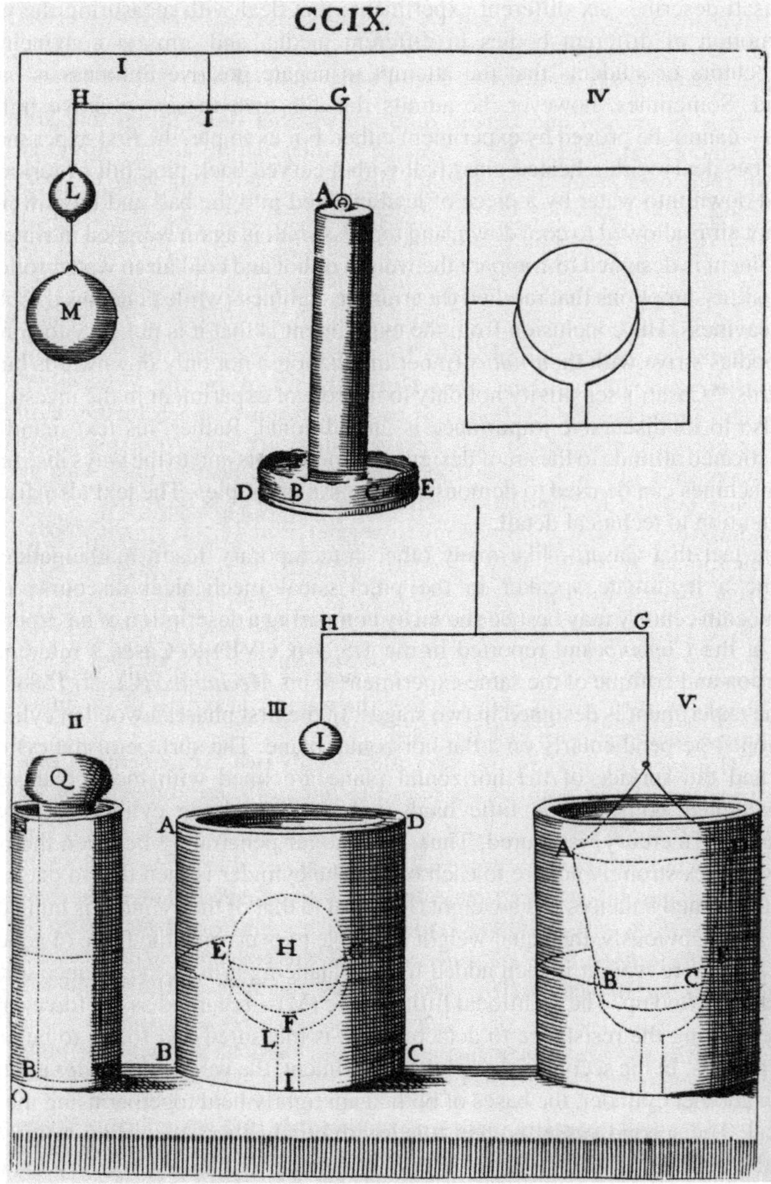

Fig. 9.1 Academicians' experiment Saggi CCII (Middleton 1971, 222)

motion down and motion up. According to them, the hypothesis is well anchored in tradition, in fact in the very old text of Plato's *Timaeus* [62c–63c] where it is declared that "Plato… maintained that heavier things are able to displace less heavy ones upwards" (Middleton 1971, 221). The academicians thus find an ancient authority to justify their thesis, which they quote in the introduction presenting their experiment. The experiment is designed to show that the heaviness of quicksilver pushes the much lighter wooden cylinder upwards in the second phase, and that the measure of the force is equal to the force needed to keep the cylinder down in quicksilver. In the following interpretation of their experiment, the academicians argue that had it indeed been the case that lightness pushes the cylinder up, then the force of lightness—5 pounds measured in the second phase of the experiment—would have overcome the force of adhesion of 3 pounds (weight M) and lift the cylinder immediately and spontaneously. Their conclusion is that something else—not lightness—causes the cylinder to move up in quicksilver. Significantly enough, they do not explicitly say that the heaviness of quicksilver actually causes the motion up. However, in light of the introduction containing the argument from Plato's *Timaeus*, it is reasonable to suppose that for them heaviness is the only force that moves bodies down or up. Also, the academicians do not really distinguish between mechanical and natural motions. For them, motion in a medium and the motion of the balance arms are discussed within one common conceptual and discursive framework.

In his *Mechanica*, Casati accurately repeats the narration of the academicians' experiment while modifying only one significant detail—the need to "free" the cylinder from adhesion in the second phase of the experiment. The slightly modified account leads him to offer a different conceptualization of the event, followed by a more elaborate calculation of the forces acting on the experimental stage. Finally, his calculation seems to confirm his conclusion that there are two kinds of forces operating on the wooden cylinder—heaviness and lightness—which eventually cancel each other in a state of equilibrium, while the body—partly sunk in mercury, partly in air—reaches a state of equilibrium (Fig. 9.2).

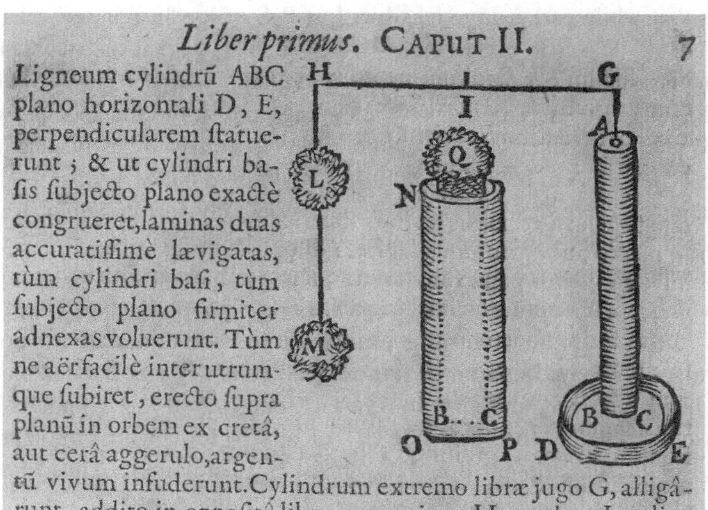
Liber primus. CAPUT II. 7

Ligneum cylindrũ ABC plano horizontali D, E, perpendicularem ſtatuerunt ; & ut cylindri baſis ſubjecto plano exactè congrueret, laminas duas accuratiſſimè lævigatas, tùm cylindri baſi, tùm ſubjecto plano firmiter adnexas voluerunt. Tùm ne aër facilè inter utrumque ſubiret, erecto ſupra planũ in orbem ex cretâ, aut cerâ aggerulo, argentũ vivum infuderunt. Cylindrum extremo libræ jugo G, alligârunt, addito in oppoſitâ libræ extremitate H pondere L cylin-

Fig. 9.2 Casati's reconstruction of academicians's experiment (Casati 1684, 7)

Casati begins his interpretation by invoking a principle that guides his explanation and measurement of the force by which a light body—relative to a medium—is pushed by a heavier one. Certainly, when two bodies of different specific heaviness gravitate in the same medium, he says, the heavier tends to push the lighter to move more quickly. And, he adds, the same is true for two light bodies "levitating" in the same medium. The principle does not contradict Archimedes' law of floating bodies, even though for Archimedes, there is no lightness. Thus, there is no need for a Jesuit to choose between Aristotle and Archimedes as representing two contradictory conceptual frameworks. Moreover, following Benedetti, Galileo and others, Casati uses Archimedes' mathematical vocabulary—such as ratios of specific heaviness or differences between specific heaviness of two bodies of different materials—to develop a causal explanation of motion of bodies in media, which for him means all motions, both natural and violent. Yet, according to Casati, not all motion is caused by extrusion (or push: *impulsio*), in contradiction to the "modern" Archimedeans like Galileo and his disciples. More specifically for the experiment under discussion, not all the motion that occurs during the experiment is due to the quicksilver, Casati claims. Rather, the wood has forces of descent or ascent of its own.

Now he applies his principles to the case of the wooden cylinder in quicksilver, part of it sunk in the liquid, the other part outside the liquid, surrounded by air. From the point of view of their bulk (*moles*), he claims, the ratio of the wood in quicksilver to the wood in air is like the ratio of the specific heaviness of wood to the difference between the specific weights between quicksilver and wood (Casati 1684, 8).[35] Hence follows the calculation of forces, first by means of algebra. Assuming the part sunk in quicksilver to be A, and the part in air to be A-S, he describes the proportion: S/R-S:A/S-A; and in the form of an eq. S(S-A)/A(R-S). Now he explains his notion of "difference between the specific weights of wood." His presupposition—which he claims to have proven in his *Terra machinis mota* (Casati 1658, diss. 105)—is that while part A in quicksilver lost R-S of its "heaviness" relative to S, part S gained—R + S to its "lightness" relative to A, so that the two cancel each other out. Thus the forces acting on the cylinder are: A(R-S) + A(−R + S) = AR − AS − AR + AS = 0. He continues by assigning numbers to the specific weight of wood (5) and quicksilver (13), calculating the respective forces of heaviness and lightness acting on A to be 5/13x8 = 40/13; and on S to be 8/13 X 5 = 40/13. In his calculation, therefore, heaviness and lightness actually fuse together, leaving the difference R-S to determine the part sunk in mercury.

Casati's presentation of this experiment comes to its close with his critique of the academicians' insufficient technical expertise in designing the experiment with the balance. In rising up, he claims, the arm of the balance draws an arc, so that the cylinder does not move in a straight line perpendicular to the horizontal plane of the balance but in an octagonal direction. In fact, one side of the cylinder's base (C) must have risen before the other side (B), creating a lever whose "potential" is at point A, the fulcrum at point B and the "resistance" at point C. Thus M = 3 could

[35] Original reads: "Quid enim partis immersae ad extantem (si moles spectetur) ea ratio est, quae specificae gravitatis ligni ad differentiam gravitatum ligni, atque mercurij."

not have been the measure of the resistance, which had to be somewhat greater. Therefore, one cannot conclude from the experiment that there is no positive lightness on the basis of assessing that the 5 pounds measuring the force of the cylinder in its motion up is much more than the 3 pounds of resistance.

Obviously, Casati's "mental" model—like that of the Academicians—for explaining and calculating motions up and down in media remains the balance. Casati, however, tries to explain away their differences with his technical critique of the experiment's design, followed by a proposal to use a "trochlea" rather than a balance. But having built up the expectations of the reader for a traditional defense of *positive* lightness, his conclusions following the presentation of his series of experiments are surprising.

> Nevertheless I would not dare to conclude a double force in a stone that acquires heaviness in water and air, but lightness in mercury or liquid lead.[36]

There is but one force, says Casati: the force of impetus which creates (*efficiens*) the movement of bodies up or down: "Ascent and descent can result from one and the same principle, depending on the varying state of the adjacent body," (Casati 1684, 11)[37] he asserts. Thus, Casati seems to reduce movements up and down to the efficacy of the impetus and the conditions of the environment. He seems to substitute a *neutral* impetus for the qualities of heaviness and lightness as the causes of motion up and down:

> Since motion accomplished in the same line clearly includes similar ubications successively acquired, be it ascent or descent, the order of their attainment being the only difference; for this reason, since ascent differs from descent in this one thing, that the stone finally obtains that ubication after the other ubications toward the end of the movement […] To change this order it does not seem that a dissimilitude of the moving force is necessary: indeed nothing dissimilar is produced. But if someone were to say that the variety of conditions suffices, he would perhaps say nothing unsound.[38]

Thus Casati undermines the whole debate about *positive lightness*, claiming it to be terminological rather than substantial. One can negate positive heaviness as well as positive lightness. In fact, nothing, concludes Casati, should be said to be *heavy* or *light* in an absolute simple manner (with the exception of the very extremes which will be discussed later). *Heavy* is always heavy in relation to something and in a certain medium. It is the movements of bodies, which are called gravitating or levitating according to their direction. And their direction is toward their natural—

---

[36] "Non ausim tamen in lapide, qui gravitatem in aqua & aere, levitatem in mercurio, aut plumbo liquente obtinet, duplicem statuere virtutem …" (Casati (1684, 11).

[37] "…ab uno eodemque principio, pro varia contigui corporis conditione, ascensum, descensumve prodire posse."

[38] "Quandoquidem motus, qui in eadem lineâ perficitur, similes planè includit ubicationes successivè acquisitas, sivè ascensus sit, sivè descensus, ordine tantùm in earum adeptione, commutato. Quare cum ascensus à descensu hoc uno differat, quòd quam ubicationem lapis demùm obtineret post alias propè finem motûs, [.. .] ad ordinem hunc permutandum non videtur necessaria virtutis motricis dissimilitudo; nihil quippe producitur dissimile. Sed si quis sufficere dicat conditionum varietatem, nihil absonum fortè loquatur." Casati 1(684, 11).

meaning physical/geometrical—place, which is not arbitrary but geometrically determined in the center or the periphery of the sphere, depending on their density or rarity. The use of the term *ubicatio* instead of *place* removes any qualitative difference between ascent and descent. From the qualitative point of view, it is the same thing to move from point A to point B as it is to move from point B to point A.[39]

Up till now we have witnessed Casati gradually modifying the Aristotelian discourse of heaviness and lightness, adapting it to an Archimedean universe where the physical, qualitative dimensions of matter are reduced to density and rarity: the one entailing approximation to the center, the other to the periphery. In this universe, motion toward the center and away from it are caused by one force: the impetus. Hence, further clarification of the notion of impetus—without delving into the full account of Casati's theory of motion—is necessary to complete our discussion of heaviness and lightness.

9.4 The Nature of Impetus

Impetus is the subject matter of two chapters in the second book of Casati's *Mechanica*. The point of departure is the assumption that moving bodies always require a cause for their motion. Bodies move when they are situated outside their natural place, defined as the place where bodies are surrounded by other bodies similar to them in their specific heaviness.[40] In that place no motion occurs and bodies are in stasis, namely in equilibrium. Thus, *gravitas* and *levitas* are the *expressions* of the inner propensity of the body, but they are expressed only when the body is in a place that is not its own, according to the circumstances. This enables Casati to conserve the idea that bodies have an inner nature or inclination to their proper place. But the inclination is not even the closest cause for movement. What really enacts the inclination is the impetus, that which is found between "nature" (namely the inclination to move) and the motion itself: "*aliquid naturae ipsi ac motui, interjectum […] (Impetum nominamus)*" (Casati 1684, 136).

Secondly, bodies move when some force is impressed on them from the outside. In both cases, Casati distinguishes between the moving body, recognized by its heaviness, the motion—continuous displacement from one place to another (Casati 1684, 138)—recognized by its velocity and by the cause which is one for all motions, namely impetus. In both cases, the impetus is responsible for the beginning of motion, as well as for its intensification—the addition of velocity in the case of a falling stone—or its retardation—the loss of velocity of arrows continuing their motion without contiguity with the throwing hand. While developing the notion of

[39] See also Des Chene (1995, 245): "To 'ubicate' something is to change its *Ubi*, that accident of a thing which consists in its being *here* or *there*."

[40] "quandiu enim corpus ita simili secundum gravitatem corpore circumfun ditur, ut naturali in loco consistere dicendum sit" Casati (1684, opening of Book II, Chap. 3, 146).

impetus as the cause of all motions: up and down, natural and violent, Casati strengthens the insight gained in the first chapter and expresses in the long title of the book itself, that there is no need to multiply forces to explain different types of motions. Nevertheless, Casati distinguishes two kinds of natural impetus:

1. Innate impetus (*impetus innatus*), which reacts to the natural propensity of substance to motion and is identical with the *gravitatio*.[41]
2. Acquired impetus (*aquisitus*), which is a kind of force—according to the most probable conjecture—added to the first impetus in small quantities of moment: "For as the nature of the descending stone remains the same [...] what could cause it to increase velocity while descending unless another impetus joins that which produced it first and then another, every single moment"[42]

While Casati stresses that the quantity of innate impetus is fixed ("[...] *certa est impetus mensura*; [...] *pari impetus producto connitatur*") he also emphasizes that the real effect of impetus depends on a variety of factors. A body acquiring the same impetus does not always move with the same velocity. Rather, the effect depends on the resistance of the medium in which the body moves. Thus Casati warns his reader not to think that the impetus of the same stone descending in air or in water is the same. Impetus is measured in accordance with the dividing medium in which the movement ends.[43] Likewise, the effect of impetus depends on the resistance of another body's *gravitas* or on its velocity. As a result, a body can be moved by the same impetus with more speed when it meets a smaller or a slower constraint (Casati 1684, 141).

Moreover, whereas the innate impetus is immediately destroyed when the body arrives at its own natural place—namely in equilibrium—the acquired impetus is only destroyed by resistance—or counter-action. Thus, Casati not only blurs the qualitative or absolute distinction between heaviness and lightness—motion downwards and upwards—but also that between natural and violent motion. Outside their natural place, all bodies acquire impetus. If the impetus is too weak in the face of the resistance and hence incapable of immediately initiating motion, or when the body continuously adds to its innate impetus a second and third impetus, in both cases the acquired impetus is not destroyed, and neither does it disappear. Rather, it is conserved in the body, either slowly growing with time until motion is effected, or immediately effecting a motion that counteracts the natural propensity of the body:

[41] "Impetum hunc, qui naturali se movendi facultati respondet, & est ipsa gravitatio, seu naturalis ad descensum propensio" Casati (1684, 146).

[42] "Cum enim eadem descendentis lapidis natura perseveret [....] qui fieri possit, ut descendens velocitatem augeat, nisi ei, quem primum produxit, alium atque alium momentis singulis impetum adjiciat" Casati (1684, 136).

[43] "Cave tamen existimes parem esse lapidis ejusdem, & in aëre, & in aquâ descendentis impetum: natura scilicet ex medio dividendo, in quo perficiendus est motus, metitur impetûs modum" Casati (1684, 138).

> Therefore the acquired impetus perishes not all at the same time; but it weakens little by little; and this for no other reason than that by the proportion a motion is impeded, for impediments may rise after all from whatever source. For when an impetus has no contrary impetus, if indeed the precise nature of impetus is inspected, it remains that it perishes by reason of those that either block the motion of the body or direct it in some other way.[44]

And in the same vein he states:

> What certainly seems to be beyond any controversy is that a heavy nature [substance, body] does not rise by its own will; therefore, what is it that forces an ivory sphere to jump back on a rock after its has already descended [...] if not because the impetus conceived by nature keeps generating the motion, while nature itself resists as much as possible.[45]

These examples point out that not only does impetus initiate motion in proportion to the body's resistance, that its accumulation or dissipation cause motion to be intensified (accelerated) or retarded, but that impetus is also the principle or the cause which enables motion to be conserved.

Our short enquiry into Casati's notion of impetus illustrates how, by adopting a concept that was not part of the original Aristotelian vocabulary but became part of the Aristotelian tradition, Casati was able to mediate between two seemingly incompatible conceptual frameworks. The unintended consequences of such a move were clear. It undermined two major dichotomies that constituted the traditional discourse: the dichotomy between heaviness and lightness, and between natural and violent motion. Heaviness and lightness could no longer be considered as *contrary* motions, while impetus became the cause both of natural as well as of violent motion.

9.5 Back to the Discourse on Heaviness and Lightness

Any attempt to understand the nature of Casati's project should take into account that his voice represents one position in a wider discursive field. Within such a field, positions crystallized in reaction to each other as well as in response to the inner logic of the body of knowledge, and to constraints imposed by institutional affiliations. Borelli's chapter on "*Positivam levitatem in rerum naturam non dari*" (Borelli 1670, 129) offers a clue to the structure of the field in which Casati attempted to carve a niche for himself and for his fellow Jesuit mathematicians in their attempts to teach and develop their version of an Aristotelian-Archimedean synthesis. There,

---

[44] "Perit igitur Acquisitus impetus non totus simul; sed sensim extenuatur; idque non aliâ ratione, quàm quâ proportione impeditur motus, quocumque tandem ex capite impedimenta oriantur. Cum enim impetus contrarium impetum non habeat, si præcisa quidem impetûs natura spectetur [...] reliquum est, ut pereat pro ratione eorum, quæ aut motui corporis obsistunt, aut illud aliò quoquomodo dirigunt."

[45] "Illud certe extra omnem controversiam positum videtur, naturam gravem sponte sua non ascendere; quid ergo illud est., quod eburneum globulum, in subjectam rupem delapsum resilire cogit [...] nisi quia conceptus natura procurante impetus pergit motum efficere, ipsa etiam natura quantum potest, obsistente" Casati (1684, 136–137).

in over one hundred pages Borelli piled up arguments, together with reconstructions and reinterpretations of several experiments, which he aimed at an unnamed interlocutor whom he addressed as his *adversarius* and called a *peripateticus in signis*. One *adversarius* in Borelli's life has been identified by historian of science Luciano Boschiero as Carlo Rinaldini (1615–1698), a senior professor of Philosophy at Pisa (Boschiero 2007, 16). According to Middleton, Rinaldini had served as a military engineer for the forces of Urban VIII and Innocent X before being appointed professor, and taught mathematics to the young prince Cosimo (later Cosimo III). Middleton also claims that Rinaldini was the first to lecture on Galileo's works and on Gassendi's atomistic philosophy in Pisa in the 1650s and 1660s.[46] Nevertheless, Boschiero has shown that in the context of the academy Rinaldini campaigned against corpuscularism and represented the position of a traditional Aristotelian. In the same vein, he supported the idea of positive lightness, though we cannot say for sure at this point whether he was the very one Borelli chose for the duel. A look at Borelli's discussion reveals many more levels in the disputation about positive lightness. Borelli answered many arguments in favor of positive lightness that cannot be detected in Casati's text. Some examples are found in the very comprehensive disputations regarding the right interpretation of the pyramid figure of a flame of fire and the figure of an air bubble ascending in water. Casati is silent on these *areas* of the discursive field. However, as we have already seen, he refers explicitly to the experiment presented to the Duke and criticizes it as being technically flawed and hence unable to prove the designers' thesis. And indeed, we find in Borelli's book of 1670 that a very similar critique is mentioned and treated.[47]

Reconstructing the positions of Borelli, his adversary and Casati as they are reflected in the texts they have left behind—the unpublished diary of the academy (Abetti and Pagnini 1942; Rinaldini 1681)[48] and the *Saggi* (1667), Borrelli's book on natural motion (1670) and Casati's *Mechanica* (1684)—can throw light on the dynamics of the field of early modern mechanics in mid-seventeenth-century Italy.[49] Borelli was an uncompromising Galileist in his drive to understand all natural phenomena in mathematical and mechanical terms and in his open hostility to Aristotelian natural philosophy and its representatives. Boschiero emphasizes three prominent features that characterize Borelli's work: his insistence on banishing explanations of cold and heat, and heaviness and lightness by means of inherent qualities; the substitution of qualitative explanations by corpuscularian, or at least quantitative mathematical explanations; and the understanding of dynamical phenomena by means of static mechanics. Throughout the chapter on "positive

[46] In 1667 Rinaldini left for Padua, see Middleton (1971, 34–35).

[47] Because Casati thinks that a trochlea will solve the problem of a perpendicular ascendance, and that the measuring was not accurate, the position Borelli refers to is that it is not possible to elevate the cylinder in this way, using either the balance or the trochlea.

[48] The unpublished Cimento diary can be found amongst the Galilean manuscripts held at the Biblioteca Nazionale Centrale di Firenze (BNCF) in the folio labeled Gal. 262.

[49] Although unable to offer a full analysis in this format, we shall try to delineate schematically the main contours of the field.

lightness" Borelli's tone is very polemical. Like his great master before him, he chose to dedicate long passages at the beginning of his text to the refutation of ancient authorities, first and foremost Aristotle, whose position vis-à-vis that of his predecessors is analyzed in detail and subsequently rejected. Later on, he deals with modern Peripatetics. Finally, Borelli betrays his impatience with his adversaries, an impatience that is often expressed as open hostility.[50] He consistently tries to portray a dichotomy in which philosophy and Aristotle were strongly opposed to mathematics, to the atomists and to Archimedean theory, a portrayal that is scarcely perceptible in Casati's text.

But does Borelli's portrayal really reflect the field as seen through the eyes of historians today? It seems to fit the picture emerging from up-to-date historiography, according to which in the mid-seventeenth century the framework of Aristotelian natural philosophy and various versions of mechanical questions were still predominant in universities and Jesuit colleges, and were hence central to the cultural field (Meli 2006, 649). Thus, Boschiero's reading of the diary and the letters exchanged between the academicians testify that, in contrast to the façade presented by the *Saggi*, according to which the academy's project was purely experimental and remote from issues of natural philosophy, beneath the surface the academy provided an arena for the ongoing battle between the mechanistic-corpuscularian position of Borelli and Viviani and the Aristotelian positions of Marsili and Rinaldini. Boschiero argues that the experiments on positive lightness were intended to show that fire was not characterized by the qualities of heat and positive lightness but rather a substance made of corpuscules of heat causing rarefaction. However, Rinaldini was not convinced by these experiments and claimed that the abstractness of mathematics and geometry could not possibly provide a valid basis for studies in physics (Boschiero 2007, Chaps. 5, 7, 8, 9).

If we do accept Boschiero's portrayal, Casati's position emerges as a third option between the uncompromising mechanists such as Borelli on the one hand and intransigent Aristotelians on the other. There is no doubt that the physico-geometrical cosmological picture presented by Casati at the opening of his work enables him to integrate the quantitative Archimedean laws into a semi-Aristotelian discourse and to explain quantitatively various phenomena related to the motion of bodies in media. Gradually, however, Aristotelian elements are being pushed aside and the most constitutive among Aristotle's categorical distinctions—those between heaviness and lightness and between natural and violent motion—are neglected and actually eroded. Finally, the reader is left with one explanatory principle—impetus—that for Casati anchors Galileo's mechanical theorems (the ratio between moments of heaviness and moments of velocity along the perpendicular and along different inclined planes) in a causal discourse on motion. Thus he is free to show his students the art of calculating the forces of various machines, which he does with endless

[50] In a letter to Paolo del Buono, October 10, 1657 he wrote of a "rotten and mouldy peripatetic" among the academicians. In Abetti and Paganini (1942, DIS, 20, note 2); Middleton (1971, fn. 83, 34).

digressions on a variety of other relevant topics such as the discussion of elastic forces, percussion, *remora* (Casati 1684, Book IV, X; XIV) and so on.

Like the early Galileo before him, Casati operated within an Aristotelian conceptual framework, assuming that all motion requires a cause—namely a force both to initiate it (an *innate impetus* creating the first movement) and to increase its *intensity* and conserve it (*acquired impetus*). Like the early Galileo, Casati believed that the Archimedean tradition could provide the toolbox necessary to mathematize the Aristotelian discourse on forces. Motion, however, was the most fundamental concept for him since he succumbed to the relativization of the notions of heaviness and lightness. In spite of the priority he gave to dynamical phenomena, however, he still believed that Archimedean statics was the only explanatory model to which all phenomena could be reduced. But in this, too, he did not differ much from the Galileists who constituted the backbone of the *Accademia del Cimento*.

Like other Galileists of his time and place, Casati adopted the experimental jargon that was fashionable in his environment. Although he did not avoid quoting examples from experience understood in a traditional way as common knowledge, his discourse opens up a large space for experiments as unique, historical events, performed in specific circumstances (Dear 2006, 106–131)). There is no doubt that his acumen was sufficient at least to design his own experiments and to offer justified technical critique of the experiments of others. Moreover, in many places Casati expressed his ideas concerning the possibilities inherent in experiments as a source of factual knowledge, but also concerning their limitations, albeit in a subtle, non-polemical manner.[51] Experiments can have rival but equally plausible interpretations. In addition, there are things—such as positive lightness—that raise serious doubt about the very possibility of proof or refutation by experiment.

Unlike the Galileists, however, Casati had a completely different attitude toward the Aristotelian tradition. His point of departure, which he never explicitly stated but which operates as a basic assumption underlying his text, is that the Aristotelian framework should be preserved despite the many modifications and adaptations it still required. Whether this was the result of actual pressure exerted on all Jesuits, as some historians have argued (Hellyer 2005), or it represented Casati's personal belief is not clear to us, and may even be unknowable. However, the case of Melchior Cornaeus makes Casati's position even more interesting, since Cornaeus had chosen another manner with which to deal with this complicated situation. In his *Curriculum philosophiae Peripateticae* (Cornaeus 1657), he denied the existence of positive lightness. Yet, in view of the explicit proscription against such a conclusion, he added that he had taught such an opinion for many years but now "because the

[51] Here is a quotation from Casati's *Terra machinis mota* where he contends that the experiment is incapable of deciding the matter of the weight of air and the connection to the existence or non-existence of void: "Quare nihil est in toto hoc experimento; quod vel leuem inferat Vacui suspicionem; in oculos semper incurrit corpus, tenue illud quidem ac rarum, à quo vasis spatia occupantur. Hoc itaque experimento *non satis probari nostri aëris grauitatem absolutè,* mihi certissimum est; sed illud vnum ex inæquali recipientis pondere antè ac post aëris extractionem vi antliæ, confici potest, quod non inficior, scilicet communem hunc nostrum aërem in aëre alio magis raro grauitare posse." Casati (1658, 171).

authority of my superiors commands something else, I say that it is probable that heaviness and lightness are two positive qualities [...] and because authority commands that we subscribe to this opinion, I subscribe and I approve it" (Feingold 2007, 226).

Unlike Cornaeus, Casati rejected the vacuum hypothesis both in a book written by him on the subject and in other places, and adhered to the concept of *positive lightness*. Still, our analysis has shown that formally reproducing the Aristotelian terminology did not necessarily mean acceptance of all the conceptual contents implied in it. The Aristotelian-Archimedean framework developed by Casati throughout the eight hundred pages of his *Mechanica* thus seems worthy of further analysis in order to understand in detail its contribution to preclassical mechanics.

References

Abetti, Griorgio, and Pietro Pagnini, eds. 1942. *L'Accademia del Ciemento: parte prima. Vol. 1. Le opere dei discepoli di Galileo Galilei*. Florence: Barbèra.

Anstey, Peter R., and A. Schuster. 2005. *The science of nature in the seventeenth century: Patterns of change in early modern natural philosophy*. Dordrecht: Springer.

Aristotle. 1585. *Quaestiones mechanicae; De miraculis naturae; Physiognomica; De lineis insecabilibus, haec & alia sequens pagina indica*. Venetijs: Bruniolo.

Aristotle, and W.K.C. Guthrie. 1939. *Aristotle in twenty-three volumes. Vol. 6. On the heavens*. Cambridge: Harvard University Press. The Loeb Classical Library.

Baldini, Ugo. 1992a. Archimed en el seicento Italiano. In *Archimede: Mito tradizione scienza (Siracusa-Catania, 9–12 ottobre 1989)*, ed. Corrado Dollo, 237–289. Florence: Olschki.

———. 1992b. *Legem impone subactis: Studi su filosofia e scienza dei Gesuiti in Italia 1540–1632*. Rome: Bulzoni.

———. 2000. *Saggi sulla cultura della Compagnia di Gesù (secoli XVI–XVIII)*. Padua: CLEUP.

Bardi, Giovanni di. 1614. *Eorum quae vehuntur in aquis experimenta*. Rome: Zannetti.

Benedetti, Giovanni Battista. 1553. *Resolutio omnium Euclidis problematum aliorumque ad hoc necessario inventorum una tantummodo circini data aperture*. Venezia, Apud Bartholomaeum Caesanum.

———. 1585. Book of various mathematical and physical ideas [Diversarum speculationum mathematicarum, et physicarum]. In *Mechanics in sixteenth-century Italy. Selections from Tartaglia, Benedetti, Guido Ubaldo, & Galileo*, ed. S. Drake and I.E. Drabkin. Turin/Madison: The University of Wisconsin Press.

Beretta, Marco, Antonio Clericuzio, and Lawrence Principe, eds. 2009. *The Accademia del Cimento and its European context, proceedings of a conference held in Florence, 14–15 December 2007*. Sasgamore Beach, MA: Science History Publications.

Blair, Ann. 1999. The Problemata as a natural philosophical genre. In *Natural particulars: Nature and the disciplines in renaissance Europe*, ed. A. Grafton and N.G. Siraisi. Cambridge, MA: MIT Press.

Borelli, Giovanni Alfonso. 1670. *De motionibus naturalibus a gravitate pendentibus*. Bologna: Dominici Ferri.

Boschiero, Luciano. 2003. Natural philosophical contention inside the Accademia del Cimento: The properties and effects of heat and cold. *Annals of Science* 60 (4): 329–349.

———. 2007. *Experiment and natural philosophy in seventeenth-century Tuscany: The history of the Accademia del Ciemento*. Dordrecht: Springer.

Buonamici, Francesco. 1591. *De motu*. Vol. 5. Florence: Sermartellius.

Casati, Paolo. 1658. *Terra machinis Mota*. Rome: Iganitij de Lazaris.
———. 1684. *P. Casati, Mechanicorum libri octo, in quibus uno eodemque principio vectis vires physice explicantur & geometrice demonstrantur, atque machinarum omnis generis componendarum methodus proponitur*. Lyon: Apud Anissonios.
Cornaeus, Melchior. 1657. *Curriculum philosophiae Peripateticae*. Herbipoli: sumptibus & typis Eliae Michaelis Zinck.
Damerow, Peter, Gideon Freudenthal, Peter McLaughlin, and Jürgen Renn, eds. 2004. *Exploring the limits of preclassical mechanics: A study of conceptual development in early modern science: Free fall and compounded motion in the work of Descartes, Galileo, and Beeckman*. 2nd ed. New York: Springer.
Dear, Peter. 1995. *Discipline and experience: The mathematical way in the Scientific Revolution*. Chicago: The University of Chicago Press.
———. 2006. The meanings of experience. In *The Cambridge history of science: Early modern science*, ed. Katherine Park and Lorraine Daston, 106–131. New York: Cambridge University Press.
Del Monte, Guidobaldo. 1577. *Guidobaldi e Marchionibus Montis mechanicorum liber*. Pisauri: Apud Hyeronimum Concordiam.
Des Chene, Dennis. 1995. *Physiologia: Natural philosophy in late Aristotelian and Cartesian thought*. Ithaca: Cornell University Press.
Dollo, Corrado. 1990. Galilei e la fisica del Collegio Romano. *Giornale Critico della Filosofia Italiana* 71: 161–201.
———, ed. 1992. *Archimede: Mito tradizione scienza*. Florence: Olschki.
Drabkin, I.E., and Stillman Drake, eds. 1960. *Galileo Galilei on motion and on mechanics*. Madison: University of Wisconsin Press.
Drake, Stillman. 1981. *Cause, experiment, and science*. Chicago: The University of Chicago Press.
Drake, Stillman, and I.E. Drabkin, eds. 1969. *Mechanics in sixteenth-century Italy. Selection from Tartaglia, Benedetti, Guido Ubaldo, & Galileo*. Madison: The University of Wisconsin Press.
Duhem, Pierre Maurice Marie. 1905. *Les origines de la statique*. Paris: Hermann.
Elazar, Michael. 2011. *Honoré Fabri and the concept of impetus: A bridge between conceptual frameworks*. Dordrecht: Springer.
Favaro, Antonio, ed. 1968. *Le opere di Galileo Galilei, nuovaristampa della edizione nazionale 1890–1909*. Vol. 20. Florence: Barbèra.
Feingold, Mordechai. 2007. Review: Marcus Hellyer. Catholic phyics: Jesuit natural philosophy in early modern Germany. *Renaissance Quarterly* 60 (1): 225–227.
Feldhay, Rivka. 2006. On wonderful machines: The transmission of mechanical knowledge by Jesuits. *Science and Education* 15 (2–4): 151–172.
Galilei, Galileo. 1587–1588. Theoremata circa centrum gravitatis solidorum. In *Le opere di Galileo Galilei: nuovaristampa della edizione nazionale 1890–1909*, ed. Antonio Favaro. Florence: Barbèra.
Galluzzi, Paolo. 1981. L'Accademia del Cimento: "Gusti" del principe, filosofia e ideologia dell'esperimento. *Quaderni Storici* 16 (48): 788–844.
———, ed. 2001. *Scienzati a corte. L'arte della sperimentazione nell'Accademia Galileana del Cimento (1657–1667)*. Firenze: Galleria degli Uffizi. 18 marzo–18 giugno 2001. Livorno: Sillabe.
Gatto, Romano. 2002. *Le mecaniche: Edizione critica e saggio introduttivo di Romano Gatto*. Florence: Olschki.
Gavagna, Veronica. 1999. Il carteggio Casati (1642–1695). *Bollettino di storia delle scienze matematiche* 18: 3–15.
Gorman, Michael John. 1998. *The scientific counter-revolution: Mathematics, natural philosophy and experimentalism in Jesuit culture, 1580–1670*. Ph.D. Dissertation. Florence: European University Institute.
———. 2003. Mathematics and modesty in the society of Jesus: The problems of Christoph Grienberger. In *The new science and Jesuit science: Seventeenth-century perspectives*, ed. M. Feingold, 1–121. Dordrecht: Springer.

Hellyer, Marcus. 2005. *Catholic physics: Jesuit natural philosophy in early modern Germany*. Notre Dame: University of Notre Dame Press.
Meli, Domenico Bertoloni. 2006. Mechanics. In *Early modern science*, ed. Katharine Park and Lorraine Daston, 632–672. Cambridge: Cambridge University Press.
Middleton, William Edgar Knowles. 1971. *The experimenters: A study of the Accademia del Cimento*. Baltimore: Johns Hopkins Press.
Mugler, Charles. 1951. *Archimède répliquant à Aristote*. Paris: Les Belles-lettres.
Palmerino, Carla Rita. 2003. Two Jesuit responses to Galileo's science of motion: Honoré Fabri and Pierre Le Cazre. In *The new science and Jesuit science: Seventeenth-century perspectives*, ed. M. Feingold, 187–227. Dordrecht: Springer.
Phillips, Edward C. 1939. The correspondence of father Christopher Clavius S. I. Preserved in the archives of the Pont. Gregorian university. *Archivum historicum societatis Iesu* VIIII (2): 193–222.
Renn, Jürgen, and Peter Damerow. 2012. *The equilibrium controversy: Guidobaldo del Monte's critical notes on the mechanics of Jordanus and Benedetti and their historical and conceptual background*. Berlin: Edition Open Access. http://www.edition-open-sources.org/sources/2/index.html.
Rinaldini, Carlo. 1681. *Philosophia rationalis, naturalis atque moralis*. Vol. 1. Padua: Frambotti.
Scaduto, Mario. 1949. Il matematico Francesco Maurolico e i Gesuiti. *Archivum historicum societatis Iesu* XVIII (35): 126–141.
Shea, William R. 1968. Galileo's discourse on floating bodies: Archimedean and Aristotelean elements. In *Actes du XIIe Congrès International d'histoire des Sciences*. Paris: Blanchard.
Schuster, John A. 2013. What was the relation of baroque culture to the trajectory of early modern natural philosophy? In *Science in the age of baroque*, ed. O. Gal and R. Chen-Morris, 13–45. Dordrecht: Springer.
Valleriani, Matteo. 2010. *Galileo engineer*. Dordrecht: Springer.
Wallace, William A. 1977. *Galileo's early notebooks: The physical questions*. Notre Dame: University of Notre Dame Press.
———. 1981. *Preludeto Galileo*. Dordrecht: Reidel Publishing Company.
———. 1984. *Galileo and his sources: Heritage of the Collegio Romano in Galileo's science*. Princeton: Princeton University Press.

Chapter 10
Jesuit Conceptions of Impetus After Galileo: Honoré Fabri, Paolo Casati, and Francesco Eschinardi

Michael Elazar and Rivka Feldhay

Abstract The fourteenth-century concept of impetus denotes an impressed force and was used to explain the continuation of the motion of projectiles and the acceleration of falling bodies. This chapter deals with the use of this concept in the period between Galileo's death (1642) and the publication of Newton's *Philosophiæ naturalis principia mathematica* (1687). Focusing on three major figures among contemporary Jesuit thinkers, the French Honoré Fabri (1608–1688) and the Italians Paolo Casati (1617–1707) and Francesco Eschinardi (1623–1703), this chapter shows how these Jesuits employed the concept of impetus in their own versions of preclassical mechanics.

Keywords Impetus · Jesuit science · Galilean science · Aristotle's physics · Honoré Fabri · Paolo Casati · Francesco Eschinardi

10.1 Introduction

As has been discussed in previous chapters, between circa 1580 and 1680 problems of motion became an important subject for mathematicians and natural philosophers all over Europe, giving rise to preclassical mechanics. This interest drew its inspiration from newly translated and published sources, both Aristotelian (especially *Mechanical Questions*) and Archimedean, as well as from the accumulated shared knowledge of contemporary practitioners and engineers, shaped by the ever growing influence of "challenging objects" that "triggered the transformation of scholastic physical concepts toward classical mechanics."[1] Galileo's work on motion

[1] Renn (2001, 7); see also Dollo (1989); Laird (1986); Laird and Roux (2001).

M. Elazar (✉)
Minerva Humanities Center, Tel-Aviv University, Tel-Aviv, Israel
e-mail: mikielazar@gmail.com

R. Feldhay
Cohn Institute for the History and Philosophy of Science and Ideas
& Minerva Humanities Center, University of Tel Aviv, Tel Aviv-Yafo, Israel
e-mail: feldhay@post.tau.ac.il

R. Feldhay et al. (eds.), *Emergence and Expansion of Preclassical Mechanics*,
Boston Studies in the Philosophy and History of Science 270,
https://doi.org/10.1007/978-3-319-90345-3_10

emerged from the knowledge shared by the community of scholars drawing on the resources of both of these traditions. In their exchanges, these scholars tended to discuss problematic issues stemming from one of these traditions—most significantly, free fall (the topic discussed in this chapter)—using the conceptual means of the Archimedean theory of buoyancy. With this framework as his starting point, Galileo succeeded in achieving some important mathematically formulated solutions for particularly difficult and unresolved physical problems such as the rate of acceleration in free fall and the parabolic shape of a projectile's trajectory. According to the historian of science John Heilbron the outcome of Galileo's effort "was a somber, limited, exact science, and a few striking results, advertised as more exact than they were, to serve as a replacement for one chapter of the vast, colorful, diffuse library of Aristotelian philosophy" (Heilbron 2010, 127).

Indeed, from the point of view of Newtonian science, Galileo's contribution may be perceived as just some scattered results devoid of any clear set of foundational principles. However, throughout most of the seventeenth century things appeared differently. In fact, in spite of the elaborate rhetoric in favor and against the traditional science of motion and in favor of and against Aristotle, preclassical mechanics provided a heterogeneous yet encompassing conceptual framework combining Aristotelian and Archimedean concepts and tools. Placing Galileo's achievements in this context makes it possible to perceive continuities streching from his predecessors to the investigations of Jesuit scientists in the aftermath of his work.

This chapter illustrates the way this precarious, yet functioning Aristotelian-Archimedean conceptual framework of preclassical mechanics provided an intellectual environment in which integration, diffusion, and—no less important—teaching of Galilean results were performed in the period between Galileo and Newton. As an example of the way traditional conceptual frameworks were transformed and eventually dissolved in the process of transformation, we will focus on the concept of impetus as articulated by three Jesuits in the course of their elaboration of Galileo's mathematical analysis of free fall and his odd-numbers rule for falling bodies. Close reading and analysis of seventeenth-century physico-mathematical Latin texts, written by three educators in Jesuit colleges in France and Italy: Honoré Fabri (1608–1688); Paolo Casati (1617–1707) and Francesco Eschinardi (1623–1703) will be at the center of this chapter.

Already in the high Middle Ages, the forerunner of impetus—the Arabic concept *mayl*—was used by philosophers to account for the increasing velocity of falling bodies (Clagett 1959, 510–514). In the fourteenth century Jean Buridan employed the concept of impressed force (*vis impressa*) to explain the acceleration of falling bodies as well as projectile motion.

Galileo, just as other representatives of preclassical mechanics, attempted to give a causal explanation of acceleration with the help of concepts compatible with the fundamental model of Aristotelian dynamics according to which motion must be caused by a force.[2] In his early work, he had explained acceleration in free fall by the gradual weakening of an impressed force. The progress he had achieved in the

[2] See Chap. 1 in this volume.

detailed mathematical account of free fall acceleration in his published work made it difficult for him to present a comprehensive dynamical explanation based on this idea. Nevertheless, he speculated on the causes of acceleration in his *Discorsi,* considering the case of free fall of a projectile after a preceding upward motion caused by an impetus:

> The diminutions of this alien impetus then continuing, and in consequence the advantage passing over to the side of the heaviness, descent commences, though slowly because of the opposition of the impressed force, a good part of which still remains in the movable.[3]

Alternatively, Galileo conceptualized free fall in terms of another Aristotelian mental model, that of natural motion taking place spontaneously without the intervention of a force:[4]

> A heavy body has from nature an intrinsic principle of moving toward the common center of heavy bodies (that is, of our terrestrial globe) with continually accelerated movement, and always equally accelerated, so that in equal times there are added equal new momenta and degrees of speed. (Galileo 1989, 77)

The causal explanation of acceleration was a problem that Galileo left open. He expressed himself even rather sceptically about the possibilities of finding such an explanation. It is therefore no wonder that Galileo's result inaugurated a lively debate over the explanation of free fall—called by historian of science Paolo Galluzzi "the new Galileo affair."[5] It is not surprising to find quite a number of Jesuit responses to Galileo's achievement. The best known of these was perhaps the rather successful attempt by Giovanni Battista Riccioli to both validate Galileo's odd numbers rule—which stated that the spaces traversed in equal consecutive times increased according to the series 1, 3, 5, 7, etc.—and measure the actual rate in which falling bodies increase their velocities.

All three Jesuit theories discussed here react to the view presented by Galileo, making use of his law of odd numbers. Since Fabri's theory has been thoroughly examined elsewhere,[6] it will only be touched upon briefly, so as to allow for a comparison with the approaches of his two colleagues.

Section 10.2 will analyze the ways in which the three Jesuits define impetus; Section 10.3 will discuss their opinions concerning the conservation of impetus and motion; Section 10.4 will discuss the manners in which the three Jesuits conceived "natural impetus," that is, the kind of impetus involved in naturally accelerated motion; Section 10.5 will analyze the way in which each of the Jesuits assimilated Galileo's law of odd numbers.

---

[3] Galileo EN VIII, 201/157 f. Translation from Damerow et al. (2004).

[4] See Chap. 1 in this volume.

[5] See Galluzzi (2001).

[6] See Elazar (2011), and also Drake (1974, 1975); Lukens (1979).

10.2 Defining Impetus

10.2.1 Fabri

Honoré Fabri was perhaps the most important among the first Jesuits who tried to assimilate Galileo's results into their physico-mathematical account of motion in a systematic way. During the 1640s the French Jesuit Fabri taught logic, metaphysics, astronomy, mathematics, and natural philosophy in *Collège de la Trinité* at Lyon, and his physical doctrine appears in a collection of his lectures, compiled by one of his students Pierre Mousnier—entitled *Tractatus physicus de motu locali*.[7] Whether intentionally or not, Fabri transmits his message right from the beginning by choosing the title *Tractatus de motu locali*—an accurate replication of the fictitious treatise on local motion read by Sagredo, Salviati and Simplicius during the third and fourth day of Galileo's *Two New Sciences*. Like Galileo, Fabri isolates certain aspects of motion—natural, violent and mixed—as the topic of his treatise. In contradistinction, however, Fabri stresses the "physical" nature of his treatise, and according to the title page, his demonstrations are declared to be "physical." His subject matter is not defined as a new science in the title, but rather is contained in his *physica*, which is fully mathematical, despite the distinction usually considered by historians to prevail between contemporary Jesuit *physica* and *mathesis*.[8] The precise contents with which Fabri will fill up his *physica*, however, will soon transpire as he unfolds the cornerstone of this old-new kind of discourse, namely the concept of "impetus."

"Since impetus," he explains, "is the immediate cause of motion," and since "a thing cannot be known as a reasoned fact (*propter quid*) unless its reason is known,"[9] therefore "there cannot be any doubt but that a discussion about impetus should come first, so that afterwards the properties of motion itself will be proved by its cause."[10] But Fabri does not stop here, and sees fit to add: "indeed I would dare say that not only motion itself but also the whole of physics depends on the knowledge of impetus alone."[11] Fabri in fact wishes to erect a whole new science of motion,

---

[7] The full title is *Tractatus physicus de motu locali, in quo effectus omnes, qui ad impetum, motum naturalem, violentum, & mixtum pertinent, explicantur, & exprincipiis physicis demonstrantur; auctore Petro Mousnerio Doctore Medico; cuncta excerpta ex praelectionibus R. P. Honorati Fabry, Societatis Iesu*. In any case, in later years Fabri treats this work as his own.

[8] See, e.g., Baldini (1999, 256–260).

[9] See Wallace (1972–1974, 29–47) and Mancosu (1996, 10–12), for the difference between an explanation quod (or *quia*), i.e. "of the fact," and a demonstration *propter quid* ("of the reasoned fact"), concepts originating from Aristotle's *Posterior Analytics*.

[10] "Cum enim impetus sit causa immediata motus, ut fuse demonstrabimus infra; & cum propter quid sit res cognosci non possit, nisi eius causa cognoscatur; dubium esse non potest, quin praemittenda sit tractatio illa, quae est. de impetu, ut deinde affectiones ipsius motus per causam eiusdem demonstrentur" Fabri (1646, 1). See also Elazar (2011, 3).

[11] "Immo ausim dicere ex unius impetus cognitione, non modo motum ipsum, verum etiam totam rem physicam pendere" Fabri (1646, 1).

based on an Aristotelian epistemology (indicated by the "propter quid" term), but which nevertheless fully utilizes recent achievements, such as Galileo's analysis of free fall.[12] In fact, Fabri assigns a major role to the concept of impetus not only in physics, but also beyond the profane sciences—he regards it also as a key concept for explaining the mystery of the Eucharist (Elazar 2011, Part IV).

According to the medieval understanding of impetus, developed and established in the work of Jean Buridan,[13] it is on the one hand "a certain motive force (*vis motiva*) of the moving body," which acts "in the direction toward which the mover was moving the moving body, either up or down, or laterally, or circularly"; on the other hand it is a thing of permanent nature (*res naturae permanentis*), namely a "quality (*qualitas*) naturally present and predisposed for moving a body in which it is impressed" (Clagett 1959, 534, 537). Impetus was sometimes presented as a "permanent" quality, that is, existing as a whole for a definite duration and tending to be conserved (in the absence of resisting factors), while motion was generally presented as a "successive" entity—representing states (i.e. places or velocities) which "succeed one another but do not proceed from one another" (Maier 1982, 160). In many cases, however, impetus was regarded as inherently transient, self-dissipating even in the absence of any obstructing factor.[14]

Fabri regards impetus, in line with Buridan's view, as a "permanent quality,"[15] and defines it, in the beginning of *De impetu* (the first book of his *De motu locali*) in the following way:

> Definition III: impetus is a quality exacting (*exigens*) motion, or spatial flow (*fluxus localis*), of its subject; or [the quality] which is the proximate cause of the motion of that mobile in which it is, i.e. in that way in which it can be a cause of motion.[16]

Fabri's more detailed definition of motion is "the transition of a mobile from one place to another in a continual fluxus (*Motus localis est transitus mobilis e loco in locum continuo fluxu*)" (Fabri 1646, 1). Fabri deems motion not a real *ens*, but a "composition of an entity and a non-entity" (*concretum quoddam ex ente & non ente*), which therefore cannot be produced (*producitur*).[17] It is not surprising therefore that Fabri chooses to put the "permanent" impetus, the cause of motion, in the

[12] This claim has been demonstrated in Elazar (2011) and will be shown here only in brief for the sake of comparison with Casati and Eschinardi.

[13] This key concept has of course a much longer history, which will not be dealt with in this article. Concerning the possible ancient origins of impetus, ranging from Hipparchus of Nicaea (second century BC) to John Philoponus (ca. 490–570 AD), see Franco (2003, 525), as well as Clagett (1959, 508–510, 543–547).

[14] Franciscus Marchia, for example, adopted such "self-expending" *vis derelicta*, and Nicole Oresme also supported a similar kind of impetus, Clagett (1959, 529); so did Galileo in his early *De motu antiquiora* (ca. 1590).

[15] Fabri (1646, lib. 1, th. 12): "Impetus est. qualitas permanens."

[16] "Definitio III: impetus est. qualitas exigens motum, seu fluxum localem sui subiecti, vel quae est. causa proxima motus illius mobilis, cui inest, eo scilicet modo, quo potest esse causa motus" (Fabri 1646, lib. 1, def. 3, p. 2). The verb "exacting" (*exigere*) means "requiring," or "compelling."

[17] Elazar (2008, 7–8). The *ens* and *non ens* to which Fabri refers signify a place just acquired and a place already left during motion.

center of his physics, rather than motion. Contrary to Buridan, Fabri takes pains to carefully distinguish between the force that a projector applies to the projectile (*potentia motrix*) and the impetus that once impressed in the projectile causes motion (Fabri 1646, lib. 1, th. 6, p. 13). Fabri defines the causality between impetus and motion by explaining that, while impetus itself (as a real *ens*) is "produced" (*producitur*) in an action involving efficient causality, motion is only "exacted" (*exigitur*) by impetus, which always acts as a formal cause. As explained elsewhere, Fabri's characterization of the creation of motion as involving formal (and not efficient) causality makes sure that impetus, as the cause of motion, is not dissipated while causing motion, and thus enables Fabri to conceive of a conservation of rectilinear motion (Elazar 2011, 127–133).

Fabri usually describes impetus as a quality, but occasionally refers to it as a *vis*.[18] In any case, Fabri clearly distinguishes between the external moving force (*potentia*) of a projectile and impetus, a quality which once impressed in the subject would "rejoice" in its inner "goal" (i.e. motion).[19] An important attribute of Fabri's impetus, which directly influences motion, is its inherent rectilinearity: "Since motion follows from impetus," he explains in his *Metaphysica demonstrativa*, "it is determined along one line only, whether it is simple, or mixed."[20] In his *Tractatus*, Fabri emphasizes that "impetus does not act simultaneously along a circle, rather only along the line of its motion,"[21] thus explicitly rejecting Buridan's above-mentioned claim that impetus can be directed "either up or down, or laterally, or circularly." Contrary to Aristotle—who stressed that "circular motion is prior to rectilinear," (Aristotle, *On the Heavens*, 1, 2, 268b17, 269a25; Aristotle 1953) and thus established what Alan Gabbey defined as the dynamical "tyranny of the circle" (Gabbey 1980, 291)—Fabri gives priority to rectilinear motion, and considers circular motion as merely an "impeded" straight motion.[22] It is clear, then, that in establishing rectilinear motion as primary Fabri's break with the Aristotelian framework is definitely more radical than Galileo's and was probably influenced by Descartes.

---

[18] For example, in theorem 12, Fabri (1646, 18) refers to it as *vis impressa*.

[19] "... cum enim motus sit finis intrinsecus impetus; certe si nihil impediret motum, haud dubie gauderet impetus suo fine" Fabri (1646, th. 44, p. 32). This is deeply connected to Fabri's view of motion as a "secondary formal effect" (*effectus formalis secundarius*) of impetus; see Elazar (2008, 8–9).

[20] "Cum motus sequatur ex impetu, hic ad unam tantum lineam est. determinatus, sive simplicem, sive mixtam" Fabri (1648, lib. 12, prop. 20, p. 496).

[21] "Non simul agit impetus in orbem sed tantum per lineam sui motus" Fabri (1646, lib. 1, th. 45, pp. 32–33).

[22] "Motus circularis in sublunaribus oritur ex recto impedito" Fabri (1646, "Synopsis amplior," "De motu circulari," par. 1). Accordingly, Fabri emphasizes that if a sling is released "a straight motion follows" Fabri (1646, lib. 1, th. 45, pp. 32–33).

10.2.2 Casati

The Italian Paolo Casati was a well-known Jesuit mathematician, highly regarded in his order, who was deeply involved, for example, with the conversion of queen Christiana of Sweden to Roman Catholicism.[23] Casati was responsible for several widely read physico-mathematical works, of which the most important and comprehensive was the *Mechanicorum libri octo*.[24] Although Casati was an excellent experimenter, he was not a philosopher by temperament. Accordingly he was not really interested in conceptualizing an impetus-based alternative to the Aristotelian philosophy of motion. Rather, his main purpose was to understand machines, and he emphasizes, in a book entitled *De causis motus Machinalis* which discusses their causes within his *Mechanicorum libri*, that in order to build successful machines, it is necessary to know the causes of their motions. These causes are impetus, the cause immediately effecting motion,[25] as well as resistances, that is, the factors that resist and inhibit impetus. Casati later argues that in order to better understand impetus we should study how and why bodies lose some of it (Casati 1684, 145), and hence the richer knowledge of these impetus-inhibiting resistances will enable us to better explain machines.[26] By defining impetus as a *motum proxime efficiens* or *vis motum efficiendi* Casati seems to imply an efficient causality between impetus and motion, in line with Buridan's view, and contrary to Fabri's formal causality. Nevertheless, Casati is no doubt familiar with Fabri's use of *exigere*, as we can infer from his remark that impetus "either effects, or at least exacts motion" ("*impetus autem motum aut efficit, aut saltem exigit*" (Casati 1684, 147)). This indicates his indifference as to the exact nature of the impetus-motion causality.

In line with the Aristotelian tenet based on the motion-implies-force model, the second chapter of *De causis motus Machinalis* (where Casati presents his concept of impetus), begins: "Anything that moves, of whatever kind, necessarily has a cause of motion, so that it would move—whether of its own volition, or by an alien force—from one place to another."[27] Casati further explains, in a traditional Aristotelian vein, that the first kind of motion concerns the motion of "heavy or light

[23] Feldhay (2006, 152). On Casati see also Gavagna (1997).

[24] The full title of Casati's work is *Mechanicorum libri octo, in quibus uno eodemque principio vectis vires physice explicantur & geometrice demonstrantur, atque Machinarum omnis generis componendarum methodus proponitur*.

[25] The title of the second chapter of *De causis motus Machinalis* promises to explain the nature of impetus as a cause immediately effecting motion ("Impetus motum proxime efficientis natura explicatur" Casati (1684, 136)). The translation to be found in the English translation of Duhem (1991, 570, n. 111), "The Nature of the impetus causing motion is explained in detail" is incorrect, for adverb *proxime* in the title therefore relates to *efficientis* and not to *explicatur* (furthermore, Casati states, in the next page that "impetus motum proxime effecit"; besides, *proxime* cannot mean "in detail" Casati (1684, 137).

[26] "Hinc nimirum in uberiorem ipsius resistantiae nototiam venientes ad explicandam motus machinalis causam propius accedemus" Casati (1684, 145).

[27] "Quicquid movetur, qualecumque est., causam habeat moventem necesse est., ut hoc quidem sponte sua, illud vero aliena vi ex alio in alium locum migret" Casati (1684, 136).

bodies" (*gravia aut levia*), which tend to arrange themselves according to the naturally constituted order of things, as well as that of animals—endowed by nature with an "appetite" (*appetitus*) to avoid harmful objects or circumstances and to acquire necessary ones; Casati also includes in this category bodies which following externally enforced expansion or contraction tend to return (by their elasticity) to their former state (Casati 1684, 136). The second kind includes objects that in moving do not "serve nature" (*in motu non serviunt naturae*), but rather are externally "dragged, pushed, moved circularly, projected" etc. Casati observes—again, according to scholastic tradition—that while the motion of heavy and light bodies involves an increase of speed (*celeritas*), the motion of projectiles is gradually weakened (Casati 1684, 136). "Even though we attribute," continues Casati, "the ultimate origin of motion to the Mover of Nature (*movens naturae*), which by the decrees of the Philosopher is a substance, nevertheless we justly recognize something interposing (*interjectum*) between that nature and motion (which we name 'impetus'), whose intension and remission is followed by speediness and slowness (*velocitas ac tarditas*).[28]

Casati refers then to impetus as a kind of "mediator" between (probably) the Prime Mover[29] and motion, which determines—according to the well-known terms taken from the medieval *latitudo formarum* vocabulary, *intensio* and *remissio*—the specific velocities of moving bodies. It is worth adding that Galileo's account of acceleration in his *Two New Sciences* also owes a great deal to this very tradition. Casati immediately adds that without the concept of impetus it would be impossible to account for the variation of velocities within either a falling stone (whose constant nature causes by itself constant velocity, and thus cannot explain its acceleration) or an upward bouncing heavy object (whose nature would force it to descend, rather than ascend) (Casati 1684, 136–137).

In keeping with tradition, Casati stresses the superiority of impetus over motion, which being merely a flux of *ubicationes* (i.e. places) cannot be considered as having any *efficientia*, and must be deemed inferior to the "permanent" impetus; "for who would deny," concludes Casati, that perishing things (*labentia*) are inferior—*caeteris paribus*—to durable things (*stantia*)?[30] At this stage Casati wants to make sure—like Fabri—that impetus is considered the cause of motion, and not the other way around, and relates this "superiority" of impetus to the principle that an effect cannot be more "eminent," or more "distinguished" (*praestabilior*) than the cause,

---

[28]"Etiamsi moventi naturæ, quæ ex Philosophi decretis substantia est., motus originem ultimam tribuamus, jure tamen optimo aliquid naturæ ipsi ac motui, interjectum agnoscimus (impetum nominamus) cujus intentionem ac remissionem velocitas ac tarditas consequatur" Casati (1684, 136).

[29]Which is indeed regarded as a substance by Aristotle (*Metaphysics*, 1071b13–22; Aristotle 1908).

[30]"... motui per se, quia ex alio in alium locum continuata migratio est., eficientiam aegre tribuere possumus; quippe qui, cum in fluxione consistat, ita ut locus loco, seu potius, ut scholae loquuntur, ubicatio ubicationi, priori scilicet pereunti succedat, posterior aeque fugax, inferioris notae censendus est. quam impetus natura sua aliquandiu permanens: labentia enim stantibus deteriora esse, caeteris paribus, quis neget? Casati (1684, 138).

for an effect can never have anything that it did not receive from the cause, or that the cause had lacked; Casati hence infers that "motion does not effect impetus; on the contrary—impetus effects motion."[31]

Somewhat later—also in line with tradition—Casati defines impetus both as a force which effects motion (*vis motum efficiendi*) (Casati 1684, 137) and as a quality "established (or designated) for motion" (*qualitas propter motum instituta*) (Casati 1684, 155). Casati also claims that he does not distinguish between impetus and *conatus* (Casati 1684, 155), a term already coined by Cicero to designate a natural inclination and "frequently employed in the seventeenth century to signify the action of a force, or tendency, during a short constant interval of time" (Jammer 1957, 111). In accordance with Aristotle's well-known dynamic laws from the *Physics*, Casati states that to move 5.25 pounds with a "velocity as 8," we need a *conatus* 1.5 times larger than the *conatus* required to move 4 pounds with a "velocity as 7."[32]

Unlike Fabri, Casati never emphasizes the inherent rectilinearity of impetus (and consequently motion). However, neither does he mention any impetus which causes non-rectilinear motion; furthermore, he clearly regards impetus as having a distinct direction, and explains that in order to determine the resultant motion of a body moved by two impetuses differing in direction, we should apply the old parallelogram rule (Casati 1684, 155). Thus Casati, despite not stressing (or even explaining) the inherent rectilinearity of impetus and motion, treats impetus in a way that in hindsight can be seen as corresponding to the modern concept of vector.

10.2.3 Eschinardi

The Italian Jesuit Francesco Eschinardi taught logic, natural philosophy and astronomy at the *Collegio Romano* (since 1658) and also in Florence and Perugia; he was also a member of the Physico-Mathematical Academy founded by Ciampini.[33] Eschinardi's main text concerning impetus physics is entitled *De impetu tractatus duplex*.[34] Like Casati, and unlike Fabri, Eschinardi does not wish to establish a new physical system, but to explain impetus inasmuch as it is needed for a "perfect understanding of machines" (*quantum sufficit ad perfectam cognitionem machinarum*) (Eschinardi 1684, 1). Eschinardi considers impetus as equivalent to *impul-*

[31] "Effectum autem causa praestabiliorem esse non posse ipsa originis notio suadet, ne quid effectus habeat, quod non acceperit, aut aliquid causa dederit, quo ipsa caereret. Non igitur impetum motus, sed motum impetus efficit" Casati (1684, 138).

[32] For $5.25 = 1.5 \cdot 7 \cdot 4 / 8$ ("Quare ad movendas libras 5¼ velocitate ut 8, requiritur conatus sesquialter conatus necessarii ad movendas libras 4 velocitate ut 7") Casati (1684, 142).

[33] On Eschinardi, see also Feldhay and Heyd (1989, 111); Middleton (1975); Shapiro (2008).

[34] The full title is *De impetu tractatus duplex: primus de impetu in communi, de motu locali et de machinis, secundus de fluidis in communi, de comparatione fluidorum cum solidis et de mensura aquarum currentium: additur in fine quamplurimum problematum seu quaesitorum solutio ex doctrinis praecedentibus*.

sus[35]: "impulse or impetus," he states, depicts not only "that quality which is produced by an external projector in some body, when it is projected," but also "the immediate cause of motion in heavy bodies, rarefaction, condensation, magnetic attraction, motions of animals etc."[36] Like Casati, and contrary to Fabri, Eschinardi is not especially interested in pinpointing the manner of impetus-motion causality, and does not mind claiming that impetus "produces" motion. Soon afterwards Eschinardi characterizes impetus as a "permanent quality causing motion" (*qualitas permanens causativa motus*) (Eschinardi 1684, 3).

Following his presentation of the concept of impetus, Eschinardi characterizes motion as "something fluid" (*aliquid fluens*), while impetus is permanent by nature, although it can "flow" (i.e. dissipate) *per accidens*.[37] He now claims that projectile motion proves the existence of impetus:

> But the existence of this quality, as distinct from substance and from any other accident, is proved: firstly, because a stone projected to a height would not be moved, unless a permanent quality causing that motion were produced in it by the throwing hand; for that stone, having been separated from the projected hand, seeks by its own nature to descend; and previous motion (*motus antecedens*) cannot be the efficient cause of following motion (*motus sequens*), as we will soon show, therefore etc.[38]

Eschinardi's detailed rejection of *motus antecedens* as a means of explaining motion will be presented in the section depicting his view of natural impetus (Sect. 10.4.3); in fact, it will be shown that our three Jesuits unanimously reject this explanation on similar grounds.

As far as we could see, Eschinardi does not refer to impetus as a *vis*. As we shall see later, Eschinardi's concept of impetus is intimately associated with actual motion only—not with a mere inclination for motion, nor with pressure of heavy bodies downwards: consequently contrary to his two colleagues discussed in this paper he finds no use for the concept "innate impetus" (Sect. 10.4.3 below). Like Fabri, Eschinardi emphasizes that each impetus has its "proper direction."[39] As we shall see in a later section, Eschinardi in fact explains that if a body cannot keep the

---

[35] This of course must not be confused with the modern term of impulse (the integral of force over time)—an effect of motion rather than its cause.

[36] "Nomine igitur impulsus, sive impetus, intelligimus hic universaliter non solum illam qualitatem, quae ab impellente extrinseco proucitur in aliquo corpore, cum proiicitur; sed possumus etiam intelligere quamcomque; cuiusmodi puto esse, non solum qualitatem productam in saxo a proiiciente, sed etiam causam immediatam motus in Gravibus, in rarefactione & addensatione, in attractione magnetica, in motu animali etc" Eschinardi (1684, 2).

[37] "Impetus vero est. aliquid de sua natura permanens; licet per accidens possit fluere" Eschinardi (1684, 2).

[38] "Probatur autem existentia huius qualitatis, prout distinctae a substantia, & a quocumque alio accidente. Primo, quia lapis proiectus in altum, non moveretur; nisi in eo producta fuisset a manu proijcente qualitas permanens causativa dicti motus; lapis enim ille separatus a manu proijcente, de sua natura petit descendere; nec motus antecedens potest esse causa efficiens motus sequentis, ut mox ostendemus; ergo &c" Eschinardi (1684, 3). The second part of Eschinardi's "proof" concerns naturally accelerated motion, and is described in Sect. 10.4.3.

[39] "Singuli enim impetus habent suam propriam directionem" Eschinardi (1684, 5).

motion determined by the impetus, which moves it, this impetus will not be conserved (Sect. 10.3.3 below).

10.3 Conservation of Impetus and Motion

10.3.1 Fabri

Fabri regards the conservation of impetus as a key property:

> **Theorem 147:** The impetus is conserved as long as nothing exacts (*exigit*) its destruction; because it is destroyed only according to the exigence (*exigentia*) of something, whatever it might be (to be discussed below), according to Ax. 14 no. 2; certainly it will not be destroyed, as long as there is nothing which exacts its destruction; therefore for this length of time it is conserved, according to Ax. 14 no. 3.[40]

Now in Fabri's physics, the conservation of impetus is directly translated to the resulting conservation of (unimpeded) motion, since Fabri accepts the possibility of motion through vacuum and strongly attacks Aristotle's objection to such motion.[41] Fabri describes Aristotle's well-known arguments against motion in the void: that instantaneous motion would occur; that every object would (paradoxically) fall in the same speed in vacuum; and that nothing would prevent a body to move forever in it.[42] Fabri replies by stating Galileo's argument from the *Two New Sciences*, which does not imply at all instantaneous motion, and indeed renders a constant rate of fall for all bodies whatsoever (see Sect. 10.4.1 below). As to Aristotle's third objection, Fabri simply answers: "I concede, of my own accord, that in a total void a stone once moved, would move indefinitely, unless it is restrained extrinsically."[43] Like Benedetti, Fabri accepts only a rectilinear impetus (Elazar 2008, 21–22), and claims accordingly that if innate impetus (i.e. gravity) did not exist, a projectile horizontally thrown "would obviously move along a horizontal straight line with constant motion."[44] Fabri's total acceptance of the anti-Aristotelian notion of an "ideal" motion devoid of any resistance is reflected in this statement, written after

[40] "Tamdiu conservatur impetus, quamdiu nihil exigit eius destructionem; quia destruitur tantum ad exigentiam alicuius, quidquid tandem illud sit, de quo infra, per Ax.14.num.2. certe tamdiu non destruitur, quamdiu nihil est., quod exigat eius destructionem; igitur tamdiu conservatur per Ax. 14.num. 3" Fabri (1646, lib. 1, th. 147, 68–69).

[41] It should be added though that Fabri objected to Torricelli's claim to have produced vacuum *de facto*; for a detailed description of Fabri's view, see Elazar (2011, 135–156).

[42] For Aristotle's opinion, see *Physics* IV, 215a28–215b23; *Physics*, IV, 216a11–216a20 *Physics*, IV, 8, 215a19 (Aristotle 1930).

[43] "Concedo ultro fore, ut in totali vacuo lapis semel motus, semper moveatur, nisi ab extrinseco retineatur" Fabri (1648, 564).

[44] Neglecting, of course, air resistance. "Theorema 38: si impetus naturalis non concurreret ad hunc motum, proiectum moveretur per lineam horizontalem rectam, ut constat, motu æquabili" Fabri (1646, 74).

proposing his discrete mathematical analysis of falling bodies (described in Sect. 10.5.1 below):

> If the heavy body were to descend in a vacuum, the above-mentioned proportions would be preserved very accurately; that is, because there is no impediment; but indeed if any impediment intervenes, no doubt they are not preserved accurately.[45]

Fabri also states that the conservation of impetus should be ultimately ascribed to God,[46] thus echoing Descartes' famous assertion that God "always preserves the same quantity of motion in the universe" (Descartes 1985, 240).

In theorem 13 of *De motu naturali deorsum* (the second book of his *Tractatus physicus de motu locali*), Fabri specifically employs the conservation of impetus to give a discrete account of acceleration under the condition of what he calls a "free medium," i.e. a medium offering neglegible resistence to the motion of fall:

> The impetus produced in the second instant in a free medium is conserved in the third, that produced in the third instant is conserved in the fourth, and so on, because they are not conserved by the initial productive cause… nor does anything exact its destruction: not a contrary impetus, none of which of course is applied, nor a resistance of the medium, which indeed exists to some amount; but certainly not so much as to be able to entirely impede motion; for I suppose a free medium, therefore also that it does not destroy impetus; since as long as the effect endures, so does the cause; therefore there is nothing which would exact the destruction of that impetus.[47]

10.3.2 Casati

We have already seen (Sect. 10.2.2 above) that Casati treats impetus as a "permanent" quality, that is, not a "self-expending" entity which by definition fails to conserve itself. Furthermore, Casati conveys a basic view concerning the conservation of impetus which is quite similar to Fabri's: since impetus, he explains, "is a quality established for motion (*propter motum instituta*), which is produced only in motion, thus in like manner it is conserved only in motion and with motion."[48] Casati

---

[45] "Theorema 71: in vacuo si corpus grave descenderet, prædictæ proportiones accuratissime servarentur; quia scilicet nullum esse impedimentum; at vero si aliquod intercedit impedimentum; haud dubie non servantur accurate" Fabri (1646, 115).

[46] Fabri (1646, 68, 81, 137); see also Elazar (2011, 130–131).

[47] "Impetus productus secundo instanti in medio libero conservatur tertio, & productus tertio conservatur quarto, atque ita deinceps; quia scilicet nec conservantur a causa primo productiva per Th.144.libri: nec aliquid exigit destructionem; non contrarius impetus, quia nullus est. applicatus, ut constat; non resistentia medii, quae quidem alicuius momenti est.; sed non tanti, ut impedire possit motum omnino, ut constat; nam suppono liberum medium, igitur nec destruere impetum; cum tamdiu duret causa quamdiu durat effectus, ut patet; igitur nihil est. quod exigat impetus huius destructionem" Fabri (1646, lib. 2, th. 13, 82).

[48] "Cum itaque impetus acquisitus, aut aliunde impressus, sit qualitas propter motum instituta, quae non nisi in motu producitur, ita pariter nisi in motu, & cum motu non conservatur" Casati (1684, 155).

explains that it cannot be the case that impetus is "circumscribed by a moment of time," otherwise pendulums would not have oscillated and arrows would not have been shot.[49] He then reemphasizes that impetus cannot perish in the very moment in which the external mover ceases, which explains why boats keep moving straight ahead after their sail is folded and oars restrained, and also why it is easier to keep in motion a wagon (or any moving load) which is already moving than to set it in motion from rest.[50]

And yet, while the conservation of impetus is highly emphasized in Fabri's account (see his theorem 147, note 40 above), that is not the case as far as Casati is concerned. The third chapter of *Mechanicorum libri octo*'s second book is entitled "on what ground is impetus, once received, destroyed" (*Qua ratione semel conceptus impetus pereat*). Furthermore, in the preceding chapter (i.e. *Impetus motum proxime efficientis natura explicatur*), while claiming that unlike a living creature, an inanimate object is completely under the control of the laws of nature and "omits nothing of the impetus" impressed on it, Casati warns: "beware however of thinking that the impetus of the same stone descending in air and in water is the same; for nature measures the mode of impetus by dividing the medium, in which motion is to be perfected."[51] Contrary to Fabri, Casati denies the possibility of descending without any material medium and he furthermore deems the resisting medium not a merely "accidental" property of motion, but an inherent property of motion which nature "requires" in order to measure the "mode of impetus." This attitude brings to mind Aristotle's view, which by definition could not abstract motion from resistance[52]—a notion still strongly held by the *Collegio Romano* Jesuit Muzio Vitelleschi, who according to Wallace ascribed such prominence to "resistive forces" as to place them "by on almost the same plane as motive force" (Wallace 1978, 407).

In the beginning of the third chapter Casati, in agreement with the Aristotelian framework, asserts that eternal violent motion is impossible and that straight motion must be limited by terminal points. He explains:

> It is clear enough that conceived impetus (*conceptus impetus*), by its own nature, does not last (*permanet*) forever, and that nor does it endure (*durat*) for [only] a single point of time; for whether it spontaneously flows forth (*profluit*) by seeking its place owed by nature [i.e. natural impetus], or whether—impressed by an alien force—it drives out a body of its place, it cannot be perpetual; for if it is violent, it certainly is not eternal; but if it is [a] natural [motion], to which [a] violent [motion] precedes, it is certainly bounded (*definitur*) by *termini*; for the body [having completed the preceding violent motion] is not infinitely distant from the place in which nature dictated rest, and accordingly will reach that place, and

[49] Casati (1684, 155).

[50] Casati (1684, 152–153).

[51] "Contra vero inanimum non suo arbitrio motus intentionem moderatur, sed naturae juribus obsequens nihil prætermittit impetus, & quantum eniti potest, opportunum in locum, sibique a natura constitutum, contendit. Cave tamen existimes parem esse lapidis ejusdem, & in aere, & in aqua descendentis impetum: natura scilicet ex medio dividendo, in quo perficiendus est. motus, metitur impetus modum" Casati (1684, 138).

[52] See, e.g., Baldini (2004, 135).

finally will come to a stop, and there will be no need for an eternal impetus, since the motion must cease.[53]

The difference between Fabri's and Casati's positions lies in the alternative of including or excluding the possibility of motion in a vacuum, an alternative that also shaped, for instance, Descartes' changing position to the problem of the motion of fall.[54]

In a chapter entitled "Whether bodies were endowed with gravity and levity" (*An corpora praedita sint graitate, & levitate*), Casati claims that motion "of heavy and light bodies" cannot be "exercised" (*exreceri*) in vacuum, and refers the reader to his own "*Vacuum proscriptum*, cap. 2, num. 9" (Casati 1684, 13). This book, published in 1649,[55] was a reaction to a work published 2 years earlier by the Capuchin Valeriano Magni (1587–1661), entitled *Demonstratio ocularis loci sine locato: corporis successive moti in vacuo, luminis nulli corpori inhaerentis*. Magni's work was an account of an experiment performed by him in Warsaw, in front of the king Wenceslas VII and the queen of Poland—which in fact repeated the famous "Torricellian experiment" from 1644.[56]

The second chapter of *Vacuum proscriptum*, to which Casati refers in his *Mechanicorum libri*, is indeed called "On a body successively moved in vacuum" (*De corpore successive moto in vacuo*)—a line taken from Magni's book title. Casati begins this chapter by completely rejecting the latter part of this title, by claiming the absurdity of light which "inheres in no body."[57] However, Casati does not reject altogether the possibility of a successive motion in the void. He indeed invokes Aristotle's dictum, that the absence of proportion between the resistance of a medium and zero resistance renders absurd velocities in the void, but dismisses it as a "false assumption" (Casati 1649, 7). But immediately he explains that despite this error, Aristotle was correct in rejecting motion on the void as far as natural motion is concerned. As we shall soon see, Casati rejects the possibility of an

---

[53] "Et sane conceptum impetum, natura sua, nec flabilem semper permanere, nec ad unicum temporis punctum durare, satis constat: sive enim sponte profluat ex natura debitum sibi locum quærente, sive aliena vi impressus suo loco corpus extrudat, perpetuus esse nequit; omnis scilicet motus terminum habeat necesse est.; nam si violentus quidem est., perennis utique non est.; sin autem naturalis, quem violentus præcesserit, certis definitur terminis; a loco enim, in quo quietem natura indixit, corpus infinito intervallo non abest, ac proinde ubi eum attigerit, demum conquiescet" Casati (1684, 146).

[54] See Damerow et al. (2004, ch. 1).

[55] Casati, *Vacuum proscriptum: disputatio physica authore Paulo Casato Placentino Societatis Iesu, in qua nullum esse in rerum natura vacuum ostenditur; & potissimum examinatur, an ab argento vivo descendente in fistula superne clausa vacuum relinquatur, huiusque experimenti symptomata exlpicantur.*

[56] Magni was to subsequently claim that when he published his book he had not even heard of Torricelli; be that as it may, Casati apparently—even though he lived in Rome—learned of this experiment only from Magni's published work; see Middleton (1964, 31, n. 32; 42–43).

[57] Casati (1649, cap. 2, n. 7, p. 7). Casati has already argued, in line with Aristotle's doctrine that "lumen non est. corpus" Casati (1649, cap. 1, n. 2, p. 4), therefore of course it cannot exist where no other body exists.

"innate impetus" which could drive bodies to the center of the universe; this rejection leads, according to Casati, to the impossibility of motion in the void, since he replaces the (traditionally accepted) inherent tendency to the center (or the periphery) with a "principle of contiguity" (*ratio contigui corporis*) as a mover of bodies, in a passage which is entitled "heavy and light bodies move only by a principle of a contiguous body."[58] There can be no doubt, Casati explains, that a body, in order to be put into straight simple motion from rest, "has only a principle of a contiguous body, and not of a distantly situated one"; in other words, in order to move, a heavy or light body must be touched by another one.[59] In a vacuum such a "principle of contiguity" would be of course irrelevant, as Casati immediately recognizes:

> May it therefore happen, that a body positioned in vacuum would be excited by nature to motion? For since it is not surrounded by any contiguous body, where would it move? For clearly there is no heavier or lighter body [in contact with it] on which it could exercise forces of levity or gravity.[60]

But Casati emphasizes that he has nothing against motions in the void which do not violate this "principle of contiguity." He repeats his position that a natural linear motion of heavy and light objects situated in vacuum (and therefore being non-contiguous) would be utterly impossible, but sees no reason to deny that "a man located in a vacuum can move about a sword, drag a spear, turn in a circle a sling hanging from his right arm, or carry a stone from one place to another."[61] However, this man would never be able to throw away the sword, spear or stone. This becomes clear in the fourth chapter, in which Casati introduces a natural and "very tight bind" (*nexus arctissimus*) by which bodies seize one another, and which cannot be severed "unless something else substitutes the place of the withdrawing body."[62]

From this "mutual adhesion of bodies," continues Casati soon afterwards, we can explain which motions a man located in vacuum cannot perform: he cannot walk, since there would be nothing which could take the place of his legs trying to recede

---

[58] "Gravia & levia moventur habita tantum ratione contingui corporis" Casati (1649, cap. 2, n. 11, p. 9).

[59] "At si sensu iudice res decernatur, non relinquitur ambigendi locus, an quodcunque corpus naturae impetu translatum simplici, ac recto motu cietur, contingui tantummodo, non vero procul positi, corporis rationem habeat, ut a quiete dimoveatur, & ad motum extimuletur" Casati (1649, cap. 2, n. 11, p. 9).

[60] "Qui ergo fieri poterit, ut corpus in vacuo constitutum, a natura ad motum excitetur? Nam, cum a nullo contiguo corpore ambiatur, quorsum moveatur? Nihil enim plane est. seu gravius, seu levius, in quod levitatis suae, aut gravitatis vires exerceat" Casati (1649, cap. 2, n. 11, p. 10).

[61] "Quamvis autem quae simplici, & recto gravium seu levium motu cientur, in vacuo constituta, nullique seu graviori, seu leviori corpori adhaerentia, immota prorsus consistere debeant, si sibi ipsa relinquantur, non negarim tamen hominem in vacuo collocatum posse gladium versare, hastam tractare, pendentemque ex dextera fundam in gyrum contorquere, aut saxum ex uno in alio locum transferre" Casati (1649, cap. 2, n. 13, p. 11).

[62] "Quia nimirum singulis corporibus hoc a natura comparatum est. (id quod effecta produnt) ut se mutuo nexu arctissime complectantur, non dimissura, nisi aliud in recedentis locum subrogetur" Casati (1649, cap. 4, n. 36, p. 32). This passage in entitled "quo nexu contigua invicem cohaereant."

(therefore they would not succeed in receding).[63] Although such a man could rotate a stone with his right hand, he would not be able to project it—it would cling to his hand; however, an angel could project a stone in vacuum, for it lacks matter—therefore no "chain of adhesion" (*vinculo adhesionis*) would connect it to the stone.[64] Casati's "mutual adhesion" (or *nexus arctissime*) notion brings to mind the "love of mutual conjunction and society" that the Coimbrans attributed to all bodies, which when united "operate much more strongly, and repel the injuries of their adversaries."[65] It is exactly this love of union which the formation of vacuum would impede, and it is of course intimately related to the idea of *horror vacui* (Des Chene 1996, 175–176). In fact, this "inexorable tendency to preserve and maintain material continuity "—as Edward Grant defines this property—was considered by scholastics (such as Roger Bacon) to be an essential characteristic of matter's universal nature (*natura universails*), while the tendency of a body to move to its natural place was ascribed to matter's *natura particularis* (Grant 1973, 329–330).

10.3.3 Eschinardi

Discussing the properties of impetus, Eschinardi rejects the opinion that "impetus is not produced *de novo*, when it is not possible to obtain by it a motion that is compelled by that impetus—that is, in its direction."[66] Eschinardi mentions a "certain author," who utilizes a "hit rock" (*rupes percussa*) as an example for this notion. Eschinardi claims that this author is mistaken, for the hit rock produces sound, which means that it trembles—therefore its parts must be moved, hence receive *impulsus* (i.e. impetus).[67] There is a good chance that this "certain author" Eschinardi attacks is no other than Fabri, who declares, in theorem 31 of his own *Tractatus physicus de motu locali*, that "impetus is not produced in a moveable that cannot be moved by a given applied motive force, while it could be moved by a stronger

[63] "Ex hac mutua corporum adhaesione inferre potes, quos motus exercere non posset homo in vacuo constitutus: si enim cubiculi vacui pavimento pedes insisterent, progredi nullatenus posset, aut saltus edere quia (cum in vacuo collocatus ponatur) nullum adesset corpus, quod in abeuntium pedum locum sufficeretur" Casati (1649, n. 40, p. 35).

[64] "At quamvis pendulum libere brachium circumagens dexteram saxo gravem contorqueret; non tamen saxum posse excuti, sed manui adhaereret. Angelus vero saxum in vacuo proijcere posset impresso impetu, quia cum omni corporeae molis concretione careat, nullo adhaesionis vinculo saxum cum eo connecteretur" Casati (1649, cap. 4, n. 40, p. 35–36).

[65] Des Chene (1996, 175–176). The last two citations are phrases quoted by Des Chene (apparently in his translation) from the Coimbran's commentary to Aristotle's physics.

[66] As stated above (Sect. 10.2.3), Eschinardi emphasizes that every impetus has its "proper direction." "Putant primo aliqui, non produci impetum de novo; cum non potest per illum obtineri motus exigitus ab illo impetu; hoc est. iuxta ipsius directionem (singuli enim impetus habent suam propriam directionem)" Eschinardi (1684, 5).

[67] "Et quidem aliquis auctor assert exemplum de rupe percussa, in quo videtur errare: hoc ipso enim quod rupes percussa sonat; signum est., quod tremit; adeoque movetur secundum partes; adeoque partes sunt impulsae" Eschinardi (1684, 5).

one."[68] Fabri explains that according to his conception of impetus, as a quality whose goal (*finis*) is motion,[69] when no motion exists, neither can impetus, because in such a case it would be "in vain" (*frustra*).[70] Fabri indeed claims accordingly that in a rock which stands still after we hit it no impetus will be created; he is well aware of the (highly plausible) objection Eschinardi raises, that is, that the sound the rock emits testifies to inner motion within it, for soon afterwards he tries to explain it away by claiming (à la Aristotle) that it is the motion of air which in fact produces the sound (Fabri 1646, th. 33, obj. 4, p. 26). Casati shares Fabri's view, and also claims that if the resistance of a body to motion is so great as to prevent it from being moved by a given force, then no impetus will be impressed in the body (Casati 1684, 149).

Eschinardi, denying that impetus is not *created* unless motion is created along its direction, asserts that it is not *conserved* unless such a motion is effected (Eschinardi 1684, 6). Accordingly, while Fabri can resort to innate impetus while explaining a stone resting on the pavement, Eschinardi—who as we shall later see finds no use for innate impetus—is led to maintain that while a new impetus downwards is constantly created, an equal resistance is also being created (by the pavement) and both keep destroying one another: the lack of motion does not therefore imply an absence of impetus, but rather an absence of its conservation, though the result is of course the same (Eschinardi 1684, 5–6).

Eschinardi regards impetus, like his two colleagues, as a "permanent" quality (Sect. 10.2.3 above); it is produced, he claims, and soon afterwards "destroyed since it is frustrated with respect to its end-goal" (*iam productus destruatur, cum frustratur suo fine*) (Eschinardi 1684, 4). This destruction, continues Eschinardi, occurs at "a greater or lesser extent in proportion to the greater or lesser extent that it is frustrated with respect to its end-goal."[71] In other words, impetus is destroyed when it cannot bring forth its end-goal: the motion of the subject in which it inheres. Eschinardi now conveys a doubt whether we can "adequately assign end-goals intended by nature through impetus."[72] Eschinardi's comment should be seen against the opinion adopted by many late Aristotelians, who on the one hand no longer considered end-goals (i.e. "final causes") as adequate physical explanations (unless rational agents are involved), and on the other hand could not abandon finality: "though the ends to which natural changes tend might not be proper causes of those changes," explains Des Chene, "it was inconceivable that natural change should not

[68] "Impetus non producitur in eo mobili, quod moveri non potest a potentia motrice applicata, licet a fortiori moveri possit" Fabri (1646, th. 31, p. 24).

[69] We have seen above how impetus "rejoices" while fulfilling that goal, i.e. succeeding im moving the subject in which it inheres.

[70] "Probatur, quia impetus est. tantum propter motum, qui eius effectus est., & finis, per Th. 15. & 16. Igitur ubi non est. motus, frustra est. impetus; sed quod frustra est., non est.; id est. non est., quod frustra esset, si esset per Ax. 6" Fabri (1646, th. 31, p. 24). Fabri immediately adds that innate impetus is an exception to this rule—see Sect. 10.4.1 below.

[71] "& quidem pro rata plus, minus; prout plus, minus frustratur fine" Eschinardi (1684, 4).

[72] "...difficultas est. in assignando adaequate fines intentos a natura per impetum" Eschinardi (1684, 4).

have ends" (Des Chene 1996, 169). Toletus, for example, while denying that a "natural" (i.e. non-rational) agent might have an end-goal, nevertheless maintained that such a natural agent could be "directed toward an end by some superior and more excellent cause, that is, by the artisan of Nature herself, who comprehends the ends of all things, and gives [to them] the propensity and powers to obtain those ends" (Des Chene 1996, 195). Toletus—like Thomas Aquinas before him and Francisco Suarez after him—invokes as a key example an archer impressing impetus upon an arrow which thus attains its target, while claiming that by the same token "glorious God has given to all things a nature by which even without foreseeing their ends they are led unerringly to them" (Des Chene 1996, 195).

Eschinardi prefers however (like his two Jesuit colleagues) not to discuss any end-goal which a "superior cause" might be striving for through impetus, but rather to assign impetus an end-goal of its own: simply to keep an object in which it inheres in motion. Eschinardi soon makes it clear that it is the extent to which the end-goal of impetus (i.e. rectilinear motion along its direction) is fulfilled which determines its conservation: it is conserved, he asserts, according to the measure by which the motion it generates keeps its direction.[73] Eschinardi also defines a static state of equilibrium (e.g. a stone resting on a pavement) as a case in which each of the opposing impetuses is "frustrated with respect to motion" by the other.[74]

Importantly, Eschinardi emphasizes that natural motion accelerates because impetus is both not "frustrated with respect to motion" (and thus conserved) and continually produced (Eschinardi 1684, 32); we shall soon observe the central role that the conservation of impetus plays in Eschinardi's explanation of the law of fall.

As for motion in vacuum, Eschinardi simply accepts this as a legitimate (though theoretical) possibility, without embarking on a detailed analysis, which either supports this type of motion (like Fabri) or rejects it (like Casati). He supports, for example, Galileo's (and Fabri's) assertion that in vacuum every object falls at the same rate. Eschinardi emphasizes that a heavy body exists always in "some fluid corporeal medium," therefore we experience only the relative gravity of a body, never its absolute gravity[75]—leaving the reader to infer that in the void we would experience absolute gravity. Much later in his *De impetu tractatus duplex* Eschinardi returns to this subject, and warns the reader that we cannot understand "mathematically" gravity "separated from medium," that is, absolute, not only because we do not exist in vacuum, but also because the instruments we use to weight objects always contain friction.[76] He adds that "even if we did exist in vacuum," and

[73] "... puto: Primo, quod impetus producatur de novo, quoties causa est. applicata; sed non conservetur, nisi cum potest haberi motus per directionem debitam tali impulsui; ita tamen, ut plus minus destruatur, prout plus minus habet de motu iuexta eam directionem" Eschinardi (1684, 6).

[74] "sic semper fit aequilibrium per duos impetus aequales contrarios fluentes, quorum uterque frustratur motu" Eschinardi (1684, 7).

[75] "Adverte hic, Grave in descensu semper esse in aliquo medio corporeo fluido, adeoque nunquam nos experiri eius gravitatem absolutam, sed solum comparativam" Eschinardi (1684, 23).

[76] "Moneo iam primo fieri non posse, ut sciamus mathematice gravitatem, sive pondus absolutum a medio; tum quia non sumus in vacuo, tem etiam quia semper datur impedimentum contactus hastae, sive brachii staterae" Eschinardi (1684, 333).

instruments were free of friction, we would still not understand exactly what "pound" and "ounce" mean, since these measures have been determined long ago in air (i.e. not in vacuum) to begin with.[77] In any case, Eschinardi has no problem in visualizing either objects existing or moving in a void. His notion of "absolute gravity" is directly linked to his adamant rejection of "absolute levity": in a material medium a relatively light material (e.g. wood in water) would rise because of the heavier medium pushing it upwards, but in the absence of any medium any material—even the lightest possible—could not be pushed upwards by its surroundings, and can thus exhibit its "absolute" unhampered gravity.[78]

10.4 Natural Impetus and Levity

10.4.1 Fabri

Fabri distinguishes between violent impetus, which is impressed from outside (and causes violent motion), and natural impetus—which is produced internally, and is divided into two kinds: natural innate impetus (*impetus innatus*) and natural acquired impetus (*impetus acquisitus*). Natural innate impetus, which Fabri identifies with gravitas ("*nam gravitas est ipse impetus innatus*") (Fabri 1646, lib. 2, th. 8, p. 81) is produced in the moment in which the body is formed. It exists and never diminishes as long as the body exists because it is never in vain, being responsible either for free fall (in the absence of any impediment), or for pressure downwards (on the impeding object, if such an object exists).[79] Fabri emphasizes, in his *Physics*, that *every* body "is heavy by absolute gravity" (*Omne corpus est. grave gravitate absoluta*), since every element is capable of receiving innate impetus—even fire (Fabri 1646, prop. 28, p. 254); this claim is of course intimately connected to his abolition of absolute levity,[80] which according to Aristotle is the most essential property of fire. Fabri, who ascribes to Archimedes the original idea of abolishing levity, in fact fully adopts the Galilean adaptation of Archimedean hydrostatics to dynamics, including the acceptance of a vacuum (in which "pure" absolute gravity is manifested), the contention that in vacuum all the bodies fall at the same rate and

[77] "Secundo etiamsi essemus in vacuo, & non impediret contactus; adhuc nesciremus libram, unciam & c.; nam cum istae mensurae fuerint impositae in medio aere, nec sint nota corpora, quibus tunc usi sunt antiqui; hinc sequitur non posse mathematice cognosci, quid intellexerint per libram &c" Eschinardi (1684, 333).

[78] "Quoad levia vero, quia probabilius est. nihil dari leve simpliciter, & absolute, seu positive; sed solum comparative, hinc quae dicuntur levia in tantum ascendunt, in quantum sursum impellantur a gravioribus petentibus maiori nisu locum inferiorum" Eschinardi (1684, 24).

[79] "Innatus est. qui vel a generante simul cum corpore gravi productus est., porro cum in corpore gravi duplex quasi proprietas sensibilis esse videatur, scilicet gravitas, seu pondus & motus deorsum... Ex his dicendum est. hunc impetum nativum nunquam destrui, quia nunquam est. frustra, habet enim semper aliquem effectum" Fabri (1646, lib. 2, th. 14, scholium, p. 83).

[80] "Immo ostendam infra, nullam dari levitatem absolutam" Fabri (1646, prop. 28, 254).

Galileo's quantitative treatment (from the *Two New Sciences*) of falling bodies in media.[81] Fabri also rejects the rules Aristotle formulated concerning falling bodies (e.g. the simple dependence of the speed of fall on weight) and directly blames him for failing to examine accelerated motion.[82]

Acquired impetus is created only when free fall occurs, that is, in the absence of an impediment to downward motion. As David Lukens sums up, Fabri's acquired impetus "is called 'natural' because it is produced by an intrinsic principle, and 'acquired' because it is not innate. It differs from the innate in that it can be destroyed by a resisting body, and can be directed along any line. Innate impetus is never destroyed and is always directed downwards" (Lukens 1979, 159).

Fabri criticizes not only Aristotle's rules concerning falling bodies, but also his claim that it is an external factor (the *generans* of the body) which is the cause for this kind of motion (Fabri 1646, lib.2, th. 1, p. 78). Fabri also rejects other "external" explanations for natural motion, such as the extruding medium, a magnetic force or Descartes' subtle matter (Fabri 1646, lib.2, th. 1, p. 78–79). Assuming to have invalidated every "external" explanation possible, Fabri concludes that the cause must be internal[83]; finally, after rejecting gravity (which stays constant and thus cannot cause an increasing velocity), he ascribes accelerated natural motion to the only remaining internal factor—impetus.[84] As will be shown below, Fabri determines, using a discrete analysis of a falling body's acceleration, that this impetus (i.e. natural acquired impetus) accumulates—in equal amounts of time—like the natural numbers law (i.e. 1, 2, 3, 4...), and proves that under the assumption of tiny such amounts of time this series converges to Galileo's law.

In his explanation of natural motion, Fabri rejects Buridan's account, which as mentioned above treats impetus as cause and effect at the same time: Fabri takes pains to show that the accumulation of impetus is the direct cause of the increasing velocity, while the latter can by no means be considered as causing the increase of impetus (Elazar 2008, 29–30). Within this context, Fabri also adamantly rejects the view which considers the "prior motion" (*motus praevius*) as the cause of the swifter following motion, by simply claiming that when this *motus praevius* is supposed to act (*agit*) it no longer exists (Fabri 1646, lib. 1, th, 40, p. 31 & lib. 2, th. 12, p. 82); accordingly Fabri regards as ridiculous the mere idea of conceiving of motion as

---

[81] Fabri (1646, prop. 28, p. 254); Elazar (2008, 18–19). Galileo's "formula" can be (anachronistically) reconstructed as $(v_o—v)/v_o = \rho_m/\rho_b$, where v = the speed of the falling body in medium m; v_o = the speed of fall of all bodies in a void; ρ_m = the specific weight of medium m; ρ_b = the specific weight of the falling body, Elazar (2008, 17).

[82] See Elazar (2011, 75–76).

[83] "Igitur non movetur corpus grave a causa prima... nec ab aere, nec a virtute magnetica... nec a caelo pellente, nec a vi sympathica, nec a generante proxime & immediate... nec ab ullo alio extrinseco, ut constat inductione; igitur ab aliqua vi intrinseca, quidquid sit, de qua alibi" Fabri (1646, lib.2, th. 1, p. 79).

[84] "Hinc motus naturalis est. ab impetu. Probatur; est. ab aliqua causa per Ax.8. lib.1; ab aliqua intrinseca per Th. 1; non a substantia corporis gravis per Th. 3; non a gravitate per Th. 4; igitur ab impetu, quia nihil aliud esse potest intrinsecum, a quo sit motus" Fabri (1646, lib.2, th. 5, p. 80).

composed of different "stages" (i.e. *partes motus*), each one causing the next.[85] Fabri is probably thus criticizing the idea adopted by Vitelleschi, following Jacopo Zabarella, that "the earlier part of a body's fall causes a greater velocity in the later part of its fall, because it then causes the medium to resist less" (Wallace 1978, 409).

10.4.2 Casati

Casati, in line with tradition (and like his two fellow-members), ascribes natural acceleration to a gradual accumulation of impetus during the fall of a heavy body. Since the nature of a falling stone remains the same, he claims, and since the stone cannot by its own power choose to accelerate or decelerate its motion, "could it happen," he asks, "that as it falls it increases velocity, unless it adds—in single moments—more and more impetus to that it initially created?"[86] Furthermore, like Fabri (and Eschinardi) he rejects the *prior motus* explanation, emphasizing that "prior motion" cannot be considered as the cause of acceleration, since it no longer exists in the moment in which the motion becomes faster (Casati 1684, 137).

Also like Fabri, Casati divides natural impetus into "innate" and "acquired" impetus, declaring that innate impetus is a non-additive property which corresponds to the ability of a body to "move itself" (*facultas se movendi*).[87] But Casati's innate impetus is very different from Fabri's: this self-moving ability, which the former ascribes to bodies, exists only when they are located in a medium whose density is different from their own. For example, Casati attributes no innate impetus to water within water; he thus differs from Aristotle and adopts an opinion close to Galileo's in his *De motu*.[88] According to Casati, only when water is held (in a bucket) above a lake can we feel its inner ability to move, expressed as a downwards propensity (which Casati describes as *gravitatio*) by which it endeavors (against the holding bucket) to return to its place.[89] What happens when a body reaches its natural

[85] "Aliquis forte diceret potentiam motricem exigere primam partem motus, quæ deinde secundam exigit, & secunda tertiam, tertia quartam, &c. Sed contra; quæro quid sit prima illa pars motus" Fabri (1646, lib 1, th. 5, p. 14.)

[86] "Cum enim eadem descendentis lapidis natura perseveret, nec illa in sua potestate sit, aut optione delata, ut eligat utrum velit, motum arbitrio suo incitare, aut remittere valeat; qui fieri possit, ut descendens velocitatem augeat, nisi ei, quem primum produxit, alium atque alium momentis singulis impetum adjiciat?" Casati (1684, 136–137).

[87] Casati (1684, 146). Casati's acquired impetus will be soon described.

[88] Aristotle insists that earth, water and air each has weight in its proper place, while fire is the only element that has no weight anywhere, even in its natural place (*On the Heavens*, Book 4, Ch. 4, 311b8 & 312b3; Aristotle 1953). On the other hand Galileo in his *De motu* strictly denies that any element gravitates in its proper place, see Heilbron (2010, 150).

[89] "Hinc nullum aquae impetum tribuo intra aquam consistenti; sed tunc solim cum situla plena e lacu extrahitur, ea. aquae pars impetum habet, quae supra subjectam lacus superficiem aere circumfusa motum expetit, quo suum repetat locum repugnans sustinteni. Impetum hunc, qui naturali se movendi facultati respondet, & est. ipsa gravitatio, seu naturalis ad descensum propensio, Innatum voco, & is est., cui extrinseca causa repugnat motum impediens" Casati (1684, 146).

place—for example, when the water in the bucket returns to the lake? According to Casati, in this case the innate impetus "immediately vanishes" (*statim perire automo*), as experience shows: for if we hang the above-mentioned water bucket within a deep tunnel and then fill it with water, then the bucket suddenly becomes lighter, having lost the weight of the water inside it—"therefore the innate impetus, by which the water filling the bucket struggled to descend, is destroyed."[90] Contrary to Fabri, who attributes to innate impetus not only the effect of motion but also pressure downwards, Casati ascribes to it motion only; consequently, when a body reaches its natural place and ceases to move, its innate impetus necessarily becomes "in vain" and therefore ceases to exist (Casati 1684, 147). Likewise, when an object is held or suspended by something (for example, a rope), the suspending force (e.g. the rope's tension) destroys the object's innate impetus.[91]

Interestingly, while Fabri can employ the "pressure downwards effect" to explain why a stone presses on the pavement on which it rests, Casati cannot invoke innate impetus for that purpose. Innate impetus, he insists, results only when bodies are moved from their natural place or state and enables them to return to it.[92] Eschinardi, who does not acknowledge innate impetus at all, simply claims that innate gravity (*gravitas innata*) exists inside bodies also when they rest. As mentioned above, he regards such a stone as experiencing a continual mutual destruction of a downward impetus and the resistance exerted by the pavement. Whereas Fabri's "ambitious" impetus serves him also for static explanations, it seems that Casati and Eschinardi wish to preserve it to actually moving bodies.

Be that as it may, Casati's view is better understood in the context of his interesting attitude toward levity and gravity and the relation between them, described in an earlier section of the *Mechanicorum libri*—in the first two chapters of the first book, that discuss the "center of heavy and light things" as well as "whether bodies were endowed with gravity and levity."[93] Casati claims there that "each body receives by nature a *facultas*, an innate inclination by which it remains in its proper place or, if removed by some force, moves itself towards it."[94] This *facultas* is of course nothing but the *gravitatio* felt by the water removed from the lake in the above-mentioned example. It is important to understand though that this *gravitatio* is totally equivalent to the contrary natural inclination, that is, *levitatio*. For Casati regards the spherical universe as arranged according to the *gravitas secundum speciem*, that is, specific gravity (or density), of bodies—the denser the body the closer it is to the

[90] "Periit ergo innatus impetus, quo aqua situlam replens descensum moliebatur" Casati (1684, 147).

[91] "Innatum videlicet gravitatis impetum vis ipsa suspendens aut sustentans elidit" Casati (1684, 154).

[92] "...ab iis impetum concipi statuo, quae aut loco naturali dimota, aut incongruam partium positionem nacta id repetunt, quod natura exigit" Casati (1684, 145). As far as we could see, Casati does not discuss this issue; perhaps he would have evoked here his "mutual adhesion of bodies" principle (see Sect. 10.3.2 above).

[93] Chapter 1 is entitled "Quid sit Centrum gravium, & levium" and Chap. 2 is called "An corpora praedita sint graitate, & levitate."

[94] Casati (1684, 2); see Chap. 9 in this volume.

center, while the lighter it is, the closer it is to the external surface;[95] therefore, any violation of this arrangement will trigger this inner *facultas*, whether it be downwards *gravitatio* or upwards *levitatio*.

Somewhat later Casati claims that he "would not dare to declare" that a stone which obtains *gravitas* in water or air and *levitas* in mercury or liquid lead, has a "twofold force" (*duplex virtus*), for "the impetus effecting motion is of the same nature, no matter the region of the sphere toward which the motion is directed."[96] He continues to explain that the only difference between an ascent and descent of the stone is the order in which the same *ubicationes* (i.e. places) are successively acquired, and maintains that "to change this order it seems that a dissimilitude of the moving force is not necessary, because actually nothing dissimilar happens." He adds that "should someone say that the variety of conditions is enough to explain it, it will not be without reason."[97]

Casati thus expresses a rather peculiar (and original) amalgamation of ideas: adopting the world picture described by Galileo in his early *De motu*, as well as the Archimedean principle underlying this cosmic view (i.e. the criterion of density), Casati manages to undermine the popular contemporary tendency to do away with absolute levity;[98] he thus "saved" Aristotle, but at the price of giving up also absolute gravity. Casati's most blatant attack against such an "absolute" tendency of bodies to reach the earth's center—deeming it an "error of the vulgar"—is contained in his *Vacuum proscriptum*, written in 1649. In this text Casati rejects the possibility of natural motion in vacuum; for how would a heavy body, he asks,

> be carried downwards in a void, where there is no difference of positions? Would it seek by some innate impetus the center of the universal corporeal mass (*universae corporeae molis centrum*)? But many will agree that this should be counted among the errors of the vulgar; for why would nature insert into individual heavy objects a desire for the center (*centri appetitus*) which she knows beforehand in almost all of them would be in vain? Because especially the elegance of nature would demand diverse natural places (*sedes*) of bodies endowed with unequal gravity. Therefore the motion of individual heavy objects does not strive to this end, that the body will cease its motion at the center, but to occupy the natural place owed to it in the universe, according to the type (*species*) of innate gravity, that is, when the equality of the bulk is observed.[99]

---

[95] See Chap. 9 in this volume.

[96] "Non ausim tamen in lapide, qui gravitatem in aqua & aere, levitatem in mercurio, aut plumbo liquente obtinet, duplicem statuere virtutem, quarum altera sursum, altera deorsum connitatur; cum enim impetus motus efficiens (ut infra constabit) ejudem naturae sit, in quamcunque demum orbis plagam dirigatur motus" Casati (1684, 11).

[97] Casati (1684, lib. 1, cap. 2, p. 13); see Chap. 9 in this volume.

[98] See Chap. 9 in this volume.

[99] "Sed quomodo deorsum ferretur in vacuo, in quo nulla est positionum differentia? An innato quodam impetu universae corporeae molis centrum peteret? Sed hoc multorum pace dictum sit, inter vulgi errores est. recensendum; ut quid enim singulis gravibus natura centri appetitum insereret, quem pene in omnibus frustrandum praecognosceret? Cum praesertim universi concinnitas diversas corporum dispari gravitate praeditorum sedes exigeret. Non igitur singulorum gravium motus eo tendit, ut corpus motum in centro quiescat, sed ut debitam sibi sedem in universo occupet, iuxta insitae gravitatis speciem, molis scilicet aequalitate spectata" Casati (1649, cap. 2, n. 10, 9).

As for acquired impetus, Casati explains that if innate impetus is not impeded, for example by letting a body fall in the air, it acquires (*acquirit*) "new impetus in single moments," becoming more and intense. Casati remarks that, were it not for the accelerating effect of acquired impetus, bodies would seek their place "exceedingly slowly" (*tarde nimis*) (ibid., 147). On the other hand, he asks the reader to imagine "how inconvenient" (*quantus incommodus*) it would be if bodies (such as our bones) were originally endowed with a high innate impetus, consequently strongly resisting motion (ibid., 147). He concludes that it is "suitable" (*peropportunum*) to have a moderate (*non vehemens*) innate impetus in individual bodies, which can nevertheless accumulate when it is allowed to continually effect motion (ibid., 147). Unlike innate impetus, acquired impetus is not destroyed when a body arrives where it should naturally either rest or move against the acquired impetus; for example, when a wooden ball falls to water from some height, it will first sink and only afterwards (possibly following several oscillations) will it begin floating (ibid., 147). It is this property of acquired impetus, adds Casati, which explains the continuing motion of a pendulum (ibid., 147–150). Later Casati claims that externally impressed (i.e. violent) impetus—which (totally unlike innate impetus) also exhibits a "motion preserving" behavior (see Sect. 10.3.2 above)—should be considered as similar to acquired impetus.[100]

10.4.3 Eschinardi

In the very beginning of *De impetu tractatus duplex*, even before defining impetus, Eschinardi considers three properties of "heavy bodies, for example a stone": 1. Innate gravity (*gravitas innata*), which causes bodies to have weight, and exists inside them whether they ascend, descend or rest (Eschinardi 1684, 1). Eschinardi sees *gravitas innata* as an "absolute" property, which would manifest itself purely only in vacuum; accordingly he totally and explicitly rejects (like Fabri) absolute levity. 2. Elicited gravity (*gravitas elicita*), that is, the act of gravitating, or *gravitatio*, "which a heavy body here and now exerts on other bodies resisting it." Eschinardi regards this "elicited gravity" as identical to impetus which is produced by "innate gravity" and which should be distinguished from the substance of the heavy body.[101] 3. The motion of a heavy body, which is distinguished both from impetus (i.e. elicited gravity) and from substance (Eschinardi 1684, 2).

We can therefore easily discern—as hinted earlier—a very important aspect in Eschinardi's theory of natural motion, in which he substantially differs from Fabri

[100] "Quare idem de impetu extrinsecus asumpto dicendum est., quod de acquisito" Casati (1649, 153).

[101] "Secundo. Consideranda est. gravitas elicita; sive actus gravitandi; qui dicitur etiam gravitatio; & est. illa, quam hic & nunc exercet grave in alia corpora sibi resistentia... quae gravitas elicita nihil aliud est., quam impetus productus a gravitate innata; & de isto impetu certumest, quod distinguitur a substantia corporis gravis" Eschinardi (1684, 2).

and Casati: Eschinardi sees no need to treat *gravitas innata* as a kind of impetus; in other words, he has no concept of "innate impetus." Instead, he prefers to associate impetus with "elicited gravity," that is, with a body actually moving naturally. Eschinardi's *gravitas elicita*—a term which is soon abandoned, replaced by mere "impetus"—amounts to what Fabri and Casati refer to as "acquired impetus."

Eschinardi emphasizes—not unlike his two colleagues—that impetus must be deemed responsible for natural acceleration, while rejecting other possible explanations; he claims in fact that the phenomenon of natural impetus can be regarded as a proof for the existence of impetus (the first part of this "proof," which concerns projectile motion, was discussed in Sect. 10.2.3 above):

> It [the existence of impetus] is proved again in natural motion of heavy bodies freely descending in air; for a heavy body always descends faster and faster; but the immediate and adequate cause for this swifter motion cannot be substance, or the remaining accidents; for they usually remain entirely the same; but by an invariant immediate necessary cause, the effect is not diversified (*variatur*); and it cannot be said that a less perfect (i.e. less swift) previous motion (*motus antecedens imperfectior*) is an adequate cause of a more perfect following motion (*motus sequens perfectior*); especially since it is merely a formal transference (*translatio formalis*) from one place to another.[102]

Eschinardi concludes—in a manner similar to Fabri's explanation—that impetus is the only intrinsic cause which could possibly account for natural motion. Contrary to Fabri, Eschinardi does not refute *external* explanations that have been proposed for natural motion.[103] Eschinardi, in line with Fabri and Casati, deems motion an "inferior" entity which cannot cause anything—and thus disqualifies "previous motion" as a possible explanation for natural acceleration; as we have seen above, Casati also refuses to ascribe any *efficientia* to motion, while Fabri emphasizes that in no way could it be "produced."

Eschinardi depicts in detail, in a chapter entitled "*De acceleratione motus naturalis*," his views concerning impetus (i.e. acquired impetus)—and reveals himself a zealous Galilean, at least concerning falling bodies, both in vacuum and in material media. In fact, he declares his very intention to merely present afresh, in a unified and perhaps better-explicated manner, Galileo's doctrine of falling bodies, "by dwelling on the things which one finds scattered in his Dialogues."[104] Eschinardi indeed immediately delivers a rather accurate paraphrase of Galileo's principles: all heavy bodies fall in vacuum at the same rate (Eschinardi 1684, 26), while in media

[102] "Probatur iterum in motu naturali gravium libere descendentium in aere; nam grave descendens, semper velocius & velocius movetur; sed causa immediata, et adaequata huius motus velocioris non est. substantia, caeteraque accidentia; nam haec plerumque remanent omnino eadem; invariata autem causa immediata necesaria, non variatur effectus; neque dici potest motus antecedens, impererfectior, quippe minus velox, causa adaequata motus sequentis perfectioris; praecipue cum sit mera translatio formalis de loco in locum" Eschinardi (1684, 3).

[103] Eschinardi (1684, 3); for Fabri's account, see p. 19 below Fabri (1646, lib. 2, th. 1, pp. 77–79); see also Elazar (2011, 70–77).

[104] "Ut autem habeamus totam simul methodum doctrinae Galilei de gravibus cadentibus, concinnavi sequentem discursum, inhaerendo tamen iis, quae sparsim inveniuntur praecipue in ipsius dialoi, addita aliqua maiori explicatione ubi oportebat" Eschinardi (1684, 26).

the difference between the actual velocity and velocity in a vacuum depends on the ratio of specific gravities.[105] A bullet shot down by a pistol from the height of 100 *ulnae* is less compressed (because of air resistance) than a bullet shot down from 3 *ulnae*.[106] Eschinardi also conveys his agreement concerning two issues on which Fabri, in contrast, disagreed with Galileo: 1. That the phase of natural acceleration (in a material medium) is followed by a constant terminal velocity[107]; 2. That the resistance of (material) media to falling bodies depends on the surface of the falling body.[108]

10.5 Galileo's Odd Numbers Law

10.5.1 Fabri

Having established (to his own satisfaction) that a gradually increasing acquired impetus is the cause of natural motion, Fabri embarks on a discretized mathematical treatment of this issue. Since the same necessary cause—the falling body's substance—acts repeatedly to increase impetus, and "hence in single equal instants equal new impetus is acquired," Fabri concludes that while within each given instant the velocity remains constant[109] from one instant to another "the impetus increases and becomes more intense equally… Therefore the speed of the motion increases equally in single instants" (Fabri 1646, 87).

Fabri now explains that "impetus increases according to an arithmetical progression, for single instants add equal impetus," and then points out that accordingly the speed increases in the same way, since in single instants equal additions of velocity are acquired. (Fabri 1646, 88). Relying on the simple rule that, if two uniform motions occur in equal times, the spaces traversed are to each other as the velocities, Fabri easily concludes that the increase of velocity according to an arithmetical progression entails the same rate of increase for successively traversed distances (Fabri 1646, 86, 88). He finally concludes:

---

105 "In proprtione gravitatum [specificarum] detrahitur de velocitate, ut ipse [Galileo] hic supponit: de velocitate inquam, qum haberent [descendentes] in vacuo" Eschinardi (1684, 27). Thus Eschinardi paraphrases the "formula" from the *Two New Sciences* which Fabri also adopted, as shown above.

106 "Patuit etiam, quod glans plumbea e sclopo emmisa deorsum in altitudine 100 ulnarum minus est. compressa quam in tribus ulnis, ut Galileus dixerat" Eschinardi (1684, 29).

107 Eschinardi (1684, 29); on Fabri's objection to this (valid) assertion, see Lukens (1979, 207–210).

108 Eschinardi (1684, 26). Fabri disagrees, and concludes that two cubes from the same material fall in the same rate not only in vacuum, but also in any material media, Fabri (1646, lib. 2, th. 112, p. 128); see also Elazar (2011, 113, note 14).

109 This is guaranteed by the important theorem 13 of his *De motu naturali deorsum*, discussed above in Sect. 10.2.3, p. 24.

> Hence the spaces increase in single equal instants according to an arithmetical progression, because they increase as the speeds [...] and these as the impetus [...] and these according to an arithmetical progression... Hence the spaces acquired in single instants are as the series of numbers... 1, 2, 3, 4, 5, 6, etc."[110]

Having developed his own "natural numbers" law, Fabri announces his full acceptance of Galileo's experimental results concerning free fall, stating that they were repeated since by many scientists *ad nauseam* (Fabri 1646, th. 61, p. 99). Furthermore, Fabri shows that Galileo's "mean speed rule"[111] and "double-distance rule"[112] are equally valid in his discrete analysis as well.

Most importantly, Fabri takes pains to show that, if we assume that the instants in which his "law of natural numbers" is valid are extremely small, then in sensible intervals of times—which include huge numbers of such "instants"—his law is indistinguishable from Galileo's "law of odd numbers" (i.e. the series 1, 3, 5, 7). In this "convergence" proof Fabri uses the lack of scalar invariance of his natural numbers law, that is, its dependence on time units. The first row (see Table 10.1 below) assumes a time unit of a discrete instant, to which Fabri's original rule applies. The second row assumes a time unit consisting of two instants; the basic rule is no longer valid since the series that results (3, 7, 11...) is different from the natural numbers series. We can easily discern that the more instants are included in a time unit, the closer we get (after a normalization) to Galileo's series (1, 3, 5, 7...). Fabri shows in a simple arithmetical analysis that the difference between the space traversed according to Galileo's rule and the space traversed in accordance with his own series is inversely proportional to the number of instants that constitute our time

Table 10.1 Fabri's "law of natural numbers" converges to Galileo's "law of odd numbers" as more and more instants are assumed to be contained in a given time unit

1,2,3,4,5,6,7,8,9,10,11,12,13,14,15,16,17,18,19,20	(t = 1 "instant")
3, 7, 11, 15, 19, 23, 27, 31, 35	(t = 2)
10, 26, 42, 58, 74, 90, 106, 122, 138, 154	(t = 4)
36, 100, 164, 228, 292, 356, 420, 484, 548	(t = 8)
136, 392, 648, 904, 1160, 1416, 1672, 1928	(t = 16)

110 Fabri (1646, 88). See also Elazar (2011, 85–86).

111 "The time in which a certain space is traversed by a moveable in uniformly accelerated movement from rest is equal to the time in which the same space would be traversed by the same moveable carried in uniform motion whose degree of speed is one–half the maximum and final degree of speed of the previous, uniformly accelerated, motion" Galileo (1989, 165).

112 If, following naturally accelerated motion (lasting for time AC), a body continues "to be moved with the same degree of speed BC, without accelerating further, then in the ensuing time CI it would pass a space double that which it passed in the equal time AC with degree of uniform speed EC, one–half the degree BC" Galileo (1989, 168).

unit. Therefore, this difference in effect vanishes as more and more instants are included in a time unit (Elazar 2011, 249–254; Fabri 1646, th. 61, p. 102).

In order to geometrically illustrate the convergence of his discrete analysis to Galileo's continuous approach, Fabri supplies a diagram (in Fig. 10.1 below) in which the horizontal "axis" FN represents time, and the vertical "axis" FA represents velocity. Fabri explains that the sum of quadrangles which constitutes the total distance in a discrete analysis[113] becomes closer to the triangle AFN (i.e. the total distance in Galileo's continuous analysis) as the total FN is divided to more and more instants.[114] Clearly, the more instants we employ in a discrete approach the closer we arrive at Galileo's result.

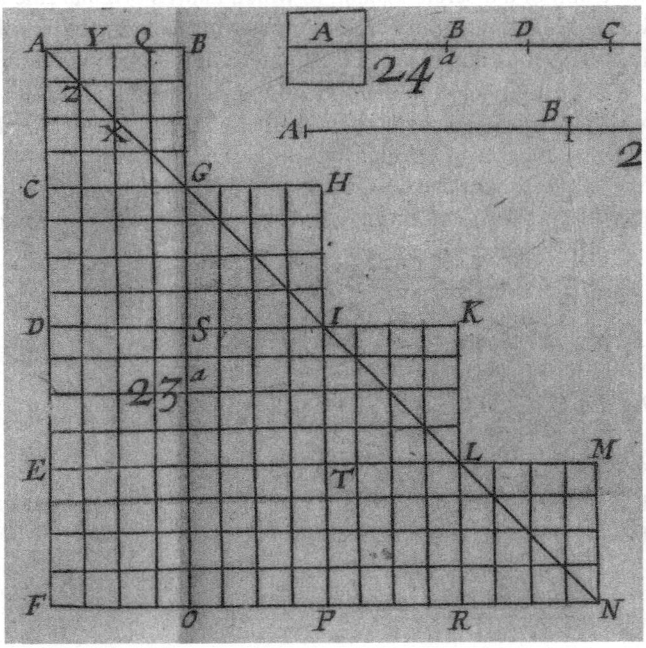

Fig. 10.1 Fabri's diagram (Plate 1, Fig. 23 from the *Tractatus de motu locali,* 1646)

[113] In figure 23: the quadrangles LMNR, IKRP, GHPO and ABOF (assuming an "instant" RN=PR = OP etc.); these areas behave of course like the natural numbers series.

[114] Fabri (1646, lib. 2, th. 61, p. 106). In triangle AFN, Consecutive areas over consecutive equal periods of time (e.g. GIPO, ILRP and LNR in figure 23) behave of course like the series 1, 3, 5 …

10.5.2 Casati

Casati's opinion on Galileo's odd numbers law appears in a chapter from the *Mechanicorum libri* that discusses the subject of *percussiones*, trying to estimate the forces with which falling objects impact the ground.[115] Casati begins this chapter by conveying a great deal of skepticism concerning the possibility of fully understanding—not to mention accurately describing mathematically—the complex behavior of bodies moving in different media: he flatly rejects not only Galileo's claim that bodies fall with the same speed, but also Aristotle's simple proportion between weight and velocity of falling bodies.[116] By completely rejecting Galileo here, Casati greatly differs also from both Fabri and Eschinardi: Fabri, although rejecting Galileo's assertion that bodies falling in media have a constant "terminal" speed, fully accepts his equal-speed principle (valid in vacuum), as well as the general relation between velocities and densities he proposed in the *Two New Sciences*. Eschinardi, for his part, accepts Galileo's assertions without any reservation. Casati rejects both Aristotle's and Galileo's opinions because he considers both as too simplistic. Casati agrees with Aristotle that different bodies fall with different velocities, and accepts Galileo's contention that the medium's resistance is related to the body's surface area (Casati 1684, 682–683). However, the latter observation immediately leads him to reject Aristotle's straightforward proportion.[117] Likewise, Casati rejects Galileo's simple proportion (which Fabri and Eschinardi adopted). Concerning bodies differing in specific gravities (i.e. made from different materials) but which weigh the same and have a similar shape, Casati is only willing to claim that the bigger will experience more resistance than the smaller. When they differ both by shape and by material, he insists that he "would not dare" to "rashly" determine the correct proportion which governs the resulting complicated behavior of resistance.[118] Casati immediately discusses specifically the behavior of air resistance, and concludes that because it depends on so many causes, it is too complicated to be successfully treated mathematically ("*ad calculos revocari*") (Casati 1684, 685).

Notwithstanding his skepticism toward previous attempts to evaluate velocities of different bodies in various media, Casati expresses absolute trust in experiments

---

[115] This chapter, entitled "Quam dispares ex motus velocitate sint percussiones," is the seventh chapter of the seventh book—*De cuneo & percussionibus*.

[116] "Hinc duae fere ex diametro oppositæ sententiæ cavendæ sunt, quarum altera gravium inæqualium motum statuit ipsorum gravitatibus analogum, ut decuplo velocius moveatur illud, quod est. decuplo gravius; altera æqualem omnibus velocitatem tribuit" Casati (1684, 682).

[117] Since, explains Casati, surfaces relate to each other as the squares of homologus sides (*in duplicata ratione laterum homologorum*), while weights—like volumes—as their cubes (*in triplicata ratione laterum homologorum*); he simply concludes that "Non igitur velocitates, quatenus ab impedimento temperantur, sunt directe gravitatibus analogae" Casati (1684, 684).

[118] "Ubi autem corpora ... si vero & dissimiles figuræ, & inæquales gravitates ponantur, ex omnibus simul compositis quodammodo conflari resistentiarum Rationem facile est. opinari; sed Rationis terminos temere definire non ausim" Casati (1684, 684).

which "a sensible man would hardly deny"—experiments that established definite ratios between distances and times. After Galileo, he explains, "such learned men" reflected upon them, with "equal outcome" and "greatest consensus," that "in all these there is present what we may call a natural complexion free of make up."[119] As an example of these proven experiments, Casati mentions an 8 ounces clay ball, which is observed passing 15 feet within one second (from rest), and 60 feet within two seconds (from rest). Many experiments of this kind, tells us Casati, were performed by Giovanni Battista Riccioli (1598–1671), and based on them it has been concluded that

> The increments of velocity are made according to the increment of the arithmetic progression of odd numbers starting with one, that is, 1, 3, 5, 7, 9, 11, 13 etc., in such a way that if a space is traversed in the first moment, which would be designated as 1, then the descending heavy body would move thrice as fast in the second moment, five times as fast in the third, seven times as fast in the fourth, and so on.[120]

As Casati immediately explains, the resulting accumulated spaces behave like the square of the times (Casati 1684, 687). Casati then—in sharp contrast to his two fellow Jesuits—prefers not to directly ascribe this experimentally proven law to Galileo, but rather to credit the "very learned" Jesuit who indeed supported it after Galileo, a support which according to Domenico Bertoloni Meli "had a sting in its tail," since Riccioli used it to oppose the possibility of the earth's motion.[121]

Exactly like Fabri and Eschinardi, Casati tries to account for the odd numbers law using the concept of impetus. He first describes the hypothesis he had formulated (so he claims) while interpreting Aristotle as a young man. According to this hypothesis, as Casati explains, he had limited the duration of any impetus to two moments only, while claiming that an impetus which is produced in a given moment reproduces itself in the next moment and perishes in the third.[122] According to this hypothesis, in the first moment of fall the first unit of impetus is created by gravity (*gravitas*); in the second moment this unit "brings forth" (*gignit*) another unit, and these two are added to the second unit created by gravity to form three units; in the

---

[119] "... spatia vero cum temporibus... sine ullo fuco veritatis color." The last sentence is copied from Cicero's *Brutus*, cap. XLIV; the translation is taken from the Loeb edition.

[120] "Plura huiusmodi experimenta (si te a tentando absterreat labor) suppeditabit Ricciolius tom.1. Almag. lib.9. Sect. 4. cap.16. ex quibus demum infertur velocitatis incrementa fieri juxta incrementum progressionis Arithmeticæ numerorum imparium ab unitate incipientis 1.3.5.7.9.11.13, &c. adeo ut, si quod spatium primo momento percurritur, statuatur ut 1, triplo velocius moveatur corpus grave descendens in secundo momento, quintuplo velocius in tertio, septuplo velocius in quarto, atque ita deinceps" Casati (1684, 686–687).

[121] As Domenico Meli explains, Riccioli claimed that "if the earth rotated, falling bodies would move with a uniform speed along a circular arc and therefore they would hit the earth regardless of their height of fall" Meli (2006, 134).

[122] "Incrementa hæc velocitatis ex concepti impetus incremento desumenda esse nullus dubito; sed operosum videri posset augescentis impetus causam exponere. Cum junior Aristotelem interpretarer, & primas curas hujusmodi rerum contemplationi impenderem, hanc excogitavi hypothesim; videlicet impetus producti diuturnitatem maximam duobus tantum momentis circumscribebam, ita ut primo momento oriretur, secundo æqualem sibi impetum gigneret, in quo superstes esset, tertio periret" Casati (1684, 687–688).

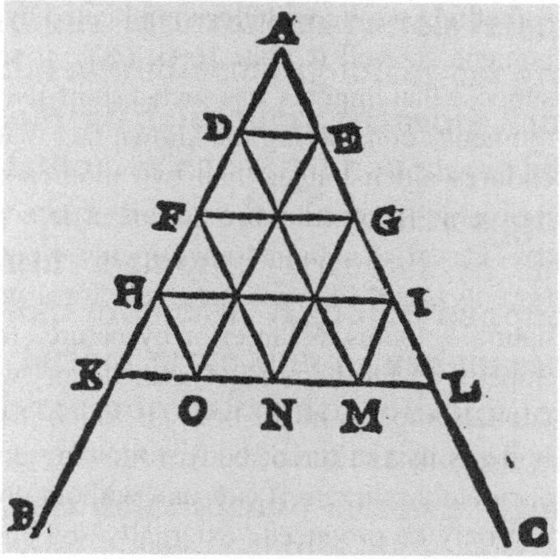

Fig. 10.2 Gassendi's *vis attrahens* and *vis impellens* approach (Gassendi 1642)

third moment the unit which was created in the first moment perishes (since no impetus unit can endure for more than two moments), while each of the two remaining units duplicates itself: a new unit formed by gravity is added to these four units to create five; in the fourth moment the two additional units from moment no. 2 perish, the remaining three multiply themselves and a new unit formed by gravity is added to create seven; in the fifth moment the three new units created in moment no. 3 perish, the four remaining duplicate themselves and if we add to these eight units the one unit newly formed by gravity we get nine.[123] We thus obtain a series which is formed by adding consecutive terms of the natural numbers series, that is, the series 1, 1 + 2, 2 + 3, 3 + 4, 4 + 5 etc., that is, the odd numbers progression. Casati's rather cumbersome attempt to come to terms with Galileo's law of fall brings to mind Gassendi's (perhaps no less cumbersome) account (in his *Epistolae de motu*, 1642), which did not employ the idea of an annihilation of any impetus older than two moments, but rather postulated the constant and simultaneous production of impetus by two forces (*vis attrahens* and *vis impellens*) and the continuous conservation of created impetuses (Palmerino 2004, 154; see Fig. 10.2).

Gassendi, in his subsequent *Epistolae de proportione* (1646), was to replace his former theory by a simpler one, which accounted for the odd numbers rule while postulating one force only (*vis attrahens*) (Palmerino 2004, 157–161). Likewise Casati—as he himself tells us in his *Mechanicorum libri*—felt that there had been

[123] "Hinc corpus descendens primo momento primum habebat impetus gradum a gravitate productum; secundo momento primus ille gradus alium gradum gignebat præter eum, qui a gravitate tunc oriebatur; quare duo novi gradus cum uno antiquo tres gradus constituebant. Tertio momento primus gradus peribat, duo secundi gradus duos pariter producebant, & gravitas suum tertium gradum; quare quinque gradus erant. Quarto momento duobus secundis gradibus pereuntibus, tres gradus tertio momento producti reliqui erant, & sibi tres alios gradus addebant, quos producebant, atque gravitas suum quartum gradum efficiebat, ut in universum essent septem gradus. Ex his septem quinto momento peribant tres tertij gradus; quatuor reliqui item alios quatuor adjiciebant quinto gradui a gravitate proficiscenti, & erant novem" Casati (1684, 688).

something wrong with his complicated hypothesis, even though it matched the phenomena so well (Casati 1684, 688). It seemed to him fictitious (*commentitia*) to suppose that impetus was such a short-lived entity, which could endure only for two moments: considering the known fact that an impetus externally impressed clearly endures much longer than two moments even when the original mover and the moveable (projectile) are no longer in contact; why would it not be the case then, asks Casati, that an internal impetus, which is never disconnected from the "origin" (gravity), will likewise persevere for a while? We may not, adds Casati, consider natural impetus as "inferior by nature" (*deterior naturae*) to external (or violent) impetus.[124] Furthermore, Casati doubts whether impetus could indeed bring forth another impetus in the body in which it inheres, while it may, of course, produce impetus in a different body, following a collision (Casati 1684, 689). Casati thus accurately reiterates (as usual, without crediting) Fabri's assertion that an impetus can only be produced "externally" by another impetus, that is, only when impact occurs.[125]

Casati therefore abandoned his initial hypothesis, looking for a "highly more probable" one, which would on the one hand "vindicate" the constantly created impetus from such a quick demise, and on the other hand "strip" from it the problematic ability to augment itself.[126] The reader might not be surprised to learn that the new hypothesis which Casati suggests is identical to Fabri's original law of fall (without mentioning any name of course): "in single moments" of descent, explains Casati, "equal single degrees of impetus" are accumulated, in such a way that the final intensity of motion (i.e. velocity) depends directly on its duration (i.e. $v \propto t$); Casati concludes that "therefore the increments of velocity take place according to the natural progression of numbers, 1, 2, 3, 4, 5 etc., for it is according to the same series that impetus—the cause of velocity—is increased, following the accumulation of single degrees in single moments."[127]

---

[124] "Huic hypothesi cum Phenomeno optime conveniebat, & ea. statuebatur impetus intensio, quæ velocitati efficiendæ par esset, servata incrementorum Ratione, quæ ex iteratis experimentis innotuerat. Verum commentitia, & fabulæ proxima videbatur tam brevis impetus vita, quam non nisi duo momenta metirentur: in iis sane, quæ vi externa moventur, & longius projiciuntur, aut in gyrum aguntur, licet extincta effectrice causa impressus impetus diutius permanet; quidni & impetus sponte sua conceptus, suæque origini cohærens aliquandiu perseveret? quippe qui aut ejusdem, aut saltem non deterioris naturæ censendus est" Casati (1684, 688–689).

[125] "Theorema 42: Impetus producitur semper ad extra ab alio impetu" Fabri (1646, 32). Accordingly, Fabri immediately concludes that "impetus agit tantum ad extra" Fabri (1646, 32); i.e. there is no way in which impetus residing in a body can increase itself.

[126] "Quare impetum a gravitate descendente productum, ex tam expedito interitu vendicandum, & virtute se novo impetu augendi spoliandum, longe probabiliore conjectura censui" Casati (1684, 689).

[127] "Singulis momentis, quibus potest descendere, singulos impetus gradus æquales prioribus adjicit, adeo ut, quot momenta motum metiuntur, tot gradus impetus postremo momento intensionem constituant, cui motus velocitas respondeat. Velocitatum igitur incrementa fiunt juxta naturalem numerorum progressionem 1.2.3.4.5, &c. nam juxta hanc eandem seriem impetus, velocitatis causa, augetur, singulis gradibus in singula momenta additis" Casati (1684, 690).

On his way to fully achieving his goal—that is, account for Galileo's odd numbers law using impetus—Casati lacks one step only: to show, exactly like Fabri before him, that when we deal with measurable magnitudes (which inevitably contain a huge number of "moments") the natural numbers law converges to the odd numbers law. For this purpose he returns to the 8 ounce clay ball example, which falls 15 feet within one second, and 60 feet within two seconds (from rest), that is, the spaces are related according to the squares of times ($4{:}1 = 2^2{:}1^2$).[128] When we are dealing with "infinitesimal" moments, for which the natural numbers law holds, this is of course not the case: the ratio between the space run through in two moments and the space acquired in a moment is $(1 + 2){:}1 = 3{:}1$. But if we consider the time required for acquiring those 15 feet as made of 10,000 moments, explains Casati, then the space acquired in this unit will be the sum 1 + 2 + 3 + ... + 10,000, that is, $10000^2/2 + 10{,}000/2 = 50{,}005{,}000$ "minimal spaces,"[129] while the space acquired in two units (20,000 moments)—that is, 60 feet—will amount to $20000^2/2 + 20{,}000/2 = 200{,}010{,}000$ minimal spaces.[130] The ratio between these spaces, $200{,}010{,}000/50{,}005{,}000 = 3\frac{49995}{50005}$, explains Casati, is only $\frac{10}{50005}$ short of 4, and who could observe, while performing this free fall experiment, such a tiny difference?[131] Casati is repeating then Fabri's analysis, concluding that the "natural numbers law" is both "agreeable to nature" (since, as we saw above, it takes into account impetus conservation and does not require a "self-producing" impetus) and after all does not defy experiment (i.e. the odd numbers law), so this is the theory we should uphold.[132]

---

[128] "Habemus experimento globum argillaceum unciarum octo percurrere uno minuto Secundo horæ pedes 15, & duobus Secundis pedes 60, hoc est. spatium quadruplum, & quia tempora sunt ut 1 ad 2, spatia sunt ut quadrata, scilicet ut 1 ad 4" Casati (1684, 691).

[129] We can easily derive the "formula" which Casati uses from the arithmetical series sum $\sum = n(a_1 + a_2)/2$. Casati calls this sum "summa mimimorum spatiorum."

[130] "Ponamus in uno Secundo esse momenta 10,000; sunt igitur ultimo momento 10,000 gradus velocitatis similes & æquales primo gradui primi momenti, & spatium ultimo hoc momento decursum, ad spatium primi momenti est. ut 10,000 ad 1. Coge igitur in summam omnia spatia incipiendo ab unitate usque ad 10,000, videlicet ultimi termini dimidiato quadrato adde ejusdem ultimi termini semissem, & prodibit omnium spatiorum summa. Ultimi termini 10,000 quadratum est. 100,000,000, cui adde ipsum ultimum terminum; & hujus summæ medietas 50,005,000 est. summa minimorum spatiorum, quibus conflantur pedes 15. In duobus Secundis erunt momenta 20,000, & similiter invenitur summa 200,010,000 minimorum spatiorum, quibus constant pedes 60" Casati (1684, 691–692).

[131] "Non sunt quidem duæ hujusmodi summæ hic inventæ 50,005,000 & 200,010,000, omnino ut 1 ad 4, sed ut 1 ad $3\frac{49995}{50005}$: verùm tantula differentia $\frac{10}{50005}$ quid officit allato experimento? an potuit observari?" Casati 1684, 692).

[132] "Cum itaque constituta hypothesis incrementi spatiorum, velocitatis, atque impetus juxta seriem naturalem numerorum sit naturæ consentanea, neque Physice repugnet experimentis, non debemus esse solliciti, ut aliam quæramus hypothesim ad statuenda incrementa exquisite juxta numeros impares" Casati (1684, 693).

10.5.3 Eschinardi

We have just seen how Eschinardi's views concerning falling bodies were shaped by Galileo's work. It is hardly surprising, therefore, that Eschinardi also adopted Galileo's law of fall. Contrary to his two Jesuit colleagues, he did so without deeming necessary for this adoption any intermediary step, such as the introduction of the natural numbers series. Eschinardi's account of the odd numbers mentions not only Galileo but also Fabri as the initiator of this intermediary step.

Already in the beginning of his book, Eschinardi clarifies that his attitude concerning falling bodies relies on the principle initially rejected—though eventually adopted—by Casati: the inherent conservation of impetus. A falling stone, explains Eschinardi, successively produces impetus, which retaining the direction of the downward falling stone is not destroyed, and therefore constantly accumulates, causing swifter motion. Eschinardi promises to explain later how this process works.[133]

Eschinardi keeps his promise, and returns to this subject in "*De acceleratione motus naturalis*." He begins by presenting the odd numbers law, illustrated by Fig. 10.1 (in Fig. 10.3 below): a ball B falls from a tower toward the ground (AC), acquiring in the first "minute of time" (*minutus temporis*) the space DE, in the second *minutus temporis* ET = 3DE and in the third *minutus temporis* TC = 5DE (Eschinardi 1684, 19–20).

Now Eschinardi proceeds to theoretically account for this law, or in his words, to supply a *ratio a priori* of this increase. He mentions those who supposed that in single moments an equal degree of impetus is created, and states that as "Mousnerius, on the basis of Father Fabri" has "well shown," the increase of space in this case would not be according to odd numbers, but rather according to the natural

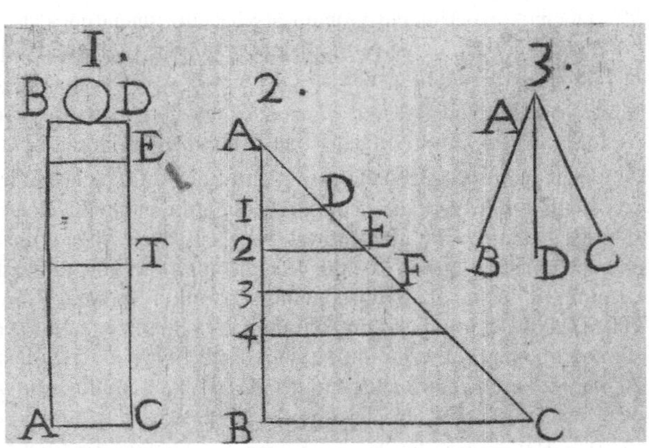

Fig. 10.3 Eschinardi's free fall analysis (1684)

[133] "... lapis, dum libere descendit, producit successive novum, & novum impetum; qui, dum obtinet intentum movendi lapidem deorsum, non destruitur, adeoque ex nova additione impetus, fit maior veluti cumulus, seu impetus intensior; adeoque causa velocioris motus: Sed hoc ipsum quomod fiat; deinde explicabimus" Eschinardi (1684, 3).

numbers.[134] He immediately adds though that he prefers to follow Galileo's opinion, by considering "physical parts of time" rather than "mathematical."[135] Eschinardi now elaborates his explanation, which relies on Fig. 10.2 (shown above, along Fig. 10.1). He begins his analysis by formulating the two following statements (numbered by Eschinardi himself):

1. The measure of impetus determines velocity, and spaces are acquired according to velocity, "for space totally depends on velocity"; furthermore, impetus is the "immediate cause" of velocity.[136]

2. Impetus is produced by a flowing point (*per punctum fluens*) just as time flows (*sicut fluit tempus*). The "production of impetus" thus "flows," and accordingly this production brings forth a line, "or rather a surface."[137]

In the second illustration of Fig. 10.3, we can see the lines and surfaces to which Eschinardi refers. Eschinardi argues that axis AB represents simultaneously both time and the production of impetus, which itself uniformly flows and corresponds to time (according to statement 2). The horizontal lines (D1, E2 etc.) represent the momentary impetuses: each of them exists in a singular moment, and they accumulate together to create ever growing surfaces (AD1, AE2 etc.). Probably following his analogy between the creation of time and the production of impetus (both depicted by the AB axis), Eschinardi assumes that A1 = D1, A2 = E2 etc. It might be the case that Eschinardi regards his first assumption, that is, that the "parts of time" are "physical" rather than "mathematical," as justifying this analogy. In any case, he now adds that the whole triangular surface is created by impetus which is "not only produced, but also successively conserved": Eschinardi imagines then the perpendicular axis AB as describing not only time but also the continuing conservation of impetus, while the horizontal lines depict the ever accelerating impetus (i.e. velocity).[138]

---

[134] "Quaeritur nunc ratio a priori huius incrementi; seu qua methodo praecisa utatur natura in hoc negotio. Putant aliqui singulis instantibus produci in singulis partibus materiae aequalem gradum, seu particulam impetus... Sed si hoc est., videtur res processura non per numeros solos impares; sed per omnnes sic 1,2,3,4,5 & c., ut bene ostendit Mousnerius ex P. Fabry pag. 105 de Motu Locali" Eschinardi (1684, 20).

[135] "Placet tamen sequi sententiam Galilei... intelligando tamen de partibus temporis physicis non mathematicis" Eschinardi (1684, 21).

[136] "Iuxta mensuram impetus est. velocitas, & iuxta velicitatem fiunt spatia (nam spatium pendet totaliter a velocitate); velocitas autem non habet aliam causam immediatam, quam impetum" Eschinardi (1684, 21).

[137] "Impetus producitur veluti per punctum fluens ita ut sicut fluit tempus; ita fluat impetus productio; ac proinde per dictam productionem sit linea, vel potius superficies, quod semper intelligas" Eschinardi (1684, 21).

[138] "Sed quia impetus productus conservatur, quando obtinet intentum motus, ut in praesenti casu; ideo in praesenti figura [Figure 2] sit AB tempus, & simul productio impetus aequaliter fluens, ac tempori correspondens; lineae vero transversae D1. E2 &c. sint impetus hic, & nunc existentes eo instanti, quo grave consideratur in ipso descensu, nempe totidem, quod fuerunt hactenus producti; ita ut D1 sit aequalis ad A1 &c.; quod si in singulis quodammodo punctis concipiantur dictae linae transversae: imo & fiat tota superficies, seu triangulum AD1, AE2 etc. impetuosae pro toto illo tempore impressae ab impetu non solum producto; sed conservato successive, figura correspondebit rei ipsi" Eschinardi (1684, 21).

Now comes the crucial part in Eschinardi's analysis, to which he devotes surprisingly little attention: proving that the area represents the distance acquired by the falling body. He first warns the reader not to regard line AB as distance, explaining that it would be too short for that.[139] He now claims that the triangular figure represents time as well as "the whole impetus (*totus impetus*), both produced, as well as conserved from the beginning until now".[140] Hence, states Eschinardi, we can deduce two "evident mathematical conclusions": 1. The distances acquired by a falling body relate to each other as the squares of the times; 2. These distances behave like the odd numbers series 1, 3, 5, 7 etc. (Eschinardi 1684, 21–22). This is how he derives the first of these two conclusions:

> The spaces achieved by a heavy body freely falling downwards are in the ratio of the squares of times (*in ratione duplicata temporum, seu quadrata temporum*) in which they take place—in relation to each other, and starting from the beginning[141]; it is proved, for they [the spaces] follow the proportion of the triangular surface described by the whole impetus (*totus impetus*), both successively produced and successively conserved; but these joint triangular parts, always taken from the beginning A, are [to each other] as the squares of times, therefore etc. It is [furthermore] proved, since they are in the ratio of the squares of the lines A2 to A1,[142] because of the similarity of the joined triangles AD1 & AE2 etc.[143]

Having proved (to his own satisfaction) conclusion 1, Eschinardi proceeds to easily prove conclusion 2—Galileo's odd numbers law: knowing that areas represent spaces, we can immediately determine that since triangle AE2 is 4 times bigger than triangle AE1 (the first space acquired), while triangle AF3 is 9 times bigger than the triangle AE1 (etc.), therefore trapezoid DE21 (the second space) must be 4–1 = 3 times bigger than triangle AE1, trapezoid EF23 (the third space) must be 9–4 = 5 times bigger than it etc.[144]

The most important question is, of course, how come Eschinardi can assume that area indeed represents space. The answer probably lies in his claim (just before deriving conclusion 1) that both time and the "whole impetus" (*totus impetus*) are

---

[139] "Adverte ne concipias lineam AB supradictam significare spatium; sed solum tempus; alioquin pro spatio deberet esse longior &c" Eschinardi (1684, 21).

[140] "Posita dicta figura, habes tempus, & productionem impetus in linea AB; totum vero impetum, tam productum, quam conservatum ab initio usque adhuc."

[141] That is, from rest.

[142] Each represents time in Figure 2.

[143] "Nempe primo spatia facta a gravi deorsum libere descendente sunt in ratione duplicata temporum: seu sicut quadrata temporum, in quibus fiunt, inter se, & incipiendo semper ab initio. Probatur, nam sequuntur proportionem superficiei triangulris descriptae a toto impetu, tum successive producto; tum successive conservato; sed istae partes communicantes trianguli, incipiendo semper ab initio A sunt ut quadrata, seu in ratione duplicata temporum, ergo &c. Probatur, nam sunt in ratione duplicata lineae A2 ad lineam A1 ob similitudinem triangulorum communicantium AD1 & AE2 &c" Eschinardi (1684, 22).

[144] "2. Singula spatia facta in descensu, crescunt per numeros impares; nam primo in triangulo AE2 cum hoc sit quadruplum trianguli AD1, trapezium DE21 est. 3 & reliquum AD1 est. unum; deinde triangulum AF3 est. novem; ergo trapezium EF23 additum est. 5 etc" Eschinardi (1684, 22).

represented by the triangular figure, together with the emphasized line within the derivation itself that repeats the concept "whole impetus." Unfortunately, Eschinardi does not clarify this important point, but it seems that his use of *totus impetus*, which is acquired as the result of impetus being both conserved and continually acquired, is akin to Galileo's "aggregate of parallels" that represent "many *momenta* of speed consumed in the accelerated motion" (Galileo 1989, 166). Like Galileo, Eschinardi seems to use an "overall velocity" concept to relate between areas (which depict *totus motus* in his case and total "momenta of speed" in Galileo's case) relying on the traditional view that when the times are equal, the spaces relate to each other as the (overall) velocities.[145] The above-mentioned statement 1, that relates space to impetus (through velocity) hints that indeed this is what Eschinardi had in mind while connecting time and *totus impetus* (which is represented by the ever increasing area formed in Fig. 10.2) to the spaces acquired. Eschinardi thus appears to follow the traditional way of identifying total velocity with space traversed.

Be that as it may, it should be added that Fabri's discrete approach, albeit rather cumbersome, had the advantage of *ipso facto* rendering surfaces equivalent to spaces—since the rectangular surfaces involved represent constant motions, which could be easily handled by contemporary mathematical means, in particular proportion theory, unlike accelerated motions.[146] We thus see that all contemporary attempts to prove the law of fall relying on contemporary means of representing change, either with the help of graphical techniques going back to the doctrine of intension and remission of qualities and motions, or relying on arithmetical techniques, are plagued with difficulties.[147] Just as the similar attempts by Galileo, Beekman, Descartes or Harriot, the treatment of the three Jesuits here discussed also display the combination of traditional conceptual tools such as the impetus concept, medieval mathematical representation techniques, the exploration of limiting procedures—all in response to the challenging problems of motion, a configuration characteristic of preclassical mechanics.[148]

References

Aristotle. 1908. *Metaphysics* (trans: Ross, W.D.). In *The works of Aristotle*, vol. 8, ed. W.D. Ross and J.A. Smith. Oxford: Oxford University Press.

———. 1930. *Physics* (trans: Hardie, R.P. and Gaye, R.K.). In *The works of Aristotle*, vol. 2, ed. W.D. Ross and J.A. Smith. Oxford: Oxford University Press.

———. 1953. *On the Heavens* (trans: Guthrie, W.K.C.). London: William Heinemann.

[145] See Damerow et al. (2004, 240–242).

[146] In other words, Fabri can easily show (and has indeed shown) that consecutive rectangles are to each other as the corresponding spaces traversed.

[147] On the problems which Galileo's view concerning the structure of the continuum posed—not only to Jesuits, but also to figures like Descartes, Gassendi and Mersenne—see Feldhay (1998); Palmerino (1999).

[148] See Damerow et al. (2004); Schemmel (2008).

Baldini, Ugo. 1999. The development of Jesuit 'physics' in Italy, 1550–1700: Structural approach. In *Philosophy in the sixteenth and seventeenth centuries: Conversations with Aristotle*, ed. Constance Blackwell, 248–279. Aldershot: Ashgate.

———. 2004. Ontology and mechanics in Jesuit scholasticism: The case of Gabriel Vasquez. In *Scientiae et artes: Die Vermittlung alten und neuen Wissens in Literatur, Kunst und Musik*, ed. Barbara Mahlmann-Bauer, 99–142. Wiesbaden: Harrassowitz Verlag.

Casati, R.P. Pauli. 1649. *Vacuum proscriptum: disputatio physica authore Paulo Casato Placentino Societatis Iesu, in qua nullum esse in rerum natura vacuum ostenditur; & potissimum examinatur, an ab argento vivo descendente in fistula superne clausa vacuum relinquatur, huiusque experimenti symptomata exlpicantur*. Genua: Peri.

———. 1684. *Mechanicorum libri octo, in quibus uno eodemque principio vectis vires physice explicantur & geometrice demonstrantur, atque Machinarum omnis generis componendarum methodus proponitur*. Lugdunum: Apud Anissonios. Joan. Posuel & Claudium Rigaud.

Clagett, Marshall. 1959. *Science of mechanics in the Middle Ages*. Madison: University of Wisconsin Press.

Damerow, Peter, Gideon Freudenthal, Peter Maclaughlin, and Jürgen Renn. 2004. *Exploring the limits of preclassical mechanics: A study of conceptual development in early modern science. Free fall and compounded motion in the work of Descartes, Galileo, and Beeckman*. Second ed. New York: Springer.

Descartes, René. 1985. *The philosophical writings of Descartes. Vol. 1. Principles of philosophy*. Trans. John Cottingham, Robert Stoothoff and Dugald Murdoch. Cambridge: Cambridge University Press

Des Chene, Dennis. 1996. *Physiologia: Natural philosophy in late Aristotelian and Cartesian thought*. Ithaca: Cornell University Press.

Dollo, C. 1989. *Archimede: Mito, tradizione, scienza, a cura di C Dollo*. Olschki: Firenze.

Drake, Stillman. 1974. Impetus theory and quanta of speed before and after Galileo. *Physis* 16: 47–75.

———. 1975. Impetus theory reappraised. *Journal of the History of Ideas* 36 (1): 27–46.

Duhem, Pierre M.M. 1991. *Origins of static: The sources of physical theory*. Dordrecht: Kluwer.

Elazar, Michael. 2008. Honoré Fabri and the Trojan horse of inertia. *Science in Context* 21 (1): 1–38.

———. 2011. *Honoré Fabri and the concept of impetus: A bridge between conceptual frameworks*. New York: Springer-Verlag.

Eschinardi, Francesco. 1684. *De impetu tractatus duplex: primus de impetu in communi, de motu locali et de machinis, secundus de fluidis in communi, de comparatione fluidorum cum solidis et de mensura aquarum currentium : additur in fine quamplurimum problematum seu quaesitorum solutio ex doctrinis praecedentibus*. Rome: Ex typographia Angeli Bernabo.

Fabri, Honoré. 1646. *Tractatus Physicus De Motu Locali: In Quo Effectus Omnes, Qui Ad Impetum, Motum naturalem, violentum, & mixtum pertinent, explicantur, & ex principiis Physicis demonstrantur*. Lugdunum: Ioannis Champion.

———. 1648. *Metaphysica demonstrativa: Sive Scientia rationum universalium*. Lugdunum: Ioannis Champion.

Feldhay, Rivka. 1998. The use and abuse of mathematical entities: Galileo and the Jesuits revisited. In *The Cambridge companion to Galileo*, ed. Peter Machamer, 80–145. Cambridge: Cambridge University Press.

———. 2006. On wonderful machines: The transmission of mechanical knowledge by Jesuits. *Science & Education* 15: 151–172.

Feldhay, Rivka, and Michael Heyd. 1989. The discourse of pious science. *Science in Context* 3: 109–142.

Franco, Abel B. 2003. Avempace, projectile motion, and impetus theory. *Journal of the History of Ideas* 64 (4): 521–546.

Gabbey, Alan. 1980. Force and inertia in the seventeenth century: Descartes and Newton. In *Descartes: Philosophy, mathematics and physics*, ed. Stephen Gaukroger, 230–320. Brighton: Harvester Press.

Galileo. 1989. *Two new sciences: Including centers of gravity and force of percussion*. Trans. Stillman Drake. Toronto: Wall & Emerson.

Galluzzi, Paolo. 2001. Gassendi and l'affaire Galilée of the laws of motion. In *Galileo in context*, ed. Jürgen Renn, 239–275. Cambridge: Cambridge University Press.

Gassendi, Pierre. 1642. *De motu impresso a motore translato: epistolae duae: in qvibus aliqvot praecipuae tum de motu universè, tum speciatim de motu terrae attributo difficultates explicantur*. Parisiis: Hevqueville.

Gavagna, Veronica. 1997. Imagini di Galileo nel 'Terra machinis mota', di Paolo Casati. *Bollettino di storia delle scienze matematiche* 17 (1): 105–148.

Grant, Edward. 1973. Medieval explanations and interpretations of the dictum that 'nature abhors a vacuum'. *Traditio* 29: 327–355.

Heilbron, John. 2010. *Galileo*. Oxford: Oxford University Press.

Jammer, Max. 1957. *Concepts of force: A study in the foundations of dynamics*. Cambridge, MA: Harvard University Press.

Laird, W.R. 1986. The scope of renaissance mechanics. *Osiris* 2: 43–68.

Laird, W.R., and S. Roux. 2001. *Mechanics and natural philosophy before the Scientific Revolution*. Dordrecht: Springer.

Lukens, David C. 1979. *An Aristotelian response to Galileo: Honoré Fabri, S.J. (1608–1688) on the casual analysis of motion*. PhD diss. Toronto University of Toronto.

Maier, Annaliese. 1982. *On the threshold of exact science: Selected writings of Anneliese Maier on late medieval natural philosophy, edited and translated by Steven D Sargent*. Philadelphia: University of Pennsylvania Press.

Mancosu, Paolo. 1996. *Philosophy of mathematics & mathematical practice in the seventeenth century*. New York: Oxford University Press.

Meli, Domenico Bertoloni. 2006. *Thinking with objects: The transformation of mechanics in the seventeenth century*. Baltimore: Johns Hopkins University Press.

Middleton, W. E. Knowles. 1964. *The history of the barometer*. Baltimore: John Hopkins Press.

———. 1975. Science in Rome, 1675–1700, and the Accademia Fisicomatematica of Giovanni Giustino Ciampini. *The British Journal for the History of Science* 8 (29): 138–154.

Palmerino, Carla Rita. 1999. Infinite degrees of speed: Marine Mersenne and the debate over Galileo's law of free fall. *Early Science and Medicine* 4 (4): 269–328.

———. 2004. Galileo's theories of free fall and projectile motion as interpreted by Pierre Gassendi. In *The reception of the Galilean science of motion in seventeenth-century Europe*, ed. C.R. Palmerino and J.M.M.H. Thijssen, 137–164. Dordrecht: Kluwer.

Renn, Jürgen. 2001. Editor's introduction. An engineer-scientist, artist, and courtier at the origins of classical science. In *Galileo in context*, ed. Jürgen Renn, 1–8. Cambridge: Cambridge University Press.

Schemmel, Matthias. 2008. *The English Galileo: Thomas Harriot's work on motion as an example of preclassical mechanics*. Dordrecht: Springer.

Shapiro, Alan E. 2008. Images: Real and virtual, projected and perceived, from Kepler to Dechales. *Early Science and Medicine* 13: 270–312.

Wallace, William A. 1972–1974. *Causality and scientific explanation. Vol. 1: Medieval and early classical science*. Ann Arbor: University of Michigan Press.

———. 1978. Causes and forces in sixteenth-century physics. *Isis* 69 (3): 400–412.

Correction to: Emergence and Expansion of Preclassical Mechanics

Rivka Feldhay, Jürgen Renn, Matthias Schemmel, and Matteo Valleriani

Correction to:
R. Feldhay et al. (eds.), *Emergence and Expansion of Preclassical Mechanics*, Boston Studies in the Philosophy and History of Science 270, https://doi.org/10.1007/978-3-319-90345-3

The book was inadvertently published with few errors in chapter 6 "**A Journey to the Center of the Earth: Cosmology and the Centrobaric Theory from Antiquity to the Renaissance**". The text has been corrected and read as follows:

1. Greek fonts were reflecting incorrectly in both the PDF and xml. These have now been corrected. An error in the production process unfortunately led to publication of this chapter prematurely, before incorporation of the final corrections. The version supplied here has been corrected and approved by the author [authors].
2. A reference in page 177 has been corrected to read as: Euripides, and D. Kovacs. 2002. Helen, Phoenician women, Orestes. Cambridge: Harvard University Press.
3. Text in page 163 has been updated from "Psalm CCCXXXVI,6" to "Psalm 103" and "Psalm XXIV,2" to "Psalm 23".

In addition to this, the text "Ribandiera" has been updated to "Ribadeneira" in line 20, page 31 of chapter 2 "**Preclassical Mechanics in Context: Practical and Theoretical Knowledge Between Sovereignty, Religion, and Science**".

The updated online version of this chapters can be found at
https://doi.org/10.1007/978-3-319-90345-3_2
https://doi.org/10.1007/978-3-319-90345-3_6
https://doi.org/10.1007/978-3-319-90345-3

R. Feldhay et al. (eds.), *Emergence and Expansion of Preclassical Mechanics*, Boston Studies in the Philosophy and History of Science 270,
https://doi.org/10.1007/978-3-319-90345-3_11

Correction to: Emergence and Expansion of Preclassical Mechanics

Rivka Feldhay, Jürgen Renn, Matthias Schemmel, and Matteo Valleriani

Correction to:
R. Feldhay et al. (eds.), *Emergence and Expansion of Preclassical Mechanics*, Boston Studies in the Philosophy and History of Science 270, https://doi.org/10.1007/978-3-319-90345-3

The book was inadvertently published with wrong volume number as 333. The same has now been corrected throughout the book. The correct volume number is 270.

The updated version of the book can be found at
https://doi.org/10.1007/978-3-319-90345-3

R. Feldhay et al. (eds.), *Emergence and Expansion of Preclassical Mechanics*,
Boston Studies in the Philosophy and History of Science 270,
https://doi.org/10.1007/978-3-319-90345-3_12

Index

A
Academia del Cimento, 61, 255, 256, 265, 269, 281
Antonio Gramsci, 64
Archimedean tradition, 140, 142, 143, 165, 262, 280
Aristotelian dynamics, 12–14, 62, 220, 225, 286
Aristotle, 7, 12, 13, 15, 18, 22, 46, 47, 86, 106, 112, 114, 122, 125, 129, 131, 133–135, 140, 143–167, 169, 171, 175, 176, 187, 189, 199, 216, 217, 237–240, 250–252, 258, 260–266, 269, 273, 279, 280, 286, 288, 290, 292, 293, 295, 297, 298, 300, 301, 303–305, 307, 313, 314
Aristotle's physics, 13, 132, 199, 262, 265, 300

B
Balance scales, 99
Benedetti, Giovanni Battista, 56, 63, 70, 181–210, 264, 273, 295
Bodin, Jean, 31–36, 40, 41
Botero, Giovanni, 31, 35, 36, 38–42

C
Casati, Paolo, 256–260, 266–271, 273–281, 285–321
Causation, 11–20
Centrobarics, 139–176
China, 82, 87–95, 103, 105, 106, 132
Classical physics, 14–20, 24, 25
Clavius, Christoph, 43, 47, 145, 167, 172–174, 203, 217–219, 262
Compilation, 116, 216
Counterweights, 85, 126–131, 133, 134

D
de coelo, 140, 144–152, 154–157, 159, 161–166, 170, 175, 223, 263

E
Educational politics, 216
Equilibrium, 22, 23, 84, 85, 87, 89, 101–103, 106, 122, 127, 131, 133, 139, 141–145, 151–159, 164, 166, 167, 169, 174–176, 188, 225, 236, 237, 273, 276, 277, 302
Eschinardi, Francesco, 285–321
Evolution of knowledge, 95–97
Exercitationes, 88, 215–252
Experience, 5, 6, 8, 9, 11, 14, 42, 48, 62, 63, 69, 92, 94, 96, 116, 117, 132, 133, 135, 152, 161, 162, 168, 172, 174, 186, 199, 207, 210, 268, 281, 302, 306, 313
External representations, 84, 86, 95–97

F
Fabri, Honoré, 48, 95, 285–321
Figure of Earth, 140

G
Galilean science, 69

R. Feldhay et al. (eds.), *Emergence and Expansion of Preclassical Mechanics*, Boston Studies in the Philosophy and History of Science 270,
https://doi.org/10.1007/978-3-319-90345-3

Galilei, Galileo, 15, 48, 56, 174, 181, 216, 218, 229, 285–321
Gravity, 17, 18, 22–24, 30, 43, 47, 48, 87, 89, 93, 102, 120, 127, 139–146, 148, 149, 152, 155, 159, 160, 165–171, 173–176, 187, 220, 225, 226, 230–232, 235–237, 239, 252, 257, 261, 262, 295, 298, 299, 302–304, 306–309, 314–316
Greece, 82, 90, 93, 113, 132–135, 145, 153
Grossman, Henryk, 57, 62–64, 186

H
Harriot, Thomas, 16–18, 62, 321
Heaviness, 47, 255–281, 287
Hessen, Boris, 57, 58, 186

I
Impetus, 14, 15, 18, 63, 182, 188, 232, 255–281, 285–321
Innovation, 81–107, 113, 182, 227
Islamicate Societies, 90

J
Jesuits, 32, 42–51, 145, 203, 204, 216–219, 221, 259, 262, 264, 265, 281, 286–288, 294, 314, 321
Jesuit science, 43, 67
Johannes de Sacrobosco, 140, 145, 159–162, 168, 172–173, 222

L
Law of the lever, 22, 24, 83–85, 87, 94, 99, 100, 103, 106, 112, 113, 121–123, 128, 130, 134, 135, 141–143, 166, 184, 232
Le Mechaniche, 264
Lever, 21–24, 82–85, 87, 90–94, 99, 100, 103, 106, 111–135, 140–143, 166, 184, 225, 226, 229–236, 239–242, 244–246, 249, 250, 257, 258, 274
Levity, 260, 298, 299, 303–310
Lightness, 47, 154, 255–281

M
Marxist historiography of science, 58
Measurement, 69, 189, 273
Mechanical problems, 22–24, 64, 86–88, 92, 93, 106, 111–119, 122–124, 126–135, 143, 167, 175, 184, 187
Mental models, 3–26, 30, 61, 84, 87, 113, 133, 134, 185, 274, 286, 287

N
Natural motion, 3, 15–18, 112, 144, 145, 147, 148, 150, 154, 155, 161, 162, 172, 174, 189, 260, 264, 268, 273, 279, 287, 288, 297, 298, 302, 304, 307–310
Neapolitan College, 216–222
Newton, Isaac, 18, 19, 24, 29, 40, 58–61, 64, 71, 72, 112, 182, 184–187, 286

P
Political epistemology, 56
Political identity of the Renaissance scientist, 182
Practical knowledge, 6, 16, 21, 30, 68, 69, 71, 92, 96, 133–135, 194, 195
Preclassical mechanics, 3, 14, 24, 29–51, 187, 259, 281, 285, 286, 321
Project, v, 46, 66, 74, 92, 101, 102, 112, 113, 132, 189, 202, 204–206, 216–218, 258, 259, 262, 278, 280, 300
Projectile trajectory, 16, 59

R
Reduction, 113, 115–123, 126, 131, 132, 184, 248
Reflection, 8–9, 13, 22, 24, 30, 37, 40, 56, 60, 63, 70, 73, 84, 86, 87, 90, 112–114, 132–135, 186
Regulative processes, 84
Roman Empire, 100, 104, 105, 127

S
Science and cultural hegemony, 66, 203
Science and technology, 61, 112, 259
Science and the court, 207
Simple machines, 21, 59, 112, 133, 134, 140, 145, 167, 184, 187, 216, 224, 225, 231, 257

Social position of the Renaissance scientist, 191–195
Socio-political history of mechanics, 181–210
Sovereignty, 29–51
Sphaera, 220, 268
Steelyard, 83, 85, 86, 88, 90, 91, 96, 99–106, 127, 128, 130, 131, 232
Syllogism, 115, 124, 125

T
Technical knowledge, 3, 4, 11, 19, 21, 25, 61, 94, 106
Theoretical knowledge, 8, 26, 29–51, 61, 63, 83–85, 87, 90, 92, 93, 95–97, 106–107, 112, 128, 134, 195, 258

W
Weighing technology, 81–107, 113

Z
Zilsel, Edgar, 56, 61, 67–73, 186, 195
Zupi, 215–252

The manufacturer's authorised representative in the EU is Springer
Nature Customer Service Centre GmbH, Europaplatz 3, 69115 Heidelberg,
Germany. If you have any concerns regarding our products, please
contact ProductSafety@springernature.com

Printed and bound by CPI Group (UK) Ltd, Croydon, CR0 4YY
30/09/2026
02225599-0012